Newnes Electrical Pocket Book

Newnes
Electrical Pocket Book

Twenty-third edition

E.A. Reeves
DFH(Hons), CEng, MIEE

Martin J. Heathcote
BEng, CEng, FIEE

Newnes

OXFORD   AMSTERDAM   BOSTON   LONDON   NEW YORK   PARIS
SAN DIEGO   SAN FRANCISCO   SINGAPORE   SYDNEY   TOKYO

Newnes
An imprint of Elsevier Science
Linacre House, Jordan Hill, Oxford OX2 8DP
200 Wheeler Road, Burlington, MA 01803

First published by George Newnes Ltd 1937
Twenty-second edition 1995
Twenty-third edition 2003

British Library Cataloguing in Publication Data
A catalogue record for this book is available from the British Library

ISBN 0 7506 4758 2

For information on all Newnes publications
visit our website at www.newnespress.com

Typeset by Laserwords Private Limited, Chennai, India.
Printed and bound in Great Britain

Contents

Preface

It is now seven years since the twenty-second edition of the Pocket Book was published, a rather longer interval than might be desirable in the rapidly moving and rapidly developing world of electrical technology. We now have a new editor and, as a result, the possibility of some differing emphasis.

Eric Reeves' name has become synonymous with the Pocket Book. He has been editor for over forty years covering some ten or more editions. He is now enjoying his 'retirement'. He has left a pocket reference work that is in good shape, but inevitably as the industry moves on, the detail is constantly subject to change.

In the UK, privatization of electricity supply was some six years consigned to history at the time of publication of the twenty-second edition. But much of the transformation of the industry, which now sees electricity traded as any other commodity like oil or coffee beans, has taken place over the last five or six years. Many of the companies that the Government set up in 1989 have now disappeared and the structure of the industry has changed beyond recognition. Changes now occur so rapidly that the details of the UK utilities as given in the previous edition have been dropped. The reader must now keep up with these developments by closely watching the business pages of his or her newspaper.

Now, if it is more profitable to sell gas than to use it to generate electricity and sell that, utilities are happy to do this. Now, the generators, transmission lines and transformers are 'assets' which assist the owners in making a profit, and the staff entrusted with the care and supervision of these are 'asset managers'. They may be more skilled in risk assessment and knowledgeable about failure rates and downtimes than their predecessors, but it is still necessary to retain a workforce who know about the plant and are able to ensure it can remain in safe and reliable operation.

Privatization of the UK electricity supply has also led to many utilities procuring equipment overseas, particularly from Europe. This has resulted in the adoption within the UK of many approaches to many aspects of electrical equipment design and specification. In a wider context this has probably provided added impetus to harmonization of standards and the acceptance of IEC and CENELEC documentation.

Today's technicians face a challenging task to keep abreast of developments even within quite narrow fields and 'continuing professional development' is a task to be pursued by all, not simply those who wish to gain advancement in their chosen field.

This is where it is hoped that this little book will remain of assistance. The danger is that it will get larger at each new edition. If it is to remain a handy pocket reference size, then to include new material it is necessary to leave out some information which has proved useful in the past. The hope is that the balance will remain about right and what Eric Reeves has achieved so successfully for many years will continue.

One chapter which might have been left out is Chapter 6 which deals with computers. These are no longer specialist tools to be used by the few; even children in primary schools are being given computing skills. There are weekly and monthly magazines by the score which can provide an introduction to computing, so its need in a work such as this might be superfluous. However, the chapter has been retained because of its relevance to electrical engineering, but it has been shortened and made less specific, hopefully in a form which will provide some useful background for those working in other branches of electrical engineering.

Chapter 4 of the twenty-second edition dealt with semiconductors as devices which have superseded valves in electronic equipment. Although many older engineers may have been introduced to semiconductors in this way, valves are no longer taught in colleges and universities. Hence the emphasis has been reversed with semiconductors introduced in their own right and some descriptions of valve devices retained because these might be encountered in special applications.

Chapter 7 has been extensively revised to include some description and theory of a.c. generators. Although few will find themselves coming into close practical contact with these, some understanding of the design and workings of the main source of electrical power is perhaps desirable for those who earn or seek to earn their livelihood in the electrical industry.

Likewise the chapter on transformers, Chapter 10, has been expanded a little to include some detail of their construction, connections, phase shifts and losses, although few in the electrical industry will encounter any but the smaller end of the size range. The section dealing with magnetic materials in Chapter 2 has also been expanded since in large transformers and generators magnetic steel is just as important a material as copper.

Since the publication of the twenty-second edition there has been a revision of BS 7671 which has brought about significant changes. A section has therefore been added to Chapter 12 detailing the changes and discussing the implications of these.

Building automatic management systems, which were highlighted in the preface to the twenty-second edition as being subject to rapid change, has seen even further development in view of the advances in computing capability. The result is that Chapter 17 has been largely rewritten to identify these developments.

Chapter 20, dealing with battery electric vehicles has been expanded a little to reflect the growth of interest in clean vehicles and particularly to describe recent developments relating to hybrid vehicles.

There have been significant changes in requirements relating to electrical equipment for use in hazardous areas in recent years as a result of two EU Directives, 94/9 relating to explosion protected equipment, and 99/92 relating to certification of the equipment. Chapter 23, which was newly written for the twenty-second edition, has, as a result, been extensively revised.

Despite what may appear a lengthy list of changes, much of what was written by Eric Reeves in the twenty-second edition remains. The hope is that readers will find both the older material that has been retained, and that which is new, of value, and that no one will feel that any vital aspect which has made Eric's formula such a successful one over so many years has been cast aside

M.J.H

Acknowledgements

Inevitably when aiming to cover as wide a spectrum of electrical engineering as does the Pocket Book, it is necessary to go to many sources in order to obtain authoritative information which can be committed to print for the benefit of readers. Many people have assisted in the preparation of the twenty-third edition, either by writing complete chapters or sections, or simply by providing constructive criticism of the editor's efforts.

The editor wishes to express grateful thanks to all those friends and colleagues, individuals and organizations who have provided assistance in this revision. In particular to my good friend W.J. (Jim) Stevens who has read most of what has been written and provided invaluable criticism and comment; to my good friend, Mike Barber, who rewrote much of Chapter 7 relating to electricity generation and the theory and practice of a.c. generators; to colleagues Bob Dodd, for the descriptions of AVRs and John Rhodes, for paragraphs on wind energy in this chapter; to Neil Pascoe for his contribution on metering transducer systems and Dan Brown for Chapter 6 on computers; Mike Rowbottom for a description of NETA, the New Electricity Trading Arrangements, in Chapter 11; to other friends and colleagues who have read and commented on specific sections and to those who have provided written contributions; Bob Bradley, TCM Tamini, on high integrity and UPS power supplies included in Chapter 7; Ian Harrison, Chloride Industrial Batteries, for much of the material for Chapter 21; Tony Martin on aluminium busbars; Terry Journeaux, Pirelli Cables, for information for Chapter 9; Paul John, Marconi Applied Technologies, for data on valves and related Marconi products. Thanks are also due to Ray Lewington of BEAMA for permission to make use of his lecture material covering the 2001 revision of BS 7671; Hugh King, Thorn Lighting, for updating Chapter 13; Steve Dalton, Johnson Industrial Control Systems, for updating Chapter 17; Simon Howard, Crompton Instruments for additions to and updating of Chapter 18; Dick Martin, CEAG Crouse Hinds, for Chapter 23. My thanks to the Institution of Electrical Engineers for permission to reproduce extracts from BS 7671, and to the many organizations who have provided the many photographs and illustrations to whom attribution is given in the text.

Finally, despite the quite extensive revision involved in the production of the twenty-third edition, the greater part of the book remains the work of Eric Reeves from the twenty-second edition, and for this due acknowledgement must be given.

Acknowledgements

Introduction

The chief function of any engineer's pocket book is the presentation in convenient form of facts, tables and formulae relating to the particular branch of engineering concerned.

In the case of electrical engineering, it is essential that the engineer should have a clear understanding of the methods by which the various formulae are derived in order that he can be quite certain that any particular formula is applicable to the conditions which he is considering. This applies with particular force in the case of alternating current work.

The first section of the Pocket Book is, therefore, devoted to the theoretical groundwork upon which all the practical applications are based. This covers symbols, fundamentals, electrostatics and magnetism.

When an engineer is called upon to deal with any particular type of electrical apparatus, for example a protective relay system, a thermostatically controlled heating system, or industrial switchgear and control gear, the first requirement is that he shall understand the principles upon which these systems operate. In order to provide this information, much space has been devoted in the various sections to clear descriptions of the circuits and principles which are used in the different types of electrical apparatus.

The inclusion of technical descriptions, together with the essential data embodied in the tables, will be found to provide the ideal combination for those engineers engaged on the utilization side of the industry, where many different types of equipment and electrical appliances, ranging from semiconductor rectifiers to electrode steam boilers, may have to be specified, installed and maintained in safe and efficient operation.

An extensive summary of the sixteenth edition of the 'IEE Regulations for Electrical Installations' (now BS 7671) is contained in Chapter 12. In 1992 when this was first issued as a British Standard, the layout and content were markedly different to the previous editions and for those personnel working in electrical contracting it is important that they obtain their own up-to-date copy of the Regulations. One of the most important changes in 1992 was the exclusion of many of the Appendices which were published as separate Guidance Notes (see page 260). Another change was the inclusion of a new Part 6, 'Special installations or locations'. Section 6 has been added to in the 2001 edition, and, in addition, in an extended Part 7, there is increased emphasis on periodic inspection and testing. More is said about these in the Preface and in Chapter 12.

1 Fundamentals and theory

Fundamentals

Current. The term 'current' is used to denote the rate at which electricity flows. In the case of a steady flow the current is given by the quantity of electricity which passes a given point in one second. (Although since 1948 the unit of current has been officially defined in terms of the electromagnetic force that it produces, see below – since this force can be most conveniently measured.) The magnitude of the current depends not only upon the electromotive force but also upon the nature and dimensions of the path through which it circulates.

Ohm's law. Ohm's law states that the current in a *direct current* (d.c.) circuit varies in direct proportion to the voltage and is inversely proportional to the resistance of the circuit. By choosing suitable units this law may be written

$$\text{Current} = \frac{\text{Electromotive force}}{\text{Resistance}}$$

The commercial units for these quantities are

Current – the ampere   (A)
Electromotive force – the volt   (V)
Resistance – the ohm   (Ω)

Using the symbols I, V and R to represent the above quantities in the order given, Ohm's law can be written

$$I = \frac{V}{R}$$

or $V = I \times R$

The law not only holds for a complete circuit, but can be applied to any part of a circuit provided care is taken to use the correct values for that part of the circuit.

Resistivity. The resistivity of any material is the resistance of a piece of the material having unit length and unit sectional area. The symbol is ρ and the unit is the ohm metre. The resistivity of a material is not usually constant but depends on its temperature. *Table 1.1* shows the resistivity (with its reciprocal, conductivity) of the more usual metals and alloys.

Resistance of a conductor. The resistance of a uniform conductor with sectional area A and length l is given by

$$R = \rho \frac{l}{A}$$

The units used must be millimetres and square millimetres if ρ is in ohm millimetre units.

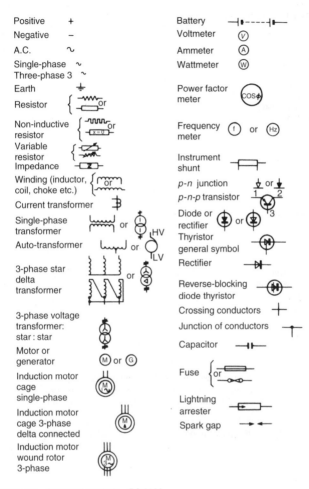

Positive	+	Battery	—┤•----┤├—
Negative	–	Voltmeter	Ⓥ
A.C.	∿	Ammeter	Ⓐ
Single-phase	∿	Wattmeter	Ⓦ
Three-phase 3	∿		
Earth	⏚	Power factor meter	cosφ
Resistor	{ —ⱲⱲⱲ— or		
Non-inductive resistor	{ ⌁⌁⌁ or [x=0]	Frequency meter	Ⓕ or Ⓗz
Variable resistor	{	Instrument shunt	⊏⊐
Impedance	—▭—	p-n junction	↓ or ↓
Winding (inductor, coil, choke etc.)	{ ⌒⌒⌒ or	p-n-p transistor	
Current transformer	⇉	Diode or rectifier	or
Single-phase transformer	or	Thyristor general symbol	
Auto-transformer	or	Rectifier	—▷⊢
3-phase star delta transformer	or	Reverse-blocking diode thyristor	
3-phase voltage transformer: star : star		Crossing conductors	+
Motor or generator	Ⓜ or Ⓖ	Junction of conductors	⊥
Induction motor cage single-phase		Capacitor	—┤├—
Induction motor cage 3-phase delta connected		Fuse	{
Induction motor wound rotor 3-phase		Lightning arrester	
		Spark gap	→• •←

Figure 1.1 Graphical symbols – BS 3939

Temperature coefficient. The resistance of a conductor at any temperature can be found as follows:

$$R_t = R_0(1 + \alpha t)$$

R_t = resistance at temperature $t°C$
R_0 = resistance at temperature $0°C$

The coefficient α is called the temperature coefficient and it can be described as the ratio of the increase in resistance per degree C rise in temperature

Table 1.1 Resistivities at 20°C

Material	Resistivity Ohm metres	Conductivity Siemens per metre
Silver	1.64 $\times 10^{-8}$	6.10×10^7
Copper (annealed)	1.72 $\times 10^{-8}$	5.8×10^7
Gold	2.4 $\times 10^{-8}$	4.17×10^7
Aluminium (hard)	2.82 $\times 10^{-8}$	3.55×10^7
Tungsten	5.0 $\times 10^{-8}$	2.00×10^7
Zinc	5.95 $\times 10^{-8}$	1.68×10^7
Brass	6.6 $\times 10^{-8}$	1.52×10^7
Nickel	6.9 $\times 10^{-8}$	1.45×10^7
Platinum	11.0 $\times 10^{-8}$	9.09×10^6
Tin	11.5 $\times 10^{-8}$	8.70×10^6
Iron	10.15 $\times 10^{-8}$	9.85×10^6
Steel	19.9 $\times 10^{-8}$	5.03×10^6
German Silver	16–40 $\times 10^{-8}$	6.3–2.5×10^6
Platinoid	34.4 $\times 10^{-8}$	2.91×10^6
Manganin	44.0 $\times 10^{-8}$	2.27×10^6
Gas carbon	0.005	200
Silicon	0.06	16.7
Gutta-percha	2 $\times 10^7$	5×10^{-8}
Glass (soda-lime)	5 $\times 10^9$	2×10^{-10}
Ebonite	2 $\times 10^{13}$	5×10^{-14}
Porcelain	2 $\times 10^{13}$	5×10^{-14}
Sulphur	4 $\times 10^{13}$	2.5×10^{-14}
Mica	9 $\times 10^{13}$	1.1×10^{-14}
Paraffin-wax	3 $\times 10^{16}$	3.3×10^{-17}

compared with the actual resistance at 0°C. The coefficient for copper may be taken as 0.004. The increase in resistance for rise of temperature is important, and for many calculations this factor *must* be taken into account.

Power. Power is defined as the rate of doing work. The electrical unit of power (P) is the *watt* (abbreviation W), and taking a steady current as with d.c.

$$1\ W = 1\ V \times 1\ A$$

or $$W = V \times A$$

or in symbols $$P = V \times I$$

(For alternating current, see page 12)
Note: 1 kW = 1000 W

Energy. Energy can be defined as power × time, and electrical energy is obtained from

$$\text{Energy} = V I t$$

where t is the time in seconds.

The unit obtained will be in joules, which is equivalent to 1 ampere at 1 volt for 1 second. The practical unit for energy is the kilowatt hour and is given by

$$\frac{\text{watts} \times \text{hours}}{1000} = \text{kWh}$$

Energy dissipated in resistance. If we pass a current I through resistance R, the volt drop in the resistance will be given by

$$V = IR$$

The watts used will be VI, therefore the power in the circuit will be $P = VI = (IR) \times I = I^2 R$.

This expression $(I^2 R)$ is usually known as the copper loss or the $I^2 R$ loss. Similarly power can be expressed as $V \times (V/R) = V^2/R$.

SI units. The SI (*Systeme Internationale*) system uses the metre as the unit of length, the kilogram as the unit of mass and the second as the unit of time. These units are defined in BS 5555 '*Specification for SI units and recommendations for the use of their multiples and of certain other units*'.

SI units are used throughout the rest of this book and include most of the usual electrical units. With these units, however, the permittivity and permeability are constants. They are:

$$\text{Permittivity } \varepsilon_0 = 8.85 \times 10^{-12} \text{ farad per metre}$$

$$\text{Permeability } \mu_0 = 4\pi \times 10^{-7} \text{ henry per metre}$$

These are sometimes called the electric and magnetic space constants respectively. Materials have relative permittivity ε_r and relative permeability μ_r hence ε_r and μ_r for a vacuum are unity.

Electrostatics

All bodies are able to become electrically charged, and this is termed static electricity. The charge on a body is measured by measuring the force between two charges, this force follows an inverse square law (i.e. the force is proportional to the product of the charges and inversely proportional to the square of the distance between them). This may be written

$$F = \frac{q_1 q_2}{4\pi \, \varepsilon_0 d^2} \text{N}$$

where q_1 and q_2 are the charges in coulombs (symbol C) and d the distance in metres – the space in between the charges being either air or a vacuum with a permittivity ε_0. N is newtons.

If the two charged bodies are separated by some other medium the force acting may be different, depending on the relative permittivity of the *dielectric* between the two charged bodies. The relative permittivity is also termed the dielectric constant.

In this case the force is given by

$$F = \frac{q_1 q_2}{4\pi \, \epsilon_r \epsilon_0 \, d^2} \, N$$

where ε_r is the constant for the particular dielectric. For air or a vacuum the value of ε_r is unity.

Intensity of field. A charged body produces an electrostatic field. The *intensity* of this field is taken as the force on unit charge.

The intensity of field at any given point due to an electrostatic charge q is given by

$$E = \frac{q}{4\pi \, \epsilon_0 \, d^2} \, V/m$$

Note: The ampere is the defined unit. Hence a coulomb is that quantity of charge which flows past a given point of a circuit when a current of one ampere is maintained for one second.

The value of the ampere, adopted internationally in 1948, is defined as that current which, when flowing in each of two infinitely long parallel conductors in a vacuum, separated by one metre between centres, causes each conductor to have a force acting upon it of 2×10^{-7} N/m length of conductor.

Dielectric flux. The field due to a charge as referred to above is assumed to be due to imaginary *tubes of force* similar to magnetic lines of force, and these tubes are the paths which would be taken by a free unit charge if acted on by the charge of the body concerned.

By means of these tubes of force we get a *dielectric flux-density* of so many tubes of force per square metre of area. For our unit we take a sphere of 1 m radius and give it unit charge of electricity. We then get a dielectric flux density on the surface of the sphere of one tube of force per square metre. The total number of tubes of force will be equal to the surface area of the sphere $= 4\pi$. For any charge q at a distance r the dielectric flux density will be

$$D = \frac{q}{4\pi r^2} \, C/m^2$$

We have seen that the intensity of field or electric force at any point is

$$E = \frac{q}{4\pi \, \epsilon_0 \epsilon_r \, r^2}$$

so that this can also be stated as $E = D/\varepsilon_r \varepsilon_0$.

Electrostatic potential. The potential to which a body is raised by an electric charge is proportional to the charge and the *capacitance* of the body – so that $C = Q/V$, where V is the potential and C the capacitance. The definition of the capacitance of a body is taken as the charge or quantity of electricity necessary to raise the potential by one volt. This unit of potential is the work done in joules, in bringing unit charge (1 coulomb) from infinity to a point at unit potential.

Capacitance. For practical purposes the unit of capacitance is arranged for use with volts and coulombs. In this case the unit is the farad (symbol F), and we get $C = Q/V$, where C is in farads, Q is in coulombs and V is in volts.

The farad is a rather large unit, so that in practice we more commonly employ the microfarad = 10^{-6} of a farad or 1 picofarad = 10^{-12} of a farad.

Capacitors

The capacitance of a body is increased by its proximity to earth or to another body and the combination of the two is termed a capacitor. So long as there is a potential difference between the two there is a capacitor action which is affected by the dielectric constant of the material in between the two bodies.

Flat plate capacitor. Flat plate capacitors (*Figure 1.2*) are usually made up of metal plates with paper or other materials as a dielectric. The rating of a plate capacitor is found from

$$C = \epsilon_r \epsilon_0 \, A/d \text{ farads}$$

where A is the area of each plate and d the thickness of the dielectric. For the multi-plate type we must multiply by the number of actual capacitors there are in parallel.

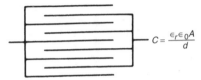

$$C = \frac{\epsilon_r \epsilon_0 A}{d}$$

Figure 1.2 Plate capacitor

Concentric capacitor. With electric cables we get what is equivalent to a concentric capacitor (*Figure 1.3*) with the outer conductor or sheath of radius r_1 m and the inner conductor of radius r_2 m. If now the dielectric has a constant of ε_r, the capacitance will be (for 1 m length)

$$C = \frac{2\pi \, \epsilon_r \epsilon_0}{\log e(r_1/r_2)} \text{ farads per metre}$$

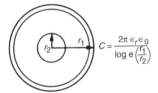

$$C = \frac{2\pi \, \epsilon_r \epsilon_0}{\log e \left(\frac{r_1}{r_2}\right)}$$

Figure 1.3 Concentric capacitor

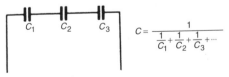

$$C = \frac{1}{\frac{1}{C_1} + \frac{1}{C_2} + \frac{1}{C_3} + \cdots}$$

Figure 1.4a Capacitors in series

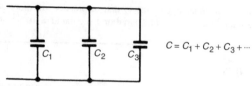

$$C = C_1 + C_2 + C_3 + \cdots$$

Figure 1.4b Capacitors in parallel

Values of ε_r for different materials

Air	1
Paper, Pressboard	2
Cotton tape (rubberized)	2
Empire cloth	2
Paper (oiled)	2
Shellac	3
Bakelite	6
Paraffin-wax	3
Mica	7
Porcelain	7
Glass	7
Marble	8
Rubber	2.5
Ebonite	2.5
Gutta-percha	4
Polyethylene	2.3
Nylon polyamide (Nomex)	3
Epoxy resin	3.4
Phenolic resin	3.5

The Magnetic Circuit

Electromagnets. Magnetism is assumed to take the form of lines of force or *magnetic flux* which flow round the magnetic circuit. This circuit may be a complete path of iron or may consist of an iron path with one or more air-gaps. The transformer iron core is an example of the former and a generator, with its combination of laminated iron stator core and rotor iron forging with an air or hydrogen filled gap between them, an example of the latter.

The lines of force are proportional to the *magneto-motive-force* of the electric circuit and this is given by

$$\text{m.m.f.} = IN \text{ ampere turns}$$

where I is the current in amperes and N the number of turns in the coil or coils carrying this current. This m.m.f. is similar in many respects to the e.m.f. of an electric circuit and in the place of the resistance we have the *reluctance* which may be regarded as the 'resistance' of the magnetic circuit to the establishing of the flux. The reluctance is found from

$$\text{Reluctance} = S = \frac{l}{A\mu_r\mu_0} \text{ ampere turns per weber (At/Wb)}$$

where l is the length of the magnetic circuit in metres, A is the cross-section in square metres and $\mu_r\mu_0$ is the *permeability* of the material. The permeability is a property of the actual magnetic circuit and not only varies with the material in the circuit but also with the number of lines of force, i.e. *flux density*, actually induced in the material if that material is a ferromagnetic material (normally iron).

The actual flux induced in any circuit is proportional to

$$\text{the ratio } \frac{\text{m.m.f.}}{\text{reluctance}} \text{ and so we get}$$

$$\text{total flux} = \phi = \frac{\text{m.m.f.}}{S} \text{ Wb}$$

The relative permeability μ_r is always given as the ratio of the number of lines of force (flux density) induced in a circuit of any ferromagnetic material compared with the number of lines induced in free space for the same conditions. The permeability of free space, μ_0, to all intents and purposes can be considered to be the same as that of air and so permeability can be taken as the magnetic conductivity compared with air.

Taking the formula for total flux given above, we can combine this by substituting values for m.m.f. and S, giving

$$\text{Total flux, } \phi = \frac{\mu_r\mu_0 I N A}{l} \text{ Wb}$$

Having obtained the total flux, we can obtain the flux density or number of lines per square metre of cross-section as follows:

$$\text{Flux density} = B = \frac{\phi}{A} \text{ tesla (T)}$$

The tesla is one weber per square metre.

In many cases the magnetic circuit (*Figure 1.5*) will have an air-gap in order that the magnetic flux can be utilized, as, for example, in the rotating armature of a motor. It is usual, in such a case, to define the flux which can be utilized as the *useful flux*. In such a situation it will be found that there is always a certain amount of 'bulging' of the flux at the edges. There will also be many lines of force which will take shorter paths remote from the air-gap so that the actual flux in the air-gap will be smaller than that produced by the coil. The ratio between these two is given by the leakage coefficient which

$$= \frac{\text{flux in air-gap}}{\text{flux in iron}}$$

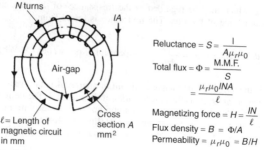

Figure 1.5 The magnetic circuit

Ampere-turns per metre (At/m). In order to deal with complex magnetic circuits such as generators, motors, etc., it is more convenient to take the various sections of the magnetic circuit separately, and for this purpose it is useful to have the ampere-turns required per metre to give a fixed flux density. Taking our complete formula above for total flux, we get

$$B = \frac{\phi}{A} = \mu_r \mu_0 \frac{IN}{l} = \mu_r \mu_0 H$$

so that the permeability and flux density are linked by the expression

$$\frac{IN}{l} = H$$

which is called the *magnetizing force* and it will be seen that this is equal to the ampere-turns per unit length (i.e. metre).

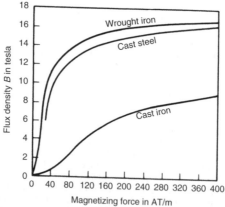

Figure 1.6 The B–H curve

The relation between B and H is usually given by means of a $B-H$ curve (*Figure 1.6*), but by using a different scale the actual value of ampere-turns per metre required can be read off. This scale is also shown in *Figure 1.6*.

Hysteresis. If a piece of iron is gradually magnetized and then slowly demagnetized it will be found that when the current is reduced to zero there is still some residual magnetism or remanence and the current has to be reversed to cancel the flux. This is shown in *Figure 1.7* where the complete curve of magnetization is shown by the circuit ABCDEF. This lagging of the flux behind the magnetizing force is termed *hysteresis* and during a complete *cycle* as shown by the figure ABCDEF energy is dissipated in the iron. Since this represents a loss to the system this is called the *hysteresis loss*. Frequency is expressed in hertz (Hz) so that 1 Hz = 1 cycle/second.

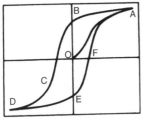

ABCDEF = Hysteresis loop
OB = Remanence
Hysteresis loss in joules/
cu cm/cycle and in watts/cycle
$W = nfB^{1.6} \times 10^{-1}$ per cu metre

Figure 1.7   Hysteresis loss

In an alternating current machine this loss is continuous and its value depends on the materials used.

$$\text{Watts loss per cubic metre} = k_1 f B_{\max}^n$$

where k_1 is a constant for any particular material. The exponent n is known as the *Steinmetz* or hysteresis exponent and is also specific for the material. Originally this was taken as 1.6 but with modern materials working at higher flux densities n can vary from 1.6 to 2.5 or higher. f is the frequency in Hz, and $B_{\max}$ is the maximum flux-density.

Almost all magnetic materials subjected to a cyclic pattern of magnetization around the hysteresis loop will also experience the flow of *eddy currents* which also result in losses. The magnitude of the eddy currents can be reduced by increasing the electrical resistance to their flow by making the magnetic circuit of thin laminations and also by the addition of silicon to the iron which increases its resistivity. The silicon also reduces the hysteresis loss by reducing the area of the hysteresis loop.

Eddy current loss is thus given by the expression

$$\text{Watts loss per cubic metre} = k_2 f^2 t^2 B_{\text{eff}}^2 / \rho$$

where k_2 is another constant for the material, t the thickness and ρ its resistivity. B_{eff} is the effective flux density which corresponds to its r.m.s. value (defined below).

When designing electrical machines it is more convenient to relate the magnetic circuit or *iron losses* to the weight of core iron used rather than its volume. This can be simply done by suitable adjustment of the constants k_1 and k_2. Typical values of combined hysteresis and eddy current losses can be from less than 1 to around 2 W/kg for modern laminations of around 0.3 mm thickness at a flux density of 1.6 tesla and a frequency of 50 Hz.

Magnetic paths in series. Where the magnetic path is made up of several different parts, the total reluctance of the circuit is obtained by adding the reluctance of the various sections. Taking the ring in *Figure 1.5* the total reluctance of this is found by calculating the reluctance of the iron part and adding the reluctance of the air-gap. The reluctance of the air-gap, of length l_0, will be given by

$$\frac{l_0}{\mu_0 A}$$

The value of $\mu_0 = 4\pi \times 10^{-7}$ H/m.

A.C. Theory

Alternating currents. Modern alternators produce an e.m.f. which is for all practical purposes sinusoidal (i.e. a sine curve), the equation between the e.m.f. and time being

$$e = E_{\max} \sin \omega t$$

where

e = instantaneous voltage
$E_{\max}$ = maximum voltage
ωt = angle through which the armature has turned from the neutral axis.

Taking the frequency as f hertz, the value of ω will be $2\pi f$, so that the equation reads

$$e = E_{\max} \sin(2\pi f)t$$

The graph of the voltage will be as shown in *Figure 1.8*.

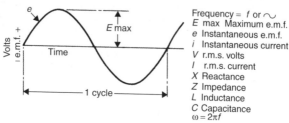

Frequency = f or $\sim$
E max Maximum e.m.f.
e Instantaneous e.m.f.
i Instantaneous current
V r.m.s. volts
I r.m.s. current
X Reactance
Z Impedance
L Inductance
C Capacitance
$\omega = 2\pi f$

Figure 1.8

Because current is generally proportional to voltage (see below) the current will also generally be sinusoidal and of the form

$$i = I_{\max} \sin[(2\pi f)t + \phi]$$

The constant ϕ represents an angular displacement between current and voltage and is further explained below.

Average or mean value. The average value of the voltage and current will be found to be 0.636 of the maximum value for a perfect sine wave, giving the equations

$$E_{ave} = 0.636E_{max} \quad \text{and} \quad I_{ave} = 0.636I_{max}$$

The mean values are only of use in connection with processes where the results depend on the current only, irrespective of the voltage, such as electroplating or battery-charging.

R.m.s. (root-mean-square) value. The values which are relevant in any circumstances involving power

$$E_{r.m.s.} = E_{max} \times \frac{1}{\sqrt{2}} = 0.707E_{max}$$

and $\quad I_{r.m.s.} = I_{max} \times \frac{1}{\sqrt{2}} = 0.707I_{max}$

are the r.m.s. values. These values are obtained by finding the square root of the mean value of the squared ordinates for a cycle or half-cycle. (See *Figure 1.8.*)

These are the values which are used for all power, lighting and heating purposes, as in these cases the power is proportional to the square of the voltage or current.

A.C. circuits

Resistance. Where a sinusoidal e.m.f. is placed across a pure resistance the current will be in phase with the e.m.f., and if shown graphically will be in phase with the e.m.f. curve (i.e. the value of ϕ in the expression above will be zero).

The current will follow Ohm's law for d.c., i.e.

$$I = V/R$$

where V is the r.m.s. value of the applied e.m.f. or voltage, and R is the resistance in ohms – the value of I will be the r.m.s. value. (See *Figure 1.9.*)

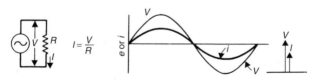

Figure 1.9

Inductance. If a sinusoidal e.m.f. is placed across a pure inductance the current will be found to be $I = V/[(2\pi f)L]$ where V is the voltage (r.m.s. value), f is the frequency and L the inductance henries, the value of I being the r.m.s. value. The current will lag behind the voltage and the graphs will

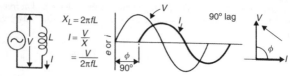

Figure 1.10

be as shown in *Figure 1.10*, the phase difference being 90° ($\phi = -90°$). The expression $(2\pi f)L$ is termed the inductive reactance (X_L).

Capacitance. If a sinusoidal e.m.f. is placed across a capacitor the current will be $I = (2\pi f)$. CV, where C is the capacitance in farads, the other values being as above. In this case the current leads the voltage by 90° ($\phi = +90°$), as shown in *Figure 1.11*. The expression $1/[(2\pi f)C]$ is termed the capacitive reactance (X_C) and the current is given by

$$I = \frac{V}{X_C}$$

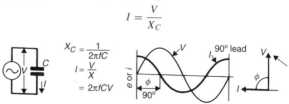

Figure 1.11

Resistance and inductance in series. In this circuit, shown in *Figure 1.12*, the current will be given by

$$I = \frac{V}{\sqrt{R^2 + X_L^2}}$$

where X_L is the reactance of the inductance ($X_L = (2\pi f)L$). The expression $\sqrt{R^2 + X_L^2}$ is called the impedance (Z), so that $I = V/Z$. The current will lag behind the voltage, but the angle of lag, ϕ, will depend on the relative values of R and X_L – the angle being such that $\tan \phi = X_L/R$ (ϕ being the angle as shown in *Figure 1.12a*).

Resistance and capacitance in series. For this circuit the current will be given by

$$I = \frac{V}{\sqrt{R^2 + X_C^2}}$$

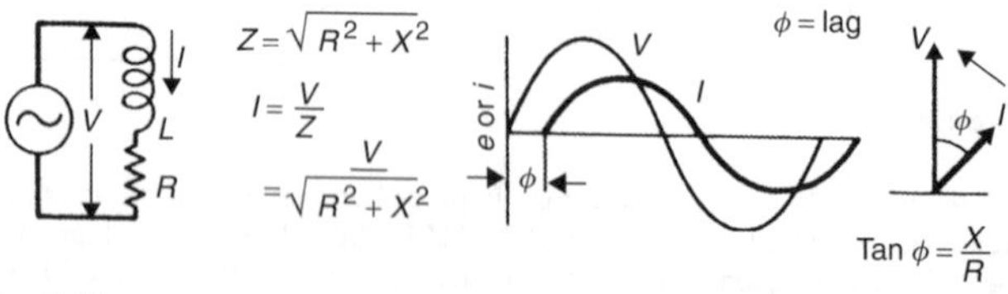

Figure 1.12a

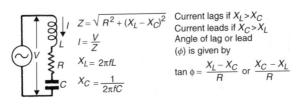

Figure 1.12b

where X_C is the reactance of the capacitance $(1/(2\pi f C))$. The current will lead the voltage and the angle of lead will be given by $\tan\phi = X_C/R$.

Resistance, inductance and capacitance in series. The impedance (Z) of this circuit will be $Z = \sqrt{R^2 + (X_L - X_C)^2}$ with $I = V/Z$, and the phase difference will be either

$$\tan\phi = \frac{X_L - X_C}{R} \quad \text{or} \quad \frac{X_C - X_L}{R}$$

whichever is the higher value. (Here $X_L =$ inductive reactance and $X_C =$ capacitance reactance.)

The IEC recommendation is that the terms inductance and capacitance can be dropped when referring to reactance, provided that reactance due to inductance is reckoned positive and to capacitance, negative.

Currents in parallel circuits. The current in each branch is calculated separately by considering each branch as a simple circuit. The branch currents are then added *vectorially* to obtain the supply current by the following method:

Resolve each branch-current vector into components along axes at right angles (see *Figure 1.13*), one axis containing the vector of the supply e.m.f. This axis is called the *in-phase* axis; the other axis at $90°$ is called the *quadrature* axis. Then the supply current is equal to

$$\sqrt{\begin{array}{c}(\text{sum of in-phase components})^2 \\ + (\text{sum of quadrature components})^2\end{array}}$$

and

$$\cos\phi = \frac{\text{sum of in-phase components}}{\text{supply current}}$$

Thus if $I_1, I_2, \ldots$, denote the branch-circuit currents, and $\phi_1, \phi_2, \ldots$, their phase differences, the in-phase components are $I_1\cos\phi_1$, $I_2\cos\phi_2$, etc. and the quadrature components are $I_1\sin\phi_1$, $I_2\sin\phi_2$, etc. Hence the line or supply current is

$$I = \sqrt{(I_1\cos\phi_1 + I_2\cos\phi_2 + \cdots)^2 + (I_1\sin\phi_1 + I_2\sin\phi_2 + \cdots)^2}$$

and

$$\cos\phi = \frac{I_1\cos\phi_1 + I_2\cos\phi_2 + \cdots}{I}$$

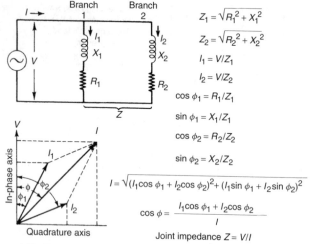

Figure 1.13 Parallel circuits

The quantities $\cos\phi_1$, $\sin\phi_1$, etc., can be obtained from the general formulae: $\cos\phi$ = resistance/impedance, and $\sin\phi$ = reactance/impedance, or $\sin\phi = [I_1\sin\phi_1 + I_2\sin\phi_2]/I$.

The equivalent impedance of the circuit is obtained by dividing the line current into the line voltage.

If the equivalent resistance and reactance of this impedance are required, they can be calculated by the formulae:

Equivalent resistance of parallel circuits = Impedance $\times \cos\phi$

Equivalent reactance of parallel circuits

$$= \sqrt{(\text{impedance})^2 - (\text{resistance})^2} = \text{impedance} \times \sin\phi$$

Current in a series-parallel circuit. The first step is to calculate the joint impedance of the parallel portion of the circuit. (*Figure 1.14*). The easiest way of doing this is to calculate the branch currents, the joint impedance, and the equivalent resistance and reactance exactly as for a simple parallel circuit. The calculations can be made without a knowledge of the voltage across the parallel portion of the circuit (which is unknown at the present stage) by assuming a value, V_1.

Having obtained the joint impedance (Z_E) of the parallel portion of the circuit, this is added vectorially to the series impedance (Z_S) to obtain the joint impedance (Z) of the whole circuit, whence the current is readily obtained in the usual manner. Thus, the joint impedance of the parallel circuits (Z_E) must be split into resistance and reactance, i.e. $R_E = Z_E\cos\phi_E$ and $X_E = Z_E\sin\phi_E$, where ϕ_E is the phase difference. This resistance and reactance is added to the resistance and reactance of the series portion of the circuit, in order to calculate the joint impedance of the whole circuit.

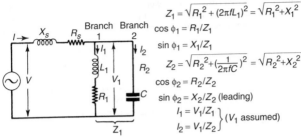

$$Z_1 = \sqrt{R_1^2 + (2\pi f L_1)^2} = \sqrt{R_1^2 + X_1^2}$$

$$\cos\phi_1 = R_1/Z_1$$

$$\sin\phi_1 = X_1/Z_1$$

$$Z_2 = \sqrt{R_2^2 + \left(\frac{1}{2\pi f C}\right)^2} = \sqrt{R_2^2 + X_2^2}$$

$$\cos\phi_2 = R_2/Z_2$$

$$\sin\phi_2 = X_2/Z_2 \text{ (leading)}$$

$$\left.\begin{array}{l} I_1 = V_1/Z_1 \\ I_2 = V_1/Z_2 \end{array}\right\} (V_1 \text{ assumed})$$

Line current I (in terms of V_1) and phase difference of parallel circuits, $\cos\phi$, as *Figure 1.13*

Joint impedance $Z_E = V_1/I$

Equivalent resistance $R_E = Z_E \cos\phi_1$

Equivalent reactance $X_E = Z_E \sin\phi$

Joint impedance of whole circuit Z

$$= \sqrt{(R_E + R_S)^2 + (X_E + X_S)^2}$$

Line current $I = V/Z$

Figure 1.14 Series-parallel circuits

Thus, the resistance term (R) of the joint impedance (Z) of the series-parallel circuit is equal to the sum of the resistance terms ($R_E + R_S$) of the separate impedances. Similarly, the reactance term (X) is equal to the sum of the reactance terms of the separate impedances, i.e. $X_E + X_S$. Hence the joint impedance of the series-parallel circuit is

$$Z = \sqrt{R^2 + X^2}$$

and line current $= V/Z$

Three-phase circuits. Three-phase currents are determined by considering each phase separately, and calculating the phase currents from the phase voltages and impedances in the same manner as for single-phase circuits. In practice, three-phase systems are usually symmetrical, the loads being balanced. In such cases the calculations are simple and straightforward. For the methods of calculations when the loads are unbalanced or the system is unsymmetrical reference should be made to the larger textbooks.

Having calculated the phase currents, the line currents are obtained from the following simple rules.

With a *star-connected* system:

$$\text{line current} = \text{phase current}$$

$$\text{line voltage} = 1.73 \times \text{phase voltage}$$

With a *delta-connected* system:

$$\text{line current} = 1.73 \times \text{phase current}$$

$$\text{line voltage} = \text{phase voltage}$$

Power in a.c. circuits. The power in a single-phase circuit is given by $W = VI\cos\phi$, where W is the power in watts, V the voltage (r.m.s.) and I the current (r.m.s.). Cos ϕ represents the power factor of the circuit, so that

$$\text{power factor} = \cos\phi = \frac{W}{VI} = \frac{\text{watts}}{\text{volt-amperes}}$$

Referring to *Figure 1.15*, I represents a current lagging by angle ϕ. This current can be split into two components, OW, the energy component, and OR, the wattless component. Only the energy component has any power value, so that the power is given by $\text{OV} \times \text{OW} = \text{OV} \times \text{OI}\cos\phi = VI\cos\phi$.

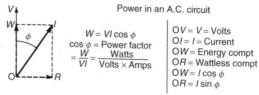

Power in an A.C. circuit

$W = VI\cos\phi$
$\cos\phi = \text{Power factor}$
$= \dfrac{W}{VI} = \dfrac{\text{Watts}}{\text{Volts} \times \text{Amps}}$

$OV = V = \text{Volts}$
$OI = I = \text{Current}$
$OW = \text{Energy compt}$
$OR = \text{Wattless compt}$
$OW = I\cos\phi$
$OR = I\sin\phi$

Figure 1.15

Three-phase working. The three windings of a three-phase alternator or transformer can be connected in two ways, as shown in *Figure 1.16*. The relations between the phase voltages and currents and the line voltages and currents are indicated in this diagram. It should be noted that with the star or Y connection a neutral point is available, whereas with the delta or mesh connection this is not so. Generators are generally star wound and the neutral point used for earthing. Motors can be either star or delta, but for low voltage small-size motors a delta connection is usually used to reduce the size of the windings.

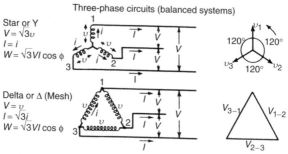

Three-phase circuits (balanced systems)

Star or Y
$V = \sqrt{3}v$
$I = i$
$W = \sqrt{3}VI\cos\phi$

Delta or Δ (Mesh)
$V = v$
$I = \sqrt{3}i$
$W = \sqrt{3}VI\cos\phi$

Figure 1.16

Power in a three-phase circuit. The total power in a three-phase circuit is the sum of the power in the three phases. Taking the star system in *Figure 1.17* and assuming a balanced system (i.e. one in which the three voltages and currents are all equal and symmetrical), the total power must by $3 \times$ power

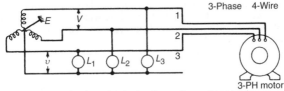

$L_1 L_2 L_3$ are single-phase loads at voltage v three-phase loads are taken from lines 1, 2 and 3 at voltage V. Note:- $V = \sqrt{3}v$

Figure 1.17

per phase. Therefore $W = 3vi \cos \phi$. Substituting the line values for phase volts and phase current, we get

$$W = 3(vi)\cos \phi = 3\left(\frac{V}{\sqrt{3}} \times I\right)\cos \phi = \sqrt{3}VI\cos \phi$$

It will be found that the same expression gives the power in a delta-connected system, and so for any balanced system the power is given by $W = \sqrt{3}VI\cos \phi$, where V and I are the line volts and line current and cos ϕ represents the power factor. For unbalanced or unsymmetrical systems the above expression does not hold good.

(Most three-phase apparatus such as motors can be assumed to form a balanced load, and calculations for current, etc., can be based on this assumption, using the above expression.)

Power in a three-phase circuit can be measured in several ways. For permanent switchboard work a three-phase wattmeter unit is used in which there are usually two elements, so that the meter will indicate both balanced and unbalanced loads. For temporary investigations either of the methods shown in *Figure 1.18* can be used. The total power = $3W$ where W is the reading on the single meter.

Power in 3-phase

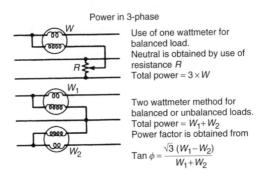

Use of one wattmeter for balanced load.
Neutral is obtained by use of resistance R
Total power = $3 \times W$

Two wattmeter method for balanced or unbalanced loads.
Total power = $W_1 + W_2$
Power factor is obtained from

$$\text{Tan } \phi = \frac{\sqrt{3}\,(W_1 - W_2)}{W_1 + W_2}$$

Figure 1.18

For an unbalanced load two units must be used, and these are connected as indicated in *Figure 1.18*. In addition to giving the total power by adding the readings on the two meters, the power factor can be obtained. It is important to note, however, that the reading of one meter will be reversed if the power factor of the system is less than 0.5. In this case the leads of one of the meters may have to be reversed in order to get a positive reading. For power factors of less than 0.5 the readings must be *subtracted* instead of added.

The power factor of the system can be obtained from

$$\tan \phi = \frac{\sqrt{3}(W_1 - W_2)}{(W_1 + W_2)}$$

which gives the tangent of the angle of lag, and the cosine can be obtained from the tables.

Power in six-phase. In a six-phase system, such as is often used for rotary converters and other rectifiers, the power of the system (assumed balanced) is given by

$$W = 6VI \cos \phi$$

where V is the phase voltage and I the phase current.

In terms of line voltage V_L and line current I_L the power equation becomes

$$\frac{3}{\sqrt{2}} V_L I_L \cos \phi$$

In both cases $\cos \phi$ is the phase angle between the phase voltage and phase current.

Three-phase 4-wire. This system (*Figure 1.16*) is now used almost universally in the UK for 400 V distribution. There are three 'lines' and a neutral.

The voltage between any one 'line' and neutral is nominally 230 V and voltage between the 'lines' is $\sqrt{3}$ times the voltage to neutral. This gives a three-phase voltage of 400 V for motors, etc. Single-phase loads are therefore taken from all 'lines' to neutral and three-phase loads from the three lines marked 1, 2 and 3. (It should be noted that the above nominal voltages are the values that have been adopted in the UK since January, 1995, in place of the values of 415 V, three-phase, and 240 V, single-phase, used previously. This is as a result of an EU Directive on voltage harmonisation – see Chapter 12).

In the distribution cable the neutral may be either equal to the 'lines' or half-size. Modern systems generally use a full-size neutral, particularly where fluorescent lighting loads predominate.

2 Properties of materials

Magnetic Materials

Low carbon steels. Low carbon steel provides the path for the magnetic flux in most electrical machines: generators, transformers and motors. Low carbon steel is used because of its high permeability, that is, a large amount of flux can be produced with the expenditure of minimal magnetizing 'effort', and it has low hysteresis thus minimizing losses associated with the magnetic field.

High levels of flux mean more powerful machines can be produced for a given size and weight.

Alternating current machines not only experience iron loss due to hysteresis, as explained in the previous chapter, but they also have losses due to circulating currents, known as eddy currents, which flow within the iron of the core. These two types of losses are present whenever a machine is energized, whether on load or not, and are together known as the no-load losses of the machine. It has been estimated that in 1987/88 the cost of no-load core losses in transformers in operation in the UK alone was £110 million. There is thus a very strong incentive to reduce this loss.

The first machines produced in the 1880s used cores made of high-grade wrought iron but around 1900 it was recognized that the addition of small amounts of silicon or aluminium greatly reduced the magnetic losses. Thus began the technology of specialized electrical steel making.

The addition of silicon reduces hysteresis, increases permeability and also increases resistivity, thus also reducing eddy current losses. It has the disadvantage that the steel becomes brittle and hard so that to retain sufficient workability for ease of core manufacture, the quantity added must be limited to about $4\frac{1}{2}\%$.

Increasing resistivity alone does not sufficiently reduce eddy currents so that it is necessary to build up the core from *laminations*. These are sheets around 0.3 mm thick, lightly insulated from each other. This greatly reduces the cross-section of the iron in the direction in which the eddy currents flow. The resistance of the eddy current path is thus increased still further. This will be explained by reference to *Figure 2.1*.

Hot-rolled steel. Electrical sheet steels from which the laminations are cut are produced by a process of rolling in the steel mill. The steels have a crystalline structure and the magnetic properties of the sheet are derived from the magnetic properties of the individual crystals or *grains*. The grains themselves are *anisotropic*. That is, their properties differ according to the direction along the crystal that these are measured. Until the 1940s the sheet steels were produced by a process of *hot-rolling* in which the grains are packed together in a random way so that the magnetic properties of the sheet have similar values regardless of the direction in which they are measured. These represent the average properties for all directions within the individual crystals. The sheet steel is therefore *isotropic*.

Grain-oriented steel. As early as the 1920s it had been recognized that if the individual steel crystals could be aligned, a steel could be produced which,

21

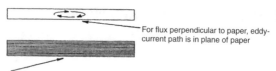

For flux perpendicular to paper, eddy-current path is in plane of paper

Laminations reduce cross-section of eddy-current path

Figure 2.1 Building core from laminations increases resistance to the flow of eddy currents. The thinner the laminations the more effective this is

in one direction, would exhibit properties related to the optimum magnetic properties of the crystals. It was not until the mid-1930s that the American, N.P. Goss developed and patented an industrial process that did this. This material is known as *cold-rolled grain-oriented steel*. It is reduced in the steel mill by a hot rolling process until it is about 2 mm thick. Thereafter it is further reduced by a series of cold reductions interspersed with annealing at around 900°C to around 0.3 mm final thickness.

In order to reduce surface oxidation and prevent the material sticking to the rolls, the steel is given a phosphate coating in the mill. This coating has a sufficiently high resistance to serve as insulation between laminations in many instances but it must generally be made good by recoating with varnish where edge burrs produced by cutting have been ground off.

Grain-oriented steel has magnetic properties in the rolling direction which are very much superior to those perpendicular to the rolling direction. To obtain maximum benefit from its use, therefore, it must be used in a machine in which the flux passes along the length of the material. This is particularly so in the case of a transformer in which the flux passes axially along the leg as shown in *Figure 2.2*. Of course the flux must cross the line of the grains at the top and bottom of the core legs where these join the yokes. As can be seen from *Figure 2.2(b)*, crossing of the grain pattern can be minimized by utilizing mitred joints at these points. Before the introduction of cold-rolled steel, leg-to-yoke joints could be simply overlapped as shown in *Figure 2.2(c)*.

High permeability steel. Cold-rolled steel as described above continued to be steadily improved until the end of the 1960s when a further step-change was introduced by the Nippon Steel Corporation of Japan. By introducing significant changes into the cold rolling process they achieved a considerable improvement in the degree of grain orientation compared with the previous grain-oriented material (most grains aligned within 3° of the ideal compared with 6° obtained previously). The steel also has a very much improved glass coating. This coating imparts a tensile stress into the steel which has the effect of reducing hysteresis loss. The reduced hysteresis loss allows some reduction in the amount of silicon which improves the workability of the material, reducing cutting burrs and avoiding the need for these to be ground off. This coupled with the better insulation properties of the coating means that additional insulation is not required. The core manufacturing process is simplified and the core itself has a better stacking factor.

Domain-refined steel. Crystals of grain-oriented steel become aligned during the grain-orientation process in large groups. These are known as *domains*. There is a portion of the core loss which is related to the size of the domains so that this can be reduced by reducing the domain size. Domain

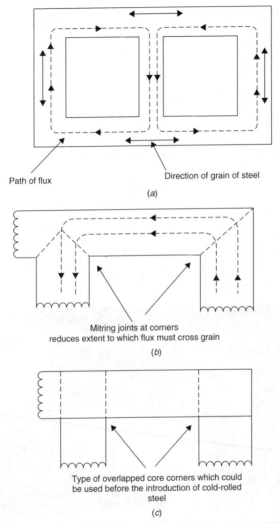

Path of flux

Direction of grain of steel

(a)

Mitring joints at corners
reduces extent to which flux must cross grain

(b)

Type of overlapped core corners which could
be used before the introduction of cold-rolled
steel

(c)

Figure 2.2 Mitring joints at intersection of limbs and yokes reduces extent of flux crossing grain orientation

size can be reduced after cold rolling by introducing a small amount of stress into the material. This is generally carried out by a process of laser etching so that this type of steel is frequently referred to as laser-etched. Improvements to the rolling process have also enabled this material to be produced in thinner sheets, down to 0.23 mm, with resulting further reduction in eddy-current loss.

Amorphous steel. Amorphous steels have developed in a totally different direction to the silicon steels described above. They were originally developed by Allied Signal Inc. Metglas Products in the USA in the early 1970s as an alternative for the steel in vehicle tyre reinforcement. It was not until the mid-1970s that the importance of their magnetic properties was recognized. Their introduction on a commercial scale is still restricted some 25 years later due to the difficulties in production and handling. Nevertheless amorphous steels offer considerable reduction in losses compared with even the best conventional steel.

Amorphous steels have a non-crystalline structure. The atoms are randomly distributed within the material. They are produced by very rapid cooling of the molten alloy which contains about 20% of a glass forming element such as boron. The material is generally produced by spraying a stream of molten alloy onto a rapidly rotating copper drum. The molten material is cooled at the rate of about 10^6 degrees C per second and solidifies to form a continuous thin ribbon. This requires annealing between 200 and 280°C to develop the required magnetic properties. Earliest quantities of the material were only 2 mm wide and about 0.025–0.05 mm thick. By the mid-1990s a number of organizations had been successful in producing strip up to 200 mm wide.

By the end of the 1980s the original developers of the material had been successful in producing a consolidated strip which could be fairly successfully built into distribution transformer sized cores. This has found more widespread use in the USA than in the UK. *Figure 2.3* shows an experimental distribution transformer manufactured in the UK using amorphous steel.

Figure 2.3 Core and windings of 200 kVA, 20/0.4 kV transformer using amorphous steel. Unfortunately very little of the core is visible, but it should be just apparent that this is of the wound construction. It will also be apparent that fairly elaborate clamping was considered necessary and that the physical size, for a 200 kVA transformer, is quite large (Alstom T&D)

Designation of core steels. Specification of magnetic materials including core steels is covered internationally by IEC 60404. This is a multi-part document covering all aspects and types of magnetic materials used in the electrical industry. In the UK this becomes BS IEC 60404-1 *Magnetic materials. Classification.* BS EN 60404 Parts 2 and 4, relating to methods of measurement of magnetic properties, have been accepted as European norms.

Permanent magnets (cast). Great advances have been made in the development of materials suitable for the production of permanent magnets. The earliest materials were tungsten and chromium steel, followed by the series of cobalt steels.

Alni was the first of the aluminium-nickel-iron alloys to be discovered and with the addition of cobalt, titanium and niobium, the Alnico series of magnets was developed, the properties of which varied according to composition. These are hard and brittle, and can only be shaped by grinding, although a certain amount of drilling is possible on certain compositions after special heat treatment. The Permanent Magnet Association (disbanded March 1975) discovered that certain alloys when heat-treated in a strong magnetic field became anisotropic. That is they develop high properties in the direction of the field at the expense of properties in other directions. This discovery led to the powerful Alcomax and Rycomax series of magnets. By using special casting techniques to give a grain-oriented structure, even better properties are obtained if the field applied during heat treatment is parallel to the columnar crystals in the magnet.

Permanent magnets (sintered). The techniques of powder metallurgy have been applied to both the isotropic and anisotropic Alnico types and it is possible to produce sintered permanent magnets which have approximately 10% poorer remanence and energy than cast magnets. More precise shapes are possible when using this method of production and it is economical for the production of large quantities of small magnets. Sintering techniques are also used to manufacture the oxide permanent magnets based on barium or strontium hexaferrite. These magnets which may be isotropic or anisotropic, have higher coercive force but lower remanence than the alloy magnets described above. They have the physical properties of ceramics, and inferior temperature stability, but their low cost makes them ideal for certain applications. Barium ferrite bonded in rubber or plastics is available as extruded strip or rolled sheet.

The newest and most powerful permanent magnets discovered to date, based on an intermetallic compound of cobalt and samarium, are also made by powder metallurgy techniques (*Table 2.1*).

Nickel-iron alloys. Nickel-iron alloy containing about 25% of nickel is practically non-magnetic, but with increased nickel content and suitable treatment some remarkably high permeability materials have been obtained. Some of the more popular alloys and their magnetic properties are shown in *Tables 2.2(a)* and *2.2(b)*.

From these tables it will be seen that there are two groups falling within the range 36–50%. The alloys with the higher nickel content have higher initial and maximum permeabilities but lower saturation inductions, remanence and coercivity.

Typical applications for these nickel-iron alloys are detailed in *Table 2.3*. From this table it will be seen that the materials are particularly suitable for high frequency applications.

Table 2.1 Properties of permanent magnets*

Material	Remanence T	Coercive force kAm^{-1}	BH_{max} kJm^{-3}	Sp. Gr.	Description
ISOTROPIC					
Tungsten steel 6%W	1.05	5.2	2.4	8.1	Rolled or forged steel
Chromium steel 6%Cr	0.98	5.2	2.4	7.8	Rolled or forged steel
Cobalt steel 3%Co	0.72	10.4	2.8	7.7	Rolled or forged steel
Cobalt steel 6%Co	0.75	11.6	3.5	7.8	Rolled or forged steel
Cobalt steel 9%Co	0.78	12.8	4.0	7.8	Rolled or forged steel
Cobalt steel 15%Co	0.82	14.4	5.0	7.9	Rolled or forged steel
Cobalt steel 35%Co	0.90	20	7.6	8.2	Rolled or forged steel
Alni	0.55	38.5	10	6.9	Cast Fe-Ni-Al
Alnico	0.75	58	13.5	7.3	Cast Fe-Ni-Al
Feroba 1 (sintered)	0.21	136	6.4	4.8	Barium ferrite
Bonded Feroba	0.17	128	5.6	3.6	Flexible strip or sheet
ANISOTROPIC					
Alcomax II	1.20	46	41	7.35	Cast Fe-Co-Ni-Al
Alcomax III	1.30	52	44	7.35	Cast Fe-Co-Ni-Al-Nb
Alcomax IV	1.15	62	36	7.35	Cast Fe-Co-Ni-Al-Nb
Columax	1.35	59	60	7.35	Grain oriented Alcomax III
Hycomax II	0.75	96	32	7.3	Cast Fe-Co-Ni-Al-Nb-Ti
Hycomax III	0.92	132	44	7.3	Cast Fe-Co-Ni-Al-Ti
Hycomax IV	0.78	160	46	7.3	Cast Fe-Co-Ni-Al-Ti
Columnar Hycomax III	1.05	128	72	7.3	Grain oriented
Feroba II	0.35	144	26.4	5.0	Barium ferrite
Feroba III	0.25	200	20	4.7	Barium ferrite
Sintered Sm Co$_5$	0.80	600	128	8.1	Cobalt-samarium

Permanent Magnets, by Malcolm McCaig, Pentech Press, 1977.

Table 2.2(a) Properties of high-permeability nickel-iron alloys (75–80% Ni-Fe alloys)

Property	Mumetal*	Nilomag*	Permalloy C*
Initial permeability	60 000	50 000	50 000
Maximum permeability	240 000	250 000	250 000
Saturation induction B_{sat} Tesla	0.77	0.70	0.80
Remanence B_{rem} Tesla	0.45	0.40	0.35
Coercivity H_c (A/m)	1.00	1.60	2.4

*Grades of higher magnetic quality, Mumetal plus Supermetal, Permalloy 'Super C' and Nilomag 771 are available.

Table 2.2(b) Properties of high-permeability nickel-iron alloys (36–50% Ni-Fe alloys)

Property	Radiometal* 50	Permalloy† B	Nilo alloy 45	Radiometal 36	Nilo alloy 36
Initial permeability	6 000	5 000	6 000	3 000	4 000
Maximum permeability	30 000	30 000	30 000	20 000	18 000
Saturation induction B_{sat} Tesla	1.6	1.6	1.2	1.2	0.8
Remanence B_{rem} Tesla	1.0	0.4	1.1	0.5	0.4
Coercivity H_c (A/m)	8.0	12.0	16.0	10.0	10.0

*Super, Hyrno and Hyrem Radiometal are derived from Radiometal 50 and offer improved permeability, electrical resistivity and remanence respectively.
†Permalloys D and F offer higher electrical and remanence respectively.
Radiometal and Mumetal are trade names are Telecon Metals Ltd; Nilo and Nilomag are trade names of Henry Wiggin Ltd; Permalloy is a trade name of Standard Telephones and Cables Ltd.

Copper and its Alloys

The electrical resistance of copper, as of all other pure metals, varies with the temperature. This variation is sufficient to reduce the conductivity of high conductivity copper at 100°C to about 76% of its value at 20°C.

$$\text{The resistance } R'_t = R_t[1 + \alpha_t(t' - t)]$$

where α_t is the constant mass temperature coefficient of resistance of copper at the reference t°C. For a reference temperature of 0°C the formula becomes

$$R_t = R_0(1 + \alpha_0 t)$$

Although resistance may be regarded for all practical purposes as a linear function of temperature, the value of the temperature coefficient is not constant

Table 2.3 Typical applications for high-permeability nickel-iron materials

Applications	% Nickel		
	75–80	45–50	36
Transformer			
Pulse	x	x	x
Audio	x	x	
Microphone	x		
Current	x		
Output		x	
Small power		x	
High frequency		x	x
Magnetic amplifiers	x	x	
Magnetic screening	x	x	
Tape recorder heads		x	
Relays			
Cores and armatures		x	x
Small motors, synchros, rotors and stators		x	
Inductors, chokes (h.f.) and filter circuits	x		

but is dependent upon, and varies with, the reference temperature according to the law

$$\alpha_t = \frac{1}{\dfrac{1}{\alpha_0} + t} = \frac{1}{234.45 + t}$$

Thus the constant mass temperature coefficient of copper referred to a basic temperature of 0°C is

$$\alpha_0 = \frac{1}{234.45} = 0.004265 \text{ per degree C}$$

At 20°C the value of the constant mass temperature coefficient of resistance is

$$\alpha_{20} = \frac{1}{234.45 + 20} = 0.00393 \text{ per degree C}$$

which is the value adopted by the IEC.

Multiplier constants and their reciprocals, correlating the resistance of copper at a standard temperature, with the resistance at other temperatures, may be obtained from tables which are included in BS 1432–1434, 4109, 7884.

Five alloys discussed below also find wide application in the electrical industry where high electrical conductivity is required. These are cadmium copper, chromium copper, silver copper, tellurium copper and sulphur copper. They are obtainable in wrought forms and also, particularly for chromium copper, tellurium copper and sulphur copper, as castings and forgings. The electrical resistivity varies from 1.71 microhm cm for silver copper in the annealed state at 20°C to 4.9 microhm cm for solution heat-treated chromium copper at the same temperature.

The main output of each alloy is determined by its major applications. For instance cadmium copper is produced as heavy gauge wire of special sections while silver copper is made generally in the form of drawn sections and strip. Much chromium copper is produced as bar and also as castings and forgings, though strip and wire forms are available.

Quantities of the five elements required to confer the differing properties on these alloys are quite small, the normal commercial ranges being: cadmium copper 0.7–1.0% cadmium; chromium copper 0.4–0.8% chromium; silver copper 0.03–0.1% silver; tellurium copper 0.3–0.7% tellurium; and sulphur copper 0.3–0.6% sulphur.

Cadmium copper, chromium copper and sulphur copper are deoxidized alloys containing small controlled amounts of deoxidant. Silver copper, like high-conductivity copper, can be 'tough pitch' (oxygen-containing) or oxygen-free, while tellurium copper may be either tough-pitch or deoxidized. 'Tough pitch' coppers and alloys become embrittled at elevated temperatures in a reducing atmosphere. Thus, when such conditions are likely to be encountered, oxygen-free or deoxidized materials should be used. Advice can be sought from the Copper Development Association which assisted in the compilation of these notes.

Cadmium copper. This material is characterized by having greater strength under both static and alternating stresses and better resistance to wear than ordinary copper. As such it is particularly suitable for the contact wires of electric railways, tramways, trolley-buses, gantry cranes and similar equipment. It is also employed for telephone wires and overhead transmission lines of long span.

Because cadmium copper retains the hardness and strength imparted by cold work at temperatures well above those at which high-conductivity copper would soften it has another field of application. Examples are electrode holders for resistance welding machines and arc furnaces, and electrodes for spot and seam welding of steel. Cadmium copper has also been employed for the commutator bars of certain types of electric motors.

Because of its comparatively high elastic limit in the work-hardened condition, cadmium copper is also used to a limited extent for small springs required to carry current. In the form of thin hard-rolled strip an important use is for reinforcing the lead sheaths of cables which operate under internal pressure. Castings of cadmium copper, though rare, do have certain applications for switchgear components and the secondaries of transformers for welding machines.

On exposure to atmosphere the material acquires the normal protective patina associated with copper. Cadmium copper can be soft soldered, silver soldered and brazed in the same manner as ordinary copper. Being a deoxidized material there is no risk of embrittlement by reducing gases during such processes.

Chromium copper. Chromium copper is particularly suitable for applications in which considerably higher strengths than that of plain copper are required. For example, for both spot and seam types of welding electrodes. Strip, and, to a lesser extent, wire are used for light springs destined to carry current. Commutator segments that are required to operate at temperatures above those normally encountered in rotating machines are another application. In its heat-treated state, the material can be used at temperatures up to about 350°C without risk of deterioration of properties.

In the solution heat-treated condition chromium copper is soft and can be machined. It is not difficult to cut in the hardened state but is not free-machining like leaded brass or tellurium copper. Chromium copper is similar to ordinary copper in respect of oxidation and scaling at elevated temperatures. Jointing methods similar to cadmium copper outlined above are applicable. As in the case of cadmium copper special fluxes are required under certain conditions, and these should contain fluorides. Chromium copper can be welded using modern gas-shielded arc-welding technology.

Silver copper. Silver copper has an electrical conductivity equal to that of ordinary high-conductivity copper, but in addition, it possesses two properties which are of practical importance. Its softening temperature, after hardening by cold work, is considerably higher than that of ordinary copper, and its resistance to creep at moderately elevated temperatures is enhanced.

The principal uses of this material are in connection with electrical machines which either run at higher than normal temperatures or are exposed to them during manufacture. Soft soldering or stoving of insulating materials are examples of the latter.

Silver copper is obtainable in the form of hard drawn or rolled rods and sections, especially those designed for commutator segments, rotor bars and similar applications. It is also available as hollow conductors and strip. It is rarely called for in the annealed condition since its outstanding property is associated with retention of work hardness at elevated temperatures.

Silver copper can be soft soldered, silver soldered, brazed or welded without difficulty but the temperatures involved in all these processes, except soft soldering, are sufficient to anneal the material if in the cold-worked condition. Because the tough pitch material contains oxygen in the form of dispersed particles of cuprous oxide, it is important to avoid heating to brazing and welding temperatures in a reducing atmosphere.

While silver copper cannot be regarded as a free-cutting material, it is not difficult to machine. This is specially true when it is in the work-hardened condition, the state in which it is usually supplied. It is similar to ordinary copper in its resistance to corrosion. If corrosive fluxes are employed for soldering, the residues should be carefully washed away after soldering is completed.

Tellurium copper. Special features of this material are ease of machining combined with high electrical conductivity, retention of work hardening at moderately elevated temperatures and good resistance to corrosion. Tellurium copper is unsuitable for welding with most procedures, but gas-shielded arc welding and resistance welding can be effected with care. A typical application of this material is for magnetron bodies, which in many cases are machined from solid blocks of the material.

Tellurium copper can be soft soldered, silver soldered and brazed without difficulty. For tough pitch, tellurium copper brazing should be carried out in an inert atmosphere (or slightly oxidizing) since reducing atmospheres are conducive to embrittlement. Deoxidized tellurium copper is not subject to embrittlement.

Sulphur copper. Like tellurium copper, sulphur copper is a high-conductivity free-machining alloy, with greater resistance to softening than high-conductivity copper at moderately elevated temperatures, and with good resistance to corrosion. It is equivalent in machinability to tellurium copper, but without the tendency shown by the latter to form coarse stringers in the structure which can affect accuracy and finish of fine machining operations.

Table 2.4 Physical properties of copper alloys

Property	Cadmium copper	Chromium copper	Silver copper	Tellurium copper	Sulphur copper
Density at 20°C (10^3 kg m^{-3})	8.9	8.90	8.89	8.9	8.9
Coefficient of linear expansion (20–100°C) (10^{-6} K^{-1})	17	17	17.7	17	17
Modulus of elasticity* (10^9 N m^{-2})	132	108	118	118	118
Specific heat at 20°C (kJ kg^{-1} K^{-1})	0.38	0.38	0.39	0.39	0.39
Electrical conductivity at 20°C (10^6 S m^{-1})					
annealed	46–53	–	57.4–58.6	56.8†	55.1
solution heat treated	–	20	–	–	–
precipitation hardened	–	44–49	–	55.7†	–
Resistivity at 20°C (10^{-8} ohm m)					
annealed	2.2–1.9	–	1.74–1.71	1.76†	1.81
solution heat treated	–	4.9	–	–	–
precipitation hardened	–	2.3–2.0	–	–	–
cold worked	2.3–2.0	–	1.78	1.80	1.85

*Solution heat treated or annealed.
†Oxygen-bearing (tough pitch) tellurium copper.

Sulphur copper finds application for all machined parts requiring high electrical conductivity, such as contacts, connectors and other electrical components. Jointing characteristics are similar to those of tellurium copper.

Sulphur copper is deoxidized with a controlled amount of phosphorus and therefore does not suffer from hydrogen embrittlement in normal torch brazing operations; long exposure to reducing atmospheres can result in some loss of sulphur and consequent embrittlement.

Aluminium and its Alloys

For many years aluminium has been used as a conductor material in most branches of electrical engineering. In addition to the pure metal, several aluminium alloys are also good conductors, combining structural strength with an acceptable conductivity. The material is lighter than copper (about one third the density) and therefore easier to handle; it is also cheaper. Another advantage is that its price is not subject to wide fluctuations as is copper. There was a sharp increase in the price of copper worldwide in the 1960s and 1970s. This led to many instances of aluminium being used in situations where copper had previously been the norm. In a few applications, for example domestic wiring and transformer foil-windings identified below, aluminium proved to be less suitable than was initially hoped, so that in the late 1990s there has been some return to copper and the use of aluminium has tended to be restricted to those applications for which it is clearly superior.

There are two groups of British Standard Specifications for aluminium, one covering aluminium for electrical purposes, which relates to high purity aluminium with emphasis on electrical properties, and the second concerning aluminium for general engineering.

Aluminium for electrical purposes covers grades with conductivities between 55% and 61% International Annealed Copper Standard (IACS) and includes pure aluminium. The following are the relevant British Standards:

BS 215 Part 1: (IEC 207) Aluminium stranded conductors for overhead power transmission purposes.

Part 2: (IEC 209) Aluminium conductors, steel-reinforced for overhead power transmission purposes.

BS 2627. Wrought aluminium for electrical purposes – wire.

BS 2897. Wrought aluminium for electrical purposes – strip with drawn or rolled edges.

BS 2898. Wrought aluminium for electrical purposes – bars, extruded round tubes and sections.

BS 3242. (IEC 208) Aluminium alloy stranded conductors for overhead power transmission.

BS 3988. Wrought aluminium for electrical purposes – solid conductors for insulated cables.

BS 6360. Specifications for conductors in insulated cables and cords.

This group of specifications include grade 1350 (formerly 1E) pure aluminium with a conductivity of 61% IACS and grade 6101A (formerly 91E) which is a heat treatable alloy with moderate strength and a conductivity of 55% IACS.

Aluminium for general engineering uses includes grades with conductivities as low as 30% IACS but with high structural strength, up to 60% of that of steel, with greater emphasis on mechanical properties. This is covered by the following British Standards:

BS 1471 Wrought aluminium and aluminium alloys – drawn tube.

BS 1472 Wrought aluminium and aluminium alloys – forging stock and forgings.

BS 1473 Wrought aluminium and aluminium alloys – rivet, bolt and screw stock.

BS 1474 Wrought aluminium and aluminium alloys – bars, extruded round tube and sections.

BS 1475 Wrought aluminium and aluminium alloys – wire.

All of the above documents are based on but not identical to ISO 209.

BS 1490 Aluminium ingots and castings (based on but not identical to ISO 3522).

BS EN 485 Aluminium and aluminium alloys – sheet, strip and plate.

This group of specifications includes grade 1050A (formerly 1B) with a conductivity of 61.6 IACS, grade 1080A (formerly 1A) also with a conductivity of 61.6 IACS, and grade 1200 (formerly 1C) with a conductivity of 59.5% IACS. These grades are generally used in sheet form, up to 10 mm thick, or plate, over 10 mm thick. Further information on aluminium grades and specifications can be obtained from the Aluminium Federation.

Busbars. Aluminium has been used for busbars for more than 60 years and from 1960 onwards is increasingly being used for a whole range of busbar applications due to its light weight and durability. Tubular aluminium is used exclusively for grid substation busbars at 275 kV and 400 kV and is increasingly being used at 132 kV for substation refurbishments and redevelopments. Aluminium is used in large industrial plants such as smelters and electrochemical plants because of the availability of large sections of cast bars (up to 600 mm × 150 mm). Aluminium is also used in switchgear and rising main systems because of its lighter weight compared with copper. A major problem with aluminium is the rapidity with which it oxidizes when the surface is prepared for bolted jointing. Much research was carried out by the former CEGB into the problem especially with the heavy currents which arise between a generator and its associated step-up transformer. This resulted in significant improvements in jointing techniques. Bolted joints in aluminium busbars which are subject to frequent dismantling are frequently electroplated using silver or tin.

Cable. Aluminium is extensively employed as the conductors over 16 mm^2 cross-sectional area for power cables up to 66 kV. Aluminium is not normally found in domestic wiring installations because of the specialized jointing and termination techniques needed to ensure longevity of trouble-free service.

Overhead lines. The a.c.s.r. (aluminium conductor steel reinforced) overhead line conductors are used worldwide for power distribution systems. A.c.a.r. (aluminium conductor aluminium alloy wire reinforced) have increasingly been used since 1960 because of the elimination of the risk of bi-metallic corrosion and improved conductivity for a given cross-section. A.c.a.r. catenary conductors for supporting the contact wire are also finding favour with railway authorities for overhead electrification schemes because of their lower weight and the reduced risk of theft in comparison with copper.

Motors. Cage rotors for induction motors often employ aluminium bars. Casings are also made from the material as are fans used for motor cooling purposes.

Foil windings. Aluminium is the norm for the windings of capacitors from the smallest types used in lighting fittings to large power capacitors. Foil windings are suitable for some transformers, reactors and solenoids. Foil thicknesses range from 0.040 mm to 1.20 mm in 34 steps. A better space factor than for a wire wound copper coil is obtained, the aluminium conductor occupying some 90% of the space as against 60% for copper wire. Heating and cooling are aided by the better space factor and the smaller amount of insulation needed for foil wound coils. Rapid radial heat transfer ensures an even temperature gradient. The disadvantage of aluminium is its poorer mechanical strength, particularly from the viewpoint of making winding end connections. The tendency, therefore has been to turn to the use of copper foil for air insulated low voltage windings. Aluminium foil is, however, almost exclusively used for the HV windings of cast resin insulated transformers as it has a thermal expansion coefficient closer to that of the resin encapsulation material than does copper which thus reduces the thermal stresses arising under load.

Table 2.5 Constants and physical properties of very high purity aluminium

Atomic number	13
Atomic volume	$10\,cm^3$/g-atom
Atomic weight	26.98
Valency	3
Crystal structure	fcc
Interatomic distance (co-ordination number 12)	2.68 kX
Heat of combustion	200 k cal/g-atom
Latent heat of fusion	94.6 cal/g
Melting point	660.2°C
Boiling point	2480°C
Vapour pressure at 1200°C	1×10^{-2} mm Hg
Mean specific heat (0–100°C)	0.219 cal/g°C
Thermal conductivity (0–100°C)	0.57 cal/cm s°C
Temperature coefficient of linear expansion (0–100°C)	23.5×10^{-6} per°C
Electrical resistivity at 20°C	2.69 microhm cm
Temperature coefficient of resistance (0–100°C)	4.2×10^{-3} per°C
Electrochemical equivalent	3.348×10^{-1} g/Ah
Density at 20°C	$2.6898\,g/cm^3$
Modulus of elasticity	$68.3\,kN/mm^2$
Modulus of torsion	$25.5\,kN/mm^2$
Poisson's ratio	0.34

Heating elements. Aluminium foil heating elements have been developed but are not widely used at present. Applications include foil film wallpaper, curing concrete and possibly soil warming.

Heatsinks. High thermal conductivity of aluminium and ease of extruding or casting into solid or hollow shapes with integral fins makes the material ideal for heatsinks. Semiconductor devices and transformer tanks illustrate the wide diversity of applications in this field. Its light weight makes it ideal for pole-mounted transformer tanks and it has the added advantage that the material does not react with transformer oil to form a sludge.

Insulating Materials

The revision in 1986 of BS 2757 (and further revision in 1994 to make it identical to IEC 60085) has introduced a different concept of insulating materials to that outlined in the same standard issued in 1956. Alteration of the title to *Method for Determining the Thermal Classification of Electrical Insulation* without reference to electrical machinery and apparatus, which appeared in the title of the edition of 1956 is indicative of this.

Thermal classes and the temperatures assigned to them are as follows:

Thermal class	*Temperature* (°C)
Y	90
A	105

E	120
B	130
F	155
H	180
200	200
220	220
250	250

Temperatures over 250°C should increase by 25°C intervals and classes designated accordingly. Use of letters is not mandatory but the relationship between letters and temperatures should be adhered to.

When a thermal class describes an electrotechnical product it normally represents the maximum temperature appropriate to that product under rated load and other conditions. Thus the insulation subjected to this maximum temperature needs to have a thermal capability at least equal to the temperature associated with the thermal class of the product. However, the description of an electrotechnical product as being of a particular thermal class does not mean, and must not be taken to imply that each insulating material used in its construction is of the same thermal capacity. It is also important to note that the temperatures in the table are actual temperatures of the insulation and not the temperature rises of the product itself.

The 1956 edition of BS 2757 gave typical examples of insulating materials and their classifications as Group Y, A, E, etc. That concept no longer exists but Table 1 of BS 5691 Part 2 (IEC 216-2) lists materials and the tests which may be appropriate for determining their thermal endurance properties. This table lists three basic classes of material which are then further subdivided. The three classes are: (a) solid insulation of all forms not undergoing a transformation during application; (b) solid sheet insulation for winding or stacking, obtained by bonding superimposed layers; and (c) insulation which is solid in its final state but applied in the form of a liquid or paste, for filling, varnishing, coating or bonding

Examples under class (a) are inorganic sheet insulation like mica, laminated sheet insulation, ceramics, glasses and quartz, elastomers, thermosetting and thermoplastic moulded insulation.

Examples under class (b) are solid sheet insulation bonded together by pressure-sensitive adhesive, heat, simple fusion and fusion combined with chemical reaction. Again mica products fall into this category as do adhesive coated films, papers, fabrics and laminates.

In the final class (c) the insulating material may be formed by physical transformation such as congealing, evaporation or a solvent or gelation. Fusible insulation materials with and without fillers, plastisols and organosols are examples. Another method is to solidify the insulation by chemical reactions such as polymerization, polycondensation or polyaddition. Thermosetting resins and certain paste materials are examples. Table II in the same standard lists available tests, the methods of carrying them out (by reference to an IEC or ISO standard), the specimen and end-point criteria.

New definitions. New definitions are now included in BS 2757 but the reader is also referred to BS 5691 and its Parts 1, 2, 3 and 4 (IEC 216 Parts 1, 2, 3 and 4).

Table 2.6 Class 2 stranded conductors for single-core and multicore cables (from BS 6360)

1	2	3	4	5	6	7	8	9	10
Nominal cross-sectional area	Minimum number of wires in the conductor						Maximum resistance of conductor at 20°C		
	Circular conductor		Circular compacted conductor		Shaped conductor		Annealed copper conductor[a]		Aluminium conductor plain or metal-clad wires
	Cu	Al	Cu	Al	Cu	Al	Plain wires	Metal-coated wires	
mm²							Ω/km	Ω/km	Ω/km
0.5	7	–	–	–	–	–	36.0	36.7	–
0.75	7	–	–	–	–	–	24.5	24.8	–
1	7	–	6	–	–	–	18.1	18.2	–
1.5	7	–	6	–	–	–	12.1	12.2	–
2.5	7	–	6	–	–	–	7.41	7.56	–
4	7	7	6	–	–	–	4.61	4.70	7.41
6	7	7	6	–	–	–	3.08	3.11	4.61
10	7	7	6	–	–	–	1.83	1.84	3.08
16	7	7	6	6	–	–	1.15	1.16	1.91
25	7	7	6	6	6	6	0.727	0.734	1.20
35	7	7	6	6	6	6	0.524	0.529	0.868
50	19	19	6	6	6	6	0.387	0.391	0.641

Nominal area							Resistance		
70	19	19	12	12	12	12	0.268	0.270	0.443
95	19	19	15	15	15	15	0.193	0.195	0.320
120	37	37	18	18	15	15	0.153	0.154	0.253
150	37	37	18	18	15	15	0.124	0.126	0.206
185	37	37	30	30	30	30	0.0991	0.100	0.164
240	61	61	34	34	30	30	0.0754	0.0762	0.125
300	61	61	34	34	30	30	0.0601	0.0607	0.100
400	61	61	53	53	53	53	0.0470	0.0475	0.0778
500	61	61	53	53	53	53	0.0366	0.0369	0.0605
630	91	91	53	53	53	53	0.0283	0.0286	0.0469
800	91	91	53	53	–	–	0.0221	0.0224	0.0367
960(4 × 240)	Number of wires not specified						0.0189	0.0189	0.0313
1000	91	91	53	53	–	–	0.0176	0.0177	0.0291
1200	Number of wires not specified						0.0151	0.0151	0.0247
1600	–	–	–	–	–	–	0.0113	0.0113	0.0186
2000	–	–	–	–	–	–	0.0090	0.0090	0.0149

*To obtain the maximum resistance of hard-drawn conductors the values in columns 8 and 9 should be divided by 0.97.

Table 2.7 Standard aluminium conductors, steel reinforced (from BS 215: Part 2)

Nominal aluminium area (mm²)	Stranding and wire diameter		Sectional area of aluminium (mm²)	Total sectional area (mm²)	Approx. overall diameter (mm)	Approx. mass per km (kg)	Calculated d.c. resistance at 20°C per km (Ω)	Calculated breaking load (kN)
	Aluminium (mm)	Steel (mm)						
25	6/2.36	1/2.36	26.24	30.62	7.08	106	1.093	9.61
30	6/2.59	1/2.59	31.61	36.88	7.77	128	0.9077	11.45
40	6/3.00	1/3.00	42.41	49.48	9.00	172	0.6766	15.20
50	6/3.35	1/3.35	52.88	61.70	10.05	214	0.5426	18.35
70	12/2.79	7/2.79	73.37	116.2	13.95	538	0.3936	61.20
100	6/4.72	7/1.57	105.0	118.5	14.15	394	0.2733	32.70
150	30/2.59	7/2.59	158.1	194.9	18.13	726	0.1828	69.20
150	18/3.35	1/3.35	158.7	167.5	16.75	506	0.1815	35.70
175	30/2.79	7/2.79	183.4	226.2	19.53	842	0.1576	79.80
175	18/3.61	1/3.61	184.3	194.5	18.05	587	0.1563	41.10
200	30/3.00	7/3.00	212.1	261.5	21.00	974	0.1363	92.25
200	18/3.86	1/3.86	210.6	222.3	19.30	671	0.1367	46.55
400	54/3.18	7/3.18	428.9	484.5	28.62	1 621	0.06740	131.9

Table 2.8 Class 1 solid conductors for single-core and multicore cables (from BS 6360)

1	2	3	4
Nominal cross-sectional area	Maximum resistance of conductor at 20°C		
	Circular, annealed copper conductors*		Aluminium conductors, circular or shaped, plain or metal-clad
	Plain	Metal-coated	
mm²	Ω/km	Ω/km	Ω/km
0.5	36.0	36.7	–
0.75	24.5	24.8	–
1	18.1	18.2	–
1.5	12.1	12.2	18.1†
2.5	7.41	7.56	12.1†
4	4.61	4.70	7.41†
6	3.08	3.11	4.61†
10	1.83	1.84	3.08†
16	1.15	1.16	1.91†
25	0.727	–	1.20
35	0.524	–	0.868
50	0.387	–	0.641
70	0.268	–	0.443
95	0.193	–	0.320
120	0.153	–	0.253
150	0.124	–	0.206
185	–	–	0.164
240	–	–	0.125
300	–	–	0.100
380 (4 × 95)	–	–	0.0800
480 (4 × 120)	–	–	0.0633
600 (4 × 150)	–	–	0.0515
740 (4 × 185)	–	–	0.0410
960 (4 × 240)	–	–	0.0313
1200 (4 × 300)	–	–	0.0250

*To obtain the maximum resistance of hard-drawn conductors the values in columns 2 and 3 should be divided by 0.97.
†Aluminium conductors 1.5 mm² to 16 mm² circular only.

Temperature index (TI). The number corresponding to the temperature in degrees Celsius derived from the thermal endurance relationship at a given time, normally 20 000 h.

Relative temperature index (RTI). The temperature index of a test material obtained from the time which corresponds to the known temperature index of a reference material, when both materials are subjected to the same ageing and diagnostics procedures in a comparative test.

Table 2.9 Class 5 flexible copper conductors for single-core and multicore cables (from BS 6360)

1	2	3	4
Nominal cross-sectional area	Maximum diameter of wires in conductor	Maximum resistance of conductor at 20°C	
		Plain wires	Metal-coated wires
mm²	mm	Ω/km	Ω/km
0.22	0.21	92.0	92.4
0.5	0.21	39.0	40.1
0.75	0.21	26.0	26.7
1	0.21	19.5	20.0
1.25	0.21	15.6	16.1
1.35	0.31	14.6	15.0
1.5	0.26	13.3	13.7
2.5	0.26	7.98	8.21
4	0.31	4.95	5.09
6	0.31	3.30	3.39
10	0.41	1.91	1.95
16	0.41	1.21	1.24
25	0.41	0.780	0.795
35	0.41	0.554	0.565
50	0.41	0.386	0.393
70	0.51	0.272	0.277
95	0.51	0.206	0.210
120	0.51	0.161	0.164
150	0.51	0.129	0.132
185	0.51	0.106	0.108
240	0.51	0.0801	0.0817
300	0.51	0.0641	0.0654
400	0.51	0.0486	0.0495
500	0.61	0.0384	0.0391
630	0.61	0.0287	0.0292

Halving interval (HIC). The number corresponding to the temperature interval in degrees Celsius which expresses the halving of the time to the end point taken at the temperature of the TI or the RTI.

Properties. The following notes give briefly the chief points to be borne in mind when considering the suitability of any material for a particular duty.

Relative density is of importance for varnishes, oils and other liquids. The density of solid insulations varies widely, e.g. from 0.6 for certain papers to 3.0 for mica. In a few cases it indicates the relative quality of a material, e.g. vulcanized fibre and pressboard.

Moisture absorption usually causes serious depreciation of electrical properties, particularly in oils and fibrous materials. Swelling, warping, corrosion

Table 2.10 Class 6 flexible copper conductors for single-core and multicore cables (from BS 6360)

1	2	3	4
Nominal cross-sectional area	Maximum diameter of wires in conductor	Maximum resistance of conductor at 20°C	
		Plain wires	Metal-coated wires
mm²	mm	Ω/km	Ω/km
0.5	0.16	39.0	40.1
0.75	0.16	26.0	26.7
1	0.16	19.5	20.0
1.5	0.16	13.3	13.7
2.5	0.16	7.98	8.21
4	0.16	4.95	5.09
6	0.21	3.30	3.39
10	0.21	1.91	1.95
16	0.21	1.21	1.24
25	0.21	0.780	0.795
35	0.21	0.554	0.565
50	0.31	0.386	0.393
70	0.31	0.272	0.277
95	0.31	0.206	0.210
120	0.31	0.161	0.164
150	0.31	0.129	0.132
185	0.41	0.106	0.108
240	0.41	0.0801	0.0817
300	0.41	0.0641	0.0654

and other effects often result from absorption of moisture. Under severe conditions of humidity, such as occur in mines and in tropical climates, moisture sometimes causes serious deterioration.

Thermal effects very often seriously influence the choice and application of insulating materials, the principal features being: melting-point (e.g. of waxes); softening or plastic yield temperature; ageing due to heat, and the maximum temperature which a material will withstand without serious deterioration of essential properties; flash point or ignitibility; resistance to electric arcs; liability to carbonize (or 'track'); ability to self-extinguish if ignited; specific heat; thermal resistivity; and certain other thermal properties such as coefficient of expansion and freezing point.

Mechanical properties. The usual mechanical properties of solid materials are of varying significance in the case of those required for insulating purposes, *tensile* strength, *transverse* strength, *shearing* strength and *compressive* strength often being specified. Owing, however, to the relative degree of inelasticity of most solid insulations, and the fact that many are quite brittle, it is frequently necessary to pay attention to such features as *compressibility*, *deformation under bending* stresses, *impact* strength and *extensibility, tearing* strength, *machinability* and ability to fold without damage.

Resistivity and insulation resistance. In the case of insulating material it is generally manifest in two forms (a) volume resistivity (or specific resistance) and (b) surface resistivity.

Electric strength (or dielectric strength) is the property of an insulating material which enables it to withstand electric stress without injury. It is usually expressed in terms of the minimum electric stress (i.e. potential difference per unit distance) which will cause failure or 'breakdown' of the dielectric under certain specified conditions.

Surface breakdown and flashover. When a high-voltage stress is applied to conductors separated only by air and the stress is increased, breakdown of the intermediate air will take place when a certain stress is attained, being accompanied by the passage of a spark from one conductor to the other.

Permittivity (specific inductive capacity). Permittivity is defined as the ratio of the electric flux density produced in the material to that produced in free space by the same electric force, and is expressed as the ratio of the capacitance of a capacitor in which the material is the dielectric, to the capacitance of the same capacitor with air as the dielectric.

Paper pressboard and wood. Before leaving the subject of solid insulation it is necessary to look in a little detail at the natural materials: paper, pressboard and wood, which are the main insulating materials used in oil-filled apparatus – primarily transformers.

Early power transformers operated in air and used asbestos, cotton, low-grade pressboard, and shellac impregnated paper. It soon became clear, however, that air insulated transformers could not match the thermal capabilities of oil-filled units. These utilized kraft paper and pressboard systems supplemented from about 1915 by insulating cylinders formed from phenol-formaldehyde resin impregnated kraft paper, or Bakelized paper, to give it its proprietary name. This material, usually referred to as s.r.b.p. (synthetic resin-bonded paper) continued to be widely used in transformers until the 1960s, and still finds many uses in locations having lower electrical stress but where high mechanical strength is required.

Paper is among the cheapest and best electrical materials known. For electrical purposes it must meet certain chemical and physical standards which in turn are dictated by the electrical requirements. The important electrical properties are:

(a) High dielectric strength.
(b) For oil-filled transformers a dielectric constant which matches as closely as possible that of oil.
(c) Low power factor (dielectric loss – discussed below).
(d) Freedom from conducting particles.

The dielectric constant for kraft paper is about 4.4, for mineral oil the figure is approximately 2.2. Kraft paper is, by definition, made entirely from unbleached softwood pulp manufactured by the sulphate process; unbleached because residual bleaching agents might hazard its electrical properties. This process is essentially one which results in a slightly alkaline residue, pH 7–9, as distinct from the less costly sulphite process commonly used for production of newsprint, for example, which produces an acid pulp. Acidic content leads to rapid degradation of the long-chain cellulose molecules and consequent loss of mechanical strength which would be unacceptable for electrical purposes. The timber is initially ground to a fine shredded texture at the location of its production in Scandinavia, Russia or Canada using carborundum or similar abrasive grinding wheels. The chemical sulphate process then removes most of the other constituents of the wood, e.g. lignin, carbohydrates, waxes, etc., to leave only the cellulose fibres. The fibres are dispersed in water which is drained to leave a wood-pulp mat. At this stage the dried mat may be transported to the mill of the specialist paper manufacturer.

The processes used by the manufacturer of the insulation material may differ one from another and even within the mill of a particular manufacturer treatments will vary according to the particular properties required from the finished product. The following outline of the type of processes used by one UK producer of specialist high quality presspaper gives some indication of what might be involved.

Presspaper by definition undergoes some compression during manufacture which increases its density, improves surface finish and increases mechanical strength. Presspaper production is a continuous process in which the paper is formed on a rotating fine mesh drum and involves building of the paper sheet from a number of individual layers. Other simpler processes may produce discrete sheets of paper on horizontal screen beds without any subsequent forming or rolling processes, but, as would be expected, the more sophisticated the manufacturing process, the more reliable and consistent the properties of the resulting product.

The process commences by repulping of the bales of dry mat using copious quantities of water, one purpose of which is to remove all residual traces of the chemicals used in the pulp extraction stage. The individual fibres are crushed and refined in the wet state in order to expose as much surface area as possible. Paper or pressboard strength is primarily determined by bonding forces between fibres, whereas the fibres themselves are stressed far below their breaking point. These physiochemical bonding forces which are known as 'hydrogen bonding' occur between the cellulose molecules themselves and are influenced primarily by the type and extent of this refining.

Fibres thus refined are then mixed with more water and subjected to intensive cleaning in multistage centrifugal separators which remove any which may not have been totally broken down or which may have formed into small knots. These can be returned to pass through the refining cycle once more. The centrifuges also remove any foreign matter such as metallic particles which could have been introduced by the refining process. The cellulose/water mixture is then routed to a wide rotating cylindrical screen. While the water flows through the screen, the cellulose fibres are filtered out and form a paper layer. An endless band of felt removes the paper web from the screen and conveys it to the forming rolls. The felt layer permits further water removal and allows up to five or six other paper plies to be amalgamated with the first before passing through the forming rolls. These then continue to extract water and form the paper to the required thickness, density and moisture content by means of heat and pressure as it progresses through the rolls. Options are available at this stage of the process to impart various special properties, for example the CLUPAK[1] process which enhances the extensibility of the paper, or impregnation with 'stabilizers' such as nitrogen containing chemicals like dicyandiamide which provide improved thermal performance. Final finish and density may be achieved by means of a calendering process in which the paper, at a controlled high moisture content, is passed through heavily loaded steel rollers followed by drying by means of heat in the absence of pressure.

The cohesion of the fibres to one another when the mat is dried is almost exclusively a property of cellulose fibres. Cellulose is a high-polymer carbohydrate chain consisting of glucose units with a polymerization level of approximately 2000. *Figure 2.4* shows its chemical structure. Hemi-cellulose molecules are the second major components of the purified wood pulp. These are carbohydrates with a polymerization level of less than 200. In a limited quantity, they facilitate the hydrogen bonding process, but the mechanical strength is reduced if their quantity exceeds about 10%. Hemi-cellulose molecules also have the disadvantage that they 'hold on' to water and make the paper more difficult to dry out.

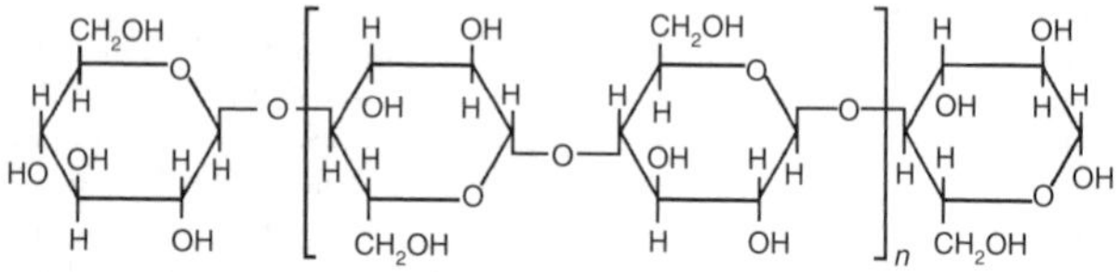

Figure 2.4 Chemical formula for cellulose

Softwood cellulose is the most suitable for electrical insulation because its fibre length of 1–4 mm gives it the highest mechanical strength. Nevertheless small quantities of pulp from harder woods may be added and, as in the case of alloying metals, the properties of the resulting blend are usually superior to those of either of the individual constituents.

Cotton cellulose. Cotton fibres are an alternative source of very pure cellulose which has been used in the UK for many years to produce the so-called 'rag' papers with the aim of combining superior electrical strength and mechanical properties to those of pure kraft paper. Cotton has longer fibres than those of wood pulp but the intrinsic bond strength is not so good. Cotton is a 'smoother' fibre than wood so that it is necessary to put in more work in the crushing and refining stage to produce the side branches which will provide the necessary bonding sites to give the required mechanical strength. This alone would make the material more expensive even without the additional cost of the raw material itself.

When first used in the manufacture of electrical paper in the 1930s the source of cotton fibres was the waste and offcuts from cotton cloth which went into the manufacture of clothing and this to an extent kept the cost competitive with pure kraft paper. In recent years this source has ceased to be an acceptable one since such cloths will often contain a proportion of synthetic fibres and other materials so that the constitution of offcuts cannot be relied upon as being pure and uncontaminated. Alternative sources have therefore had to be found. Cotton linters are those cuts taken from the cotton plant

---

[1] Clupak Inc.'s trademark for its extensible paper manufacturing process.

after the long staple fibres have been cut and taken for spinning into yarn for the manufacture of cloth. First grade linters are those taken immediately after the staple. These are of a length and quality which still renders them suitable for high quality insulation material. They may provide the 'furnish' or feedstock for a paper-making process of the type described, either alone or in conjunction with new cotton waste threads.

Cotton fibre may also be combined with kraft wood pulp to produce a material which optimizes the advantages of both constituents giving a paper which has good electrical and mechanical properties as well as maximum oil absorption capability. This latter requirement can be of great importance in paper used for high to low wraps or wraps between layers of round-wire distribution transformer high voltage windings where total penetration of impregnating oil may be difficult even under high vacuum.

Other fibres such as manila, hemp, and jute may also be used to provide papers with specific properties developed to meet particular electrical purposes, for example in capacitors and cable insulation. British Standard 5626: 1979, *Cellulosic papers for electrical purposes*, which is identical to IEC 60554, lists the principal paper types and properties. Presspapers are covered by British Standard EN 60641, *Pressboard and presspaper for electrical purposes*.

Pressboard: At its most simple, pressboard represents nothing more than thick insulation paper made by laying up a number of layers of paper at the wet stage of manufacture. *Figure 2.5* shows a diagrammatic arrangement of the manufacturing process. Of necessity this must become a batch process rather than the continuous one used for paper, otherwise the process is very similar to that used for paper. As many thin layers as are necessary to provide the required thickness are wet laminated without a bonding agent. Pressboard can, however, be split into two basic categories:

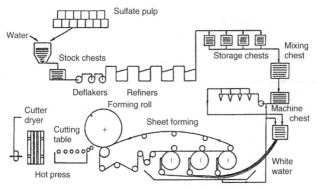

Figure 2.5 Manufacturing process for precompressed transformers board (H Weidmann AG)

(a) That built up purely from paper layers in the wet state without any bonding agent, as described above.
(b) That built up, usually to a greater thickness, by bonding individual boards using a suitable adhesive.

Pressboards and presspapers in the former category are covered by a British Standard, BS EN 60641 *Pressboard and presspaper for electrical purposes*. This is a multi-part document, Part 1 of which gives the general requirements and defines the various types. A similar multi-part document, BS EN 60763 *Laminated pressboard*, details the technical requirements for the laminated boards. As in the case of paper insulation, there are a number of variants around the theme and all the main types of material are listed in the above documents. Raw materials may be the same as for presspaper, that is all wood pulp, all cotton, or a blend of wood and cotton fibres.

Pressboard in the first of the above categories is available in thicknesses up to 8 mm and is generally used at thicknesses of around 2–3 mm for inter-winding wraps and end insulation of oil-filled transformers and 4.5–6 mm for strips used to form oil cooling ducts. The material is usually produced in three subcategories.

The first is known as calendered pressboard and undergoes an initial press-ing operation at about 55% water content. Drying by means of heat without pressure then follows to take the moisture level to about 5%. The pressboard thus produced has a density of about 0.90 to 1.00. Further compression is then applied under heavy calenders to take the density to between 1.15 and 1.30.

The second category is mouldable pressboard which receives little or no pressing after the forming process. This is dried using heat only to a moisture content of about 5% and has a density of about 0.90. The result is a soft press-board with good oil absorption capabilities which is capable of being shaped to some degree to meet the physical requirements of particular applications.

The third material is precompressed pressboard. Dehydration, compression and drying are performed in hot presses direct from the wet stage. This has the effect of bonding the fibres to produce a strong, stable, stress-free material of density about 1.25 which will retain its shape and dimensions throughout the stages of transformer manufacture and the thermal cycling in oil under service conditions to a far better degree than the two boards previously described. Because of this, high stability precompressed material is now the preferred pressboard of most transformer manufacturers for most applications.

Laminated pressboard starts at around 10 mm thickness and is available in thicknesses up to 50 mm or more. The material before lamination may be of any of the categories of unlaminated material described above but generally precompressed pressboard is preferred. This board is used in large power transformers for winding support platforms, winding end support blocks and distance pieces as well as cleats for securing and supporting leads.

Liquid dielectrics. Liquid dielectrics are used

(a) As a filling and cooling medium for transformers, capacitors and rheostats.
(b) As an insulating and arc-quenching medium in switchgear, such as cir-cuit breakers.
(c) As an impregnant of absorbent insulations, e.g. paper, pressboard, and wood, used in transformers, switchgear, capacitors and cables.

The desirable properties for these liquids are, therefore, (i) high elec-tric strength, (ii) low viscosity, (iii) high chemical stability and resistance to oxidation, (iv) high flash point, (v) low volatility.

The most important liquid dielectric in general use is mineral oil. This is specified in BS 148 : 1984. *Specification for unused mineral insulating oils for*

Table 2.11 Representative properties of typical insulating materials

Insulant	n^*	ε_r	$\tan \delta$ 50 Hz	$\tan \delta$ 1 MHz
Vacuum	∞	1.0	0	0
Air	∞	1.0006	0	0
Mineral insulating oil	11–13	2–2.5	0.0002	–
Chlorinated polyphenols	10–12	4.5–5	0.003	–
Paraffin wax	14	2.2	–	0.0001
Shellac	13	2.3–3.8	0.008	–
Bitumen	12	2.6	0.008	–
Pressboard	8	3.1	0.013	–
Ebonite	14	2.8	0.01	0.009
Hard rubber (loaded)	12–16	4	0.016	0.01
Paper, dry	10	1.9–2.9	0.005	–
Paper, oiled	–	2.8–4	0.005	–
Cloth, varnished cotton	13	5	0.2	–
Cloth, silk	13	3.2–4.5	–	–
Ethyl cellulose	11	2.5–3.7	0.02	0.02
Cellulose acetate film	13	4–5.5	0.023	–
S.R.B.P.	11–12	4–6	0.02	0.04
S.R.B. cotton	7–10	5–11	0.03	0.06
S.R.B. wood	10	4.5–5.4	–	0.05
Polystyrene	15	2.6	0.0002	0.0002
Polyethylene	15	2.3	0.0001	0.0001
Methyl methacrylate	13	2.8	0.06	0.02
Phenol formaldehyde wood-filled	9–10	4–9	0.1	0.09
Phenol formaldehyde mineral-filled	10–12	5	0.015	0.01
Polystyrene mineral-filled	–	3.2	–	0.0015
Polyvinyl chloride	11	5–7	0.1	–
Porcelain	10–12	5–7	–	0.008
Steatite	12–13	4–6.6	0.0012	0.001
Mycalex, sheet, rod	12	7	–	0.002
Mica, Muscovite	11–15	4.5–7	0.0003	0.0002
Glass, plate	11	6–7	–	0.004
Quartz, fused	16	3.9	–	0.0002

*Volume resistivity: $\rho = 10^n$ ohm-m. The value of n is tabulated. Information in the above table is taken from the 13th Edition of *Electrical Engineer's Reference Book* published by Butterworths.

use in transformers and switchgear. This document is similar but not identical to IEC 296. Mineral oil is used as insulant and coolant in virtually all outdoor transformers and in most underground cables of 132 kV and above. Oil is also used as an arc-quenching medium and insulant in much of the switchgear currently in service at voltages of 33 kV and below. For switchgear, the high maintenance requirement associated with oil has led to its being superseded first by air-break equipment and more recently by the widespread introduction of vacuum and sulphur hexafluoride (SF_6) (see below).

Mineral insulating oils are highly refined hydrocarbon oils obtained from selected crude petroleum. The refinement process is the means of removing impurities, mainly compounds containing sulphur nitrogen and oxygen, and of separating the lower viscosity hydrocarbons required for the electrical oils from the heavier lubricating and fuel-oil constituents. This is carried out by distillation, filtration and catalytic breakdown of some of the larger chain molecules.

Hydrocarbons present in mineral oil fall into three classes: naphthenes, paraffins and aromatics. Most crude oils consist of a mixture of all three types, but for electrical purposes an oil which is predominantly naphthenic is preferred. Many paraffins tend to produce wax which impedes flow at low temperatures. Aromatics are chemically less stable than the other two types and if present in large quantity would not provide the required high chemical stability. A typical electrical oil might contain 65% naphthenes, 30% paraffins and 5% aromatics.

BS 148 lists acceptable characteristics for three classes of mineral oil, Class I, II and III. Class is determined by viscosity, with Class I, which has the highest viscosity, relatively speaking, being used in transformers. The lower viscosity oil is specified for circuit breakers since this enables the oil to flow more quickly in between the parting current-interruption contacts which assists in extinguishing the arc.

Another requirement associated with long-term chemical stability is resistance to oxidation. Oxidation is more of a problem for oil in transformers than in switchgear as these operate at a higher temperature. Oil which has become oxidized is acidic and deposits a sludge in the transformer windings which reduces cooling efficiency and shortens transformer life. Selection of the correct chemical make up of the oil will assist in resisting oxidation, but the oxidation resistance may also be improved by the addition of *inhibitors*. Oil containing inhibitors is known as *inhibited oil*. In the UK there has long been a preference for oil which does not rely on inhibitors as it is not known how long they will retain their inhibiting properties. Oil without inhibitors is known as *uninhibited oil*.

Water is slightly soluble in electrical oils but for good electrical strength it is desirable that water content is kept to a minimum. Water content is measured in parts per million (p.p.m.) and solubility varies with type of oil, but typically at $20°C$ it will dissolve up to 40 p.p.m. while at $80°C$ this increases to 400 p.p.m. The greatest hazard to electrical insulation strength is the presence of free water, as undissolved droplets, in combination with contamination with minute fibres. When used in switchgear the oil will also become contaminated by carbon particles arising from arc interruption. This must be periodically removed by filtration.

BS 148 allows oil as supplied by bulk tanker to contain up to 30 p.p.m. dissolved water. When supplied in drums it may contain up to 40 p.p.m. A laboratory test (Karl Fischer test – see BS 2511) is necessary to establish water content in parts per million but an easy and convenient test for the presence of free water is the crackle test. To carry out this test a sample of the oil is heated quickly in a test tube over a silent flame. If free water is present this will boil off with an audible crackle before it is able to dissolve in the hotter oil. In this test oil shown to contain water should not be used in electrical equipment without suitable filtration and drying.

The other important test for oil quality is the *breakdown strength*. For this test oil is subjected to a steadily increasing alternating voltage between two electrodes spaced at 2.5 mm apart in a test cell, until breakdown occurs. The breakdown voltage is the voltage reached at the time of the first spark whether this is transient or total. The test is carried out six times on the same cell filling, and the electric strength of the oil is the average of the six breakdown values obtained. BS 148 specifies that breakdown strength for oil as supplied should be 30 kV minimum. For good quality oil as, for example, that taken from a high voltage transformer, this value should be easily exceeded, a figure of at least 50 or 60 kV being obtained.

Before the present breakdown voltage test was introduced in the 1972 edition of BS 148, the *electrical strength* test was used. This utilized a similar test cell having spherical electrodes 4 mm apart. The oil sample was required to withstand the test voltage of 40 kV for one minute. Any transient discharges which did not develop into an arc were ignored. To pass the test two out of three samples were required to resist breakdown. This test has not been totally abandoned. Since it is less searching than the breakdown strength test it is still accepted as a method of testing used oil and is included as such in BS 5730 *Code of practice for maintenance of insulating oil.*

A source of concern where a significant quantity of mineral oil is used in electrical equipment is the risk of fire. In a power station, for example, where a ready supply of water is available, waterspray fire protection is usually provided on all the large transformers. Where the risk of fire is not considered acceptable, for example in buildings, it may be considered preferable to use another low flammability fluid instead of mineral oil. Some of these alternatives are listed below.

Silicone fluids. Are low flammability liquid dielectrics suitable for insulation purposes and generally restricted to 66 kV and below. They have high thermal stability and chemical inertness. They have a very high flash point and in a tank will not burn below 350°C even when subjected to a flame. They are used in power and distribution transformers, small aircraft transformers, and as an impregnant for capacitors. One disadvantage of silicone fluids is that their arc-quenching properties do not make them suitable for use in on-load tapchangers.

Synthetic ester fluids. Complex esters or hindered esters are already widely accepted in the fields of high temperature lubrication and hydraulics, particularly in gas turbine applications and as heat transfer fluids generally. In these fields they have largely replaced petroleum and many synthetic oils which have proved toxic or unsuitable in some other respect. More recently a similar ester has been developed to meet the requirements of application as a high voltage dielectric for transformers and on-load tapchangers. It has very low toxicity and is biodegradable. It also possesses excellent lubrication properties enabling it to be used in forced cooled (i.e. pumped) transformers of all types.

Polychlorinated bi-phenyls (PCBs). This class of synthetic liquids – also known as askarels – should be mentioned for the sake of completeness as they were from their introduction by Monsanto in the 1940s until the late 1970s very widely used in capacitors and transformers. Outside the electrical industry they also had widespread application as a heat transfer fluid. However, due to the non-biodegradable nature of PCBs, which causes them to remain in the environment and ultimately to enter the food chain, plus their close association with a more hazardous material, dioxin, production of these liquids in most parts of the world has now ceased and their use is being phased out.

In the 1980s a number of specialist organizations developed the skills for draining askarel-filled transformers, refilling them with alternative liquids and safely disposing of the askarels. The process is, however, fraught with difficulties as legislation is introduced in many countries requiring that fluids containing progressively lower and lower levels of PCBs be considered and handled as PCBs. It is very difficult as well as costly to remove all traces of PCB from a transformer so that retrofilling in this way is tending to become a far less viable option. By the 1990s those considering the problem of what to do with a PCB-filled transformer are strongly encouraged to scrap it in a safe manner and replace it. Guidance concerning disposal is available from the Health and Safety Executive.

Gas insulation. As mentioned above, the gas sulphur hexafluoride, SF_6, has now replaced mineral oil as the most widely used insulant and arc-extinguishing medium in all classes of switchgear up to 400 kV and beyond.

SF_6 gas is stable and inert up to about 500°C, it is incombustible, non-toxic, odourless and colourless. SF_6 gas possesses excellent insulating properties when pressurized in the range 2 to 6 bar and has a dielectric strength some 2.5 to 3 times that of air at the same pressure. The gas is about five times heavier than air with a molecular weight of 146 and specific gravity of 6.14 g/l. At normal densities the gas is unlikely to liquefy except at very low operating temperatures less than −40°C and equipment may be fitted with heaters if this is likely to be a problem. Industrial SF_6 gas used in circuit breakers and busbar systems is specified to have a purity of 99.9% by weight and has impurities of SF_4 (0.05%), air (0.05% O_2 plus N_2), 15 p.p.m. moisture and 1 p.p.m. HF. Absorbed moisture leaving the switchgear housing and insulator walls leads to the moisture content of the SF_6 gas stabilizing at between 20 and 100 p.p.m. by weight when in service.

Gases at normal temperatures are good insulators but the molecules tend to dissociate at the elevated arc temperatures ($\sim$2000 K) found during the circuit breaking process and become good conductors. SF_6 gas also dissociates during the arcing process and is transformed into an electrically conductive plasma which maintains the current until the next or next but one natural power frequency current zero. SF_6 gas has proven to be an excellent arc-quenching medium. This arises not only from its stability and dielectric strength but also its high specific heat, good thermal conductivity and ability to trap free electrons. It cools very rapidly, within a few μs, and the fluorine and sulphur ions quickly recombine to form stable insulating SF_6. Such properties all assist in the removal of energy from the arc during the circuit breaking process. More will be said about SF_6 in Chapter 15.

Vacuum insulation. Vacuum is now being employed for arc-extinguishing applications in switchgear and motor control gear. Although vacuum is not a 'material' in the sense covered in this chapter, vacuum circuit breakers have characteristics which are specifically related to the arc-interruption medium. These will be discussed in Chapter 15.

Power factor and dielectric losses. When an alternating stress is applied to, say, the plates of a capacitor in which the dielectric is 'perfect', e.g. dry air or a vacuum, the current passed is a pure capacitance current and leads the voltage by a phase angle of 90°. In the case of practically all other dielectrics conduction and other effects (such as dielectric hysteresis[2]) cause a certain amount of energy to be dissipated in the dielectric, which results in the current leading the voltage by a phase angle less than 90. The value of the angle which is complementary with the phase angle is, therefore, a measure of the losses occurring in the material when under alternating electric stress.

In the phasor diagram, *Figure 2.6*, the phase angle is ϕ and the complementary angle δ is known as the loss angle. As this angle is usually quite small, the power factor ($\cos\phi$) can be taken as equal to $\tan\delta$ (for values of $\cos\phi$ up to, say, 0.1).

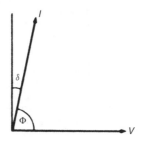

Figure 2.6 Vector diagram for a material having dielectric loss

The energy loss (in watts) is $V^2 C\omega \tan\delta$ where V is the applied voltage. C the capacitance in farads, and $\omega = 2\pi f$, where f is the frequency in hertz.

This loss, known as the dielectric loss, is seen to depend upon the capacitance, which, for given dimensions of dielectric and electrodes, is determined by the permittivity of the insulating material. The properties of the dielectric which determine the amount of dielectric losses are, therefore, power factor ($\tan\delta$) and permittivity. It is consequently quite usual practice to quote figures for the product of these two, i.e. $k \times \tan\delta$ for comparing insulating materials in this respect. It will also be noted that the losses vary as the square of the voltage.

Power factor varies, sometimes considerably, with frequency, also with temperature, values of $\tan\delta$ usually increasing with rise in temperature, particularly when moisture is present, in which case the permittivity also rises with the temperature, so that total dielectric losses are often liable to a considerable increase as the temperature rises. This is very often the basic cause of electrical breakdown in insulation under a.c. stress, especially if it is thick, as the losses cause internal temperature rise with consequent increase in the dielectric power factor and permittivity, this becoming cumulative and resulting in thermal instability and, finally, breakdown, if the heat developed in the interior cannot get away faster than it is generated.

These properties are, of course, of special importance in the case of radio and similar uses where high frequencies are involved.

---

[2] Dielectric hysteresis is a phenomenon by which energy is expended and heat produced, as the result of the reversal of electrostatic stress in a dielectric subjected to alternating electric stress.

Superconductivity

The ideal superconducting state exhibited by certain materials is characterized by two fundamental properties: (a) the disappearance of resistance when the temperature is reduced to a critical value and (b) the expulsion of any magnetic flux which may be in the material when the critical, or transition, temperature is reached. The discovery of superconductivity was made at the University of Leiden in 1911 by Professor Onnes when he was examining the relationship between the resistance and temperature of mercury. In the years that followed many other elements were found to exhibit superconductivity and theories were developed to explain the phenomenon. The transition temperatures were typically about $10\,K$ ($-263°C$) which, in practice, meant that they had to be cooled with liquid helium at $4\,K$. In general these materials were of little more than academic value because they could only support a low current density in a low magnetic field without losing their superconducting properties.

In the 1950s a new class of superconductors was discovered which were alloys or compounds and which would operate with very high current densities – typically $10^5\,A/cm^2$ – and high magnetic flux densities – typically 8 tesla. The most important materials in this class were the alloy NbTi and the compound Nb_3Sn. The consequence of these discoveries was the initiation of a significant activity worldwide on their application to many types of power equipment and magnets for research purposes. Other applications investigated were computers and very sensitive instruments.

In the UK the former CEGB (now privatized) commenced studies on the superconducting power cables and magnetohydrodynamic (MHD) power generation and an electrical machines programme was commenced by NEI International Research and Development Co. Ltd. This company designed and built the world's first superconducting motor in 1966 (now in the Science Museum) and followed this with a series of other d.c. motors and generators up to the early 1980s; they also played a leading role in the development of superconducting a.c. generators for central power stations and a fault current limiter for use in power transmission/distribution networks. The driving force for these developments was a reduction of electrical losses, in some cases the elimination of magnetic iron, and reductions in the size and weight of plant. Many other countries had similar programmes and there was a good measure of international collaboration through conferences. Milestones were the design, construction and test by NEI-IRD of a 2.44 MW low-speed superconducting motor in 1969; an 87 kVA a.c. generator by the Massachusetts Institute of Technology in the late 1960s; a very large superconducting Bubble Chamber magnet at CERN in Switzerland in the late 1960s and a superconducting levitated train by Siemens in the mid-1970s. Many other countries have made significant achievements since these dates.

One of the problems that arose during the 20 years or so of these developments was the high cost of designing to meet the engineering requirements at liquid helium temperature and, by the early 1980s, many programmes had been terminated and others were proceeding very slowly.

In late 1986, Bednorz and Mueller working in Zurich discovered that a ceramic material LaBaCuO was superconducting at 35 K (they were awarded a Nobel prize) and in 1987, Professor Chu at the University of Houston in Texas discovered that YBaCuO was superconducting at 92 K and since that time the transition temperature of other materials has crept up to over 105 K.

The enormous significance of these discoveries is that the materials will work in liquid nitrogen instead of liquid helium. The consequence has been an unprecedented upsurge of activity in every country in the world with a technology base. Much of this work is directed at seeking new superconductors with higher transition temperatures and to establishing production routes for the materials. Some of the major problems are that the new materials are brittle and, unless they are in the form of very thin films, the current density is rather low $10^3 - 10^4$ A/cm^2; indications are that they will operate at very high flux densities in excess of 50 T. However, good progress is being made with the development of materials and attention is being turned to applications. There are very significant advantages in using liquid nitrogen instead of liquid helium – for example, the efficiency of refrigeration is nearly 50 times better. Many of the organizations in different countries who were working with the earlier lower temperature materials are now re-examining their designs but based upon the use of liquid nitrogen. The impact upon industry is expected to be as important as the silicon chip and many new applications will probably be identified which will open up completely new markets.

3 Plastics and rubber in electrical engineering

Properties of Moulding Materials

Plastics have become established as very important materials for the electrical engineer especially as insulation, but also for structural parts and in some cases as replacements for metals. The term 'plastics' is an omnibus one covering a great number of substantially synthetic materials which have rapidly increasing fields of application.

Plastics can be conveniently divided into two different groups known as thermosetting and thermoplastics materials. The two groups behave differently when external heat is applied. With thermosetting materials the application of heat initially causes softening and during this period the material can be formed or moulded. Continuation of heating, however, results in chemical changes in the material generally resulting in rigid cross-linked molecules. These are not appreciably affected by further heating at moderate temperatures. Overheating, however, may cause thermal decomposition. In addition to heat, thermosetting resins may be hardened (or 'cured') by catalysts, radiation, etc.

When thermoplastics materials are heated, they soften and become less stiff: they may eventually reach a stage at which they become a viscous liquid. On cooling, such a material stiffens and returns to its former state. This cycle of softening and hardening may in theory be repeated indefinitely. Overheating can, however, result in irreversible decomposition.

A major difference between thermosetting and thermoplastics materials is that the former are seldom used without the addition of various reinforcing or filling materials such as organic or inorganic fibres or powders. Thermoplastics are more often used in unfilled form, but fibres, fillers and other additives can be added when special properties are required.

In general, many thermoplastics come into the low loss, low permittivity category whereas thermosetting materials often have higher loss and permittivity values.

Thermosetting Materials

The number of thermosetting materials available to industry is smaller than for the thermoplastics, and mention will be made only of the more important groups used in electrical engineering. These are the phenolic, aminoplastic, polyester, epoxy, silicone, polyimide and polyaralkylether/phenol resins. More details about each of these materials is given later in this section.

As already noted, thermosetting resins are seldom used untreated and the following basic processing methods may be considered as typical.

Laminating. The impregnation of fibrous sheet materials such as glass and cotton materials, cellulose paper, synthetic fibre and mica with resins and forming into sheets, tubes and other shapes by the action of heat and pressure in a press or autoclave.

Compression moulding. Complex-shaped components produced by curing filled and reinforced compounds under heat and pressure in a matched metal mould cavity.

Transfer moulding. Similar to compression moulding but involves transferring fluxed moulding material from a heated transfer pot by means of a plunger through a runner system.

Casting. Used for liquid resin-based compounds such as polyester and epoxy, to form complex shapes. Large insulators and encapsulated components are examples.

Properties. Most thermosetting materials are used as composites and the resultant characteristics are naturally highly dependent on the properties of the constituents. For example, the use of glass fabric to reinforce a particular resin system will produce a material with a higher modulus, a better impact strength, a better resistance to high temperature than a similar material using a cellulose paper as reinforcement. The variations in properties which can be produced by using combinations of the various resins and reinforcing materials is very wide and it is possible to give only a few typical examples.

Phenolic resin (phenol formaldehyde) – PF. This is the bakelite type material so well known in industry. The resin is made by reacting phenolic or cresylic materials with formaldehyde at temperatures around $90-100°C$ either with or without a catalyst. The processing is done in a digester equipped for refluxing, usually with arrangements for the removal of water formed during the reaction. Ammonia, soda or other alkaline catalysts are generally employed although sometimes acid catalysts are used. The final polymerizing or 'curing' time of the resin, which vitally affects its utility, is varied as desired by the manufacturing process. Some resins cure in a few seconds at temperatures around $150°C$ whereas others may require an hour or more. The resins are sometimes in a semi-liquid form but more usually they are solids with softening temperatures ranging from 60 to $100°C$.

Like all thermosetting materials, PF may be compounded with fillers, pigments and other ancillary materials to form moulding compounds which can be processed by several techniques. Perhaps the most widely used method involves curing under heat and pressure in metal moulds to produce the finished article. The fillers may be cheap materials used partly to economize in resin but also to improve performance and often to reduce difficulties due to effects such as mould shrinkage, coefficient of thermal expansion, etc. Examples of some typical fillers are chopped cotton cloth to produce greater strength, and graphite to produce a material that is an electrical conductor with good wearing properties. Wood flour is often used as a general purpose filler. The PF resins are relatively cheap.

As already mentioned this resin can be used to impregnate fabrics, wood veneers or sheets of paper of various types. When these are hot-pressed high strength laminates are produced. The liquid resins suitably modified can be used as adhesives and insulating varnishes.

Depending on the fillers used, PF materials may be considered as being suitable for long-term operation at temperatures in the 120°C to 140°C region although in some forms they may be suitable for even higher temperatures.

The main disadvantages of PF materials are the restriction to dark brownish colours and, from an electrical point of view, the poor resistance to tracking – when surface contaminants are present. Nevertheless, their good all-round electrical performance and low cost make them useful for a wide range of applications. They are employed extensively in appliances and in some electrical accessories.

Aminoplastic resins (ureaformaldehyde – UF – and melamine-formaldehyde – MF). These two resins are aminoplastics and are more expensive than phenolics. Resins are clear and uncoloured while compounds are white or pastel coloured, and are highly resistant to surface tracking effects. They are particularly suitable for domestic applications but should not be used where moisture might be present.

The resins are produced by reaction of urea or melamine with formaldehyde. If the condensation process for UF resins is only partially carried out, useful water-soluble adhesives are obtained. They can be hardened after application to joints by means of the addition of suitable curing agents. Hot-setting MF resins with good properties are also available.

By addition of various fillers UF and MF resins may be made into compounds which can be moulded to produce finished articles. In addition, melamine resins are employed to produce fabric laminated sheet and tubes. In decorative paper-based sheet laminates for heat resistant surfaces the core of the material is generally formed from plies of cellulose paper treated with PF resin but the decorative surfaces comprise a layer of paper impregnated with MF resin.

Depending on the fillers used, UF and MF resin-bonded materials may be considered as being suitable for long-term operation at 110–130°C.

As a result of their good electrical properties and excellent flammability resistance, UF materials are suitable for domestic wiring devices.

Alkyd and polyester resins (UP). The group known as alkyds is mainly used in paints and varnishes but slightly different materials are used for mouldings. These alkyd resins are the condensation products of polybasic acids (e.g. phthalic and maleic acids) with polyhydric alcohols (e.g. glycol and glycerol). They are substantially non-tracking and some newer types, when used in conjunction with fillers, have very high heat resistance.

The unsaturated polyester resins (UP) are usually solutions of unsaturated alkyd resins in reactive monomers, of which styrene is the most commonly used. By the addition of suitable catalysts these resins may be cured at ambient temperatures with zero pressure (contact moulding) to produce large, strong structures at comparatively low capital costs. Alternatively, preimpregnated glass fibre reinforced compounds, known as dough moulding compounds (DMC), may be rapidly cured under high temperatures and pressures to form durable and dimensionally accurate mouldings and insulating sheets with good mechanical properties and thermal stability. Such mouldings are used in contactors. As well as sheets for insulation, glass fibre polyester or reinforced plastics (RP or GRP) mouldings are used for covers and guards, line-operating poles, insulating ladder and many other applications where large, strong, complex insulated components are required.

Silicones. In addition to hard, cured resins, silicone elastomers (rubbers) are also available. These materials have outstanding heat and chemical resistance and their resistance to electrical discharges is excellent. Silicone elastomers can be applied to glass fabrics and woven sleevings to produce flexible insulants suitable for high temperature use. Filled moulding compositions, encapsulating and dielectric liquids based on silicone resins for use at elevated temperatures are also available.

Polyimide resins (PI). A fairly recent development has been the organic polyimide resins which give good performance at temperatures in the 250–300°C region. For this reason, they are generally used with glass or other high temperature fibrous reinforcements. Curing is by heat and pressure and care is needed during the processing operation if good properties are to be obtained. This is because volatile products are evolved during the process although it is hoped that this difficulty will be overcome in time. Electrical properties are also excellent.

Polyaralkylether/phenol resin. Recently developed by the Friedel Crafts route is a resin based on a condensation reaction between polyaralkylether and phenol. The material is similar in mechanical and electrical properties to some of the epoxy resins but having temperature capabilities more like those of silicone resins. At present the cost is approximately between the two. Long-term operation at 220–250°C is possible.

These resins are generally used with glass fabric reinforcements to produce laminates or tubes and with fibrous asbestos fillers to produce mouldings.

Epoxy resins. Epoxy resins are produced by a reaction between epichlorohydrin and diphenylalpropane in an alkaline medium. The electrical properties of this thermosetting material are outstanding with resistance to alkalis and non-oxidizing acids good to moderate. Water absorption is very low and stable temperature range is between about −40°C and +90°C.

Epoxy resins are widely used by the electrical industry for insulators, encapsulating media for distribution and instrument transformers and, when used with glass reinforcement, for printed circuits.

Thermoplastics Materials

The main thermoplastics materials used in electrical engineering applications are discussed below with an indication of their uses. Later their electrical properties are treated in some detail.

Polyethylene (PE). This tough resilient material was first used as an insulant for high frequency low voltage cables in radar and its low-loss properties are also exploited in high performance submarine cables, and as an insulator and sheathing for telephone cables.

Polytetrafluorethylene (PTFE). PTFE is a relatively soft flexible material which is chemically inert, can withstand continuous temperatures in the range ±250°C and has excellent insulating and non-tracking properties over a wide temperature range. It is used as a dielectric and insulator in high temperature cables, as spacers and connectors in high frequency cables, as a hermetic seal for capacitors and transformers and in valve holders.

Polyvinylchloride (PVC). Unplasticized PVC is hard, stiff and tough, and has good weathering, chemical and abrasion resistance. It is employed for conduit and junction boxes. Incorporation of plasticizers can produce PVC compounds with a wide range of flexibilities, for which the main electrical use is in low frequency cable insulation and sheathing and for moulded insulators.

Polypropylene (PP). Polypropylene combines strength, fatigue resistance, stiffness and temperature withstand and excellent chemical resistance. Its good electrical properties are exploited in high frequency low-loss cable insulation. Biaxially oriented polypropylene film is used to make power capacitors of either film-foil or film-paper-foil construction, and also high energy rapid discharge capacitors.

Thermoplastics polyester (PTP). As a biaxially oriented film, polyethylene terephthalate has high dielectric strength, high volume resistivity, flexibility, toughness, excellent mechanical strength and a high working temperature. Major electrical applications include motor insulation, cable wrapping, insulation in transformers, coils, and relays, printed circuit flat cables, and in capacitors as metallized film.

Other plastics. In film form polycarbonate (PC), polyphenylene oxide (PPO) and polysulphone are used as capacitor dielectrics. Many thermoplastics find applications in electrical engineering for mechanical rather than electrical reasons. For example, housings, casings and containers are made from ABS (acrylonitrile-butadiene-styrene), PVC (polyvinylchloride), POM (acetal), PC, PPO PP and nylon. Outdoor illuminated signs make use of CAB (cellulose acetate butyrate), PMM (acrylic) and PVC while diffusers for fluorescent light fittings employ PMM and PS (polystyrene).

Electrical properties. Most thermoplastics are good electrical insulators, some outstandingly so. Often the choice of a plastics for a particular application depends primarily on factors other than electrical properties. For example, mechanical properties such as creep, long-term strength, fatigue and impact behaviour (see BS 4618) are often the deciding factors. Corrosion resistance or thermal stability may also govern the choice. An increasingly important factor is flammability. Many thermosets, PVC, polycarbonate and nylon are inherently flame-retardant, whereas many other plastics materials, when unmodified, will support combustion.

Where electrical properties are important five characteristics are of interest: resistivity, permittivity and power factor, electrical breakdown, electrostatic behaviour and conductance.

Some plastics at room temperature (e.g. highly plasticized PVC) exhibit ohmic behaviour so that the current reaches a steady value. For most plastics, however, the resistivity depends on the time of electrification and *Table 3.3* gives data on the apparent volume resistivity after various times of electrification. Surface resistance values depend on the state of the surface of the plastics, particularly on the presence of hydrophilic impurities which may be present or as additives in the plastics. Results depend greatly on the ambient conditions, particularly on the relative humidity.

The permittivity of many non-polar plastics, e.g. PE, is essentially constant with frequency and changes with temperature may be related to changes in

Table 3.1 Typical properties of some moulding materials*

Type of thermosetting resin	Phenol formaldehyde	Urea formaldehyde	Melamine formaldehyde	Epoxy	Unsaturated polyester
Type of filler	Wood flour/cotton flock	Alpha cellulose	Alpha cellulose	Mineral	Mineral
Density, kg/m³	1320–1450	1470–1520	1470–1520	1700–2000	1600–1800
Heat distortion temperature, °C	125–170	130–140	200	105–150	90–120
Water absorption after 24 h at room temperature, %	0.3–1.0	0.4–0.8	0.1–0.6	0.05–0.2	0.1–0.5
Volume resistivity at room temp, Ωm	10^7–10^{11}	10^{10}–10^{11}	10^{10}–10^{12}	$>10^{13}$	10^{11}
Electric strength at room temp, kV r.m.s./mm	8–17	12–16	12–16	18–21	15–21
Loss tangent at 1 MHz and room temp, tan δ	0.05–0.10	0.025–0.035	0.025–0.050	0.01–0.03	0.015–0.030
Permittivity at 1 MHz and room temp	4.5–6.0	6.0–7.5	7.0–8.0	3.0–3.8	3.5–4.2
Flexural strength at room temp, MN/m²	55–85	70–110	70–110	80–110	60–100
Tensile strength at room temp, MN/m²	45–60	40–90	50–100	60–85	20–35
Elastic modulus at room temp, GN/m²	5–8	7–9	7–9	8–10	7–9
Thermal conductivity perpendicular to surface, W/m°C	0.17–0.21	0.3–0.4	0.25–0.4	0.5–0.7	0.4–0.6

*It should be noted that the values given in this table have been collected from a wide variety of sources and the test methods and specimen dimensions may therefore be such as to make direct comparisons impossible. Wide limits have been given because of the wide types and grades of materials available.

Table 3.2 Typical properties of some laminated materials*

Type of thermosetting resin	Phenol formaldehyde				Melamine formaldehyde	Unsaturated polyester	Epoxy	Silicone	Polyimide	Polyaralkydether/phenols
Reinforcement	Cellulose paper	Cotton fabric	Wood veneer†	Asbestos paper	Glass fabric	Glass fabric	Glass fabric	Glass fabric	Glass fabric	Glass fabric
Density, kg/m³	1340	1330	1300	1700	1700	1750	1770	1650	1850	1770
Water absorption after 24 h in water, %	0.5–3.0	0.5–3	1.5–3	0.1–0.5	0.5–1	0.2–0.5	0.1–0.3	0.1–0.2	0.1–0.2	0.1–0.2
Loss tangent at 1 MHz, tan δ	0.02–0.04	0.04–0.06	0.05	N/a	0.02–0.04	0.01–0.03	0.005–0.02	0.001–0.003	0.01–0.02	0.01–0.03
Permittivity at 1 MHz	5–6	5–6	4–5	N/a	6.5–7.5	4–5	4.5–5.5	3.5–4.5	4–4.5	4.8
Electric strength normal to laminate in oil at 90°C kV.r.m.s./mm	12–24	12–18	3–5	1–5	10–14	10–18	16–18	10–14	16–20	28–34
Electric strength along laminate in oil at 90°C kV.r.m.s./mm	40–50	25–35	25–30	5–15	25–35	30–50	35–45	30–40	60–80	
Flexural strength, MN/m²	140–210	105–140	105–210	140–210	200–310	280–350	350–450	105–175	350–520	520
Tensile strength, MN/m²	85–110	70–105	85–170	105–140	175–240	210–240	240–310	105–175	350–450	350–450
Elastic modulus, GN/m²	6–11	5–8	14–17	7–9	12–15	7–12	12–14	4–9	20–28	35
Coeff. of thermal expansion in plane of sheet per °C × 10⁶	10–15	17–25	8–15	15–25	10–15	10–15	10–15	10–12	5–10	N/a

*Some of the values are very dependent on the direction of the grain. N/a = No data available.

†Some of the values given in the table have been collected from a wide variety of sources and specimen dimensions therefore may be such as to make direct comparison impossible. Because of the wide variations in types and grades available wide limits have been given.

60

Table 3.3 Apparent volume resistivity at 20°C for differing times of electrification

Material	Time of electrification (sec)		
	10	100	1000
Low density polythene	10^{18}	10^{19}	$>10^{19}$
High density polythene	$>10^{16}$	$>10^{16}$	$>10^{16}$
Polypropylene	10^{17}	10^{18}	10^{18}
Flexible PVC	$>10^{14}$	$>10^{14}$	$>10^{14}$
Rigid PVC	2×10^{15}	3×10^{16}	10^{17}
Poly (methyl methacrylate)	2×10^{15}	2.5×10^{16}	10^{17}
PTFE	$>10^{18}$	$>10^{19}$	$>10^{19}$
Polyacetal	–	$>10^{14}$	–
Nylon 6	–	$>10^{12}$	–
Nylon 66	–	$>10^{12}$	–
Nylon 610	–	$>10^{13}$	–
Thermoplastics polyesters (oriented film)	10^{16}	10^{17}	10^{17}
Polyethersulphone (dried)	3×10^{16}	1.5×10^{17}	10^{18}
Polysulphone	4.1×10^{16}	5.2×10^{17}	3.2×10^{18}
Polycarbonate	–	9×10^{15}	–

density using the Clausius–Mosotti relationship. For popular materials these considerations do not apply.

Power factor (loss tangent and loss angle) data have maxima which depend on both frequency and temperature. The levels of dielectric loss can range from a few units of 10^{-6} in tan δ to values as high as 0.3. For low values of loss tangent it is usual to modify tan δ to the angular notation in microradians (1 microradian = 10^{-6} in tan δ units). Materials of high power factor such as PVC transform electrical energy into heat; at high frequencies the heat generated may lead to softening of the plastics and this is the basis of dielectric heating used commercially.

Permittivity and power factor data are best presented as contour maps of the relevant property on temperature–frequency axes, as recommended in BS 4618, but this is beyond the scope of this section. In *Table 3.4* data are given for permittivity and power factor at 20°C at frequencies in the range 100 Hz to 1 MHz. *Table 3.5* indicates the temperature dependence of these properties from −50°C to +150°C at 1 kHz.

When a dielectric/conductor combination is subject to high voltages in the absence of electrical discharges, the effect may be to induce a new set of thermal equilibrium conditions, or to produce thermal runaway behaviour resulting in electrical breakdown. However. with many plastics, failure results from the electrical (spark) discharges before the onset of thermal runaway, either by the production of conducting tracks through or on the surface of the material, or by erosion. This highlights the importance of the standard of surface finish in affecting track resistance. The designer should therefore aim for complete freedom from discharges in assemblies where long life is required.

One consequence of the high resistivities which many plastics have is the presence of electrostatic charge in or on an article. Often associated problems such as dust pick-up and build-up of charge on carpets can be resolved by

Table 3.4 Frequency dependence of permittivity and power factor at 20°C and relative humidity of 50%

Material	Permittivity					Power factor				
	100 Hz	1 kHz	10 kHz	100 kHz	1 MHz	100 Hz	1 kHz	10 kHz	100 kHz	1 MHz
Low density polythene	2.28	2.28	2.28	2.28	2.28	$<10^{-4}$	$<10^{-4}$	$<10^{-4}$	$<10^{-4}$	1.5×10^{-4}
High density polythene	2.35	2.35	2.35	2.35	2.35	$<10^{-4}$	$<10^{-4}$	$<10^{-4}$	$<10^{-4}$	$<10^{-4}$
Polypropylene	2.3	2.3	2.3	2.3	2.3	$<5 \times 10^{-4}$	$<5 \times 10^{-4}$	$<5 \times 10^{-4}$	$<5 \times 10^{-4}$	$<5 \times 10^{-4}$
Flexible PVC	5–6	5–6	4–5	3–4	3	$\sim 10^{-1}$	$\sim 10^{-1}$	$\sim 10^{-1}$	$\sim 10^{-1}$	$<10^{-1}$
Rigid PVC	3.5	3.4	3.3	3.3	–	10^{-2}	$\sim 2 \times 10^{-2}$	$\sim 2 \times 10^{-2}$	$\sim 2 \times 10^{-2}$	–
Poly (methyl methacrylate) dried	3.2	3.0	2.8	2.8	2.7	5×10^{-2}	4×10^{-2}	3×10^{-2}	2×10^{-2}	1.5×10^{-2}
PTFE	2.1	2.1	2.1	2.1	2.1	$<5 \times 10^{-5}$	$<5 \times 10^{-5}$	$<5 \times 10^{-5}$	$<5 \times 10^{-5}$	$<10^{-4}$
PCTFE	2.7	2.65	2.56	2.5	2.48	1.9×10^{-2}	2.5×10^{-2}	2×10^{-2}	1.2×10^{-2}	8×10^{-3}
Polyacetal	3.7	3.7	3.7	3.7	–	2×10^{-3}	2.5×10^{-3}	2.8×10^{-3}	3.2×10^{-3}	–
Nylon 6	–	4	–	–	3.6	–	3×10^{-3}	–	–	6×10^{-2}
Nylon 66	4.9	4.5	4.0	3.7	3.4	$\sim 5 \times 10^{-2}$	$\sim 6 \times 10^{-2}$	$\sim 6 \times 10^{-2}$	$\sim 6 \times 10^{-2}$	$\sim 5 \times 10^{-2}$
Nylon 610	3.9	3.7	3.5	3.4	3.2	$\sim 3 \times 10^{-2}$	$\sim 4 \times 10^{-2}$	$\sim 4 \times 10^{-2}$	$\sim 4 \times 10^{-2}$	5×10^{-2}
Polyesters (linear oriented film)	3	3	3	3	–	3×10^{-3}	7×10^{-3}	1.2×10^{-2}	1.5×10^{-2}	–
Polycarbonates	2.96	2.95	2.95	2.95	–	1.2×10^{-3}	1×10^{-3}	2×10^{-2}	5×10^{-3}	–
Polyethersulphone, dried	3.6	3.6	3.6	–	–	1.5×10^{-3}	2.1×10^{-3}	3.1×10^{-3}	–	–
Polysulphone	3.16	3.16	3.16	–	–	2.2×10^{-3}	9.7×10^{-4}	1.6×10^{-3}	–	–

Table 3.5 Temperature dependence °C of permittivity and power factor at 1 kHz

Material	Permittivity					Power factor				
	−50	0	50	100	150	−50	0	50	100	150
Low density polythene	2.28	2.28	2.28	2.28	2.28	$<10^{-4}$	$<10^{-4}$	$<10^{-4}$	$<10^{-4}$	–
Polypropylene	2.3	2.3	2.3	2.3	2.3	$<5 \times 10^{-4}$	$<5 \times 10^{-4}$	$<5 \times 10^{-4}$	$<5 \times 10^{-4}$	–
Flexible PVC	3	3–4	7	7–8	–	$\sim2 \times 10^{-2}$	$\sim8 \times 10^{-2}$	$\sim8 \times 10^{-2}$	$\sim2 \times 10^{-2}$	–
Rigid PVC	3.1	3.3	3.7	12	–	$\sim2 \times 10^{-2}$	$\sim2 \times 10^{-2}$	$\sim2 \times 10^{-2}$	1×10^{-1}	–
Poly (methyl methacrylate) dried	2.7	2.9	3.3	4.2	–	1×10^{-2}	3×10^{-2}	7×10^{-2}	8×10^{-2}	–
PTFE	2.1	2.1	2.05	2.0	1.95	5×10^{-5}	$<5 \times 10^{-5}$	$<5 \times 10^{-5}$	$<5 \times 10^{-5}$	$<5 \times 10^{-5}$
PCTFE	2.43	2.55	2.75	2.83	2.89	5×10^{-3}	2×10^{-2}	1×10^{-2}	3×10^{-3}	2.5×10^{-3}
Polyacetal	3.5	3.7	3.75	3.8	–	2×10^{-2}	2.5×10^{-2}	1.5×10^{-3}	2.5×10^{-2}	–
Polyester (linear) oriented film	3	3	3	3	3.3	1.2×10^{-2}	1×10^{-2}	3×10^{-5}	5×10^{-3}	8×10^{-3}
Polycarbonate	3	3	3	3	3	6×10^{-3}	1.5×10^{-3}	1×10^{-3}	1×10^{-3}	2×10^{-3}
Polyethersulphone	–	–	3.57	3.56	3.54	–	–	8×10^{-4}	1.2×10^{-3}	7.4×10^{-4}

applying a hygroscopic coating to the surface or by incorporating an anti-static agent in the plastics which will migrate to the polymer surface and reduce surface resistivity. Where a spark discharge could lead to a hazardous situation, anti-static grades of plastics should be used.

Electrically conducting plastics are usually produced by incorporating high concentrations of filamentary carbon black or certain pyrolized materials into an otherwise electrically insulating polymer matrix.

Rubber in Electrical Engineering

Rubbers are organic compounds derived from the latex plant or by synthesis. They are characterized by some unique qualities, notably high elasticity. Natural or synthetic latex is the chief material of the rubber industry and vulcanized rubber in its many forms is the main end product. To be classed as a rubber rather than a plastic, a material must meet four requirements:

1. It must consist of very long-chain molecules.
2. The long-chain molecules must be lightly cross-linked (vulcanized).
3. The glass transition temperature must be well below room temperature.
4. The melting point of the crystallites, if any, must also be below room temperature.

Vulcanized rubber meets all four requirements and finds applications in the cable industry and other electrical fields. The temperature limits in degrees Celsius for practical applications of the various rubbers is given in *Table 3.6*.

Table 3.6 Temperature limits in °C of some rubbers

Natural rubber	−50 to 60
Styrene butadiene rubber	−45 to 65
Butyl rubber	−30 to 80
Chloroprene rubber	−20 to 70
Nitrile rubber	−10 to 90
Polysulphide rubber	−50 to 100
Silicone rubber	−70 to 150
Polyurethane rubber	−55 to 90
Fluor rubber	−30 to 150
Ethylene propylene rubber	−50 to 90

The electrical properties of some rubbers are shown in *Table 3.7*. A glance at the table reveals that in general the material possesses good electrical properties with the exception of nitrile polychloropropene and polyurethane. The latter are not suitable for high frequency applications because of high dielectric losses. Ebonite is no longer used for electrical purposes as it has been replaced by plastics like polyethylene, but is included in the table for completeness.

Table 3.7 Electrical properties of rubber

Material	Volume resistance in 10^{-2} ohm-cm	Dielectric constant at 1 kHz	Tan δ at 1 kHz	Breakdown voltage in kV/cm (50 Hz after 2 hours)
Natural rubber	10^{15}–10^{17}	2.3–3.0	0.0025–0.0030	210
SBR	10^{15}	2.9	0.0030	260
Butyl rubber	10^{17}	2.1–2.4	0.0030	220
Nitrile rubber	10^{10}	13.5	0.055	165
Polychloroprene rubber	10^{11}	9.0	0.030	225
Polysulphide rubber	10^{12}	7.0–9.5	0.001–0.005	200
Silicone rubber	10^{11}–10^{17}	3.0–3.5	0.001–0.010	220
Fluor rubber	10^{18}	2.0	0.0002	250
Polyurethane rubber	10^{11}	7.0	0.017–0.09	165
Polyethylene rubber	10^{15}–10^{19}	2.3	0.0005	200
Ebonite	10^{14}–10^{16}	2.8–4.0	0.005	250

Gutta-percha, an isomeric form of natural rubber, which has been used in sea cables, has also been replaced by polyethylene.

The cable industry is a large user of rubbers for insulating and sheathing purposes. Ethylene propylene rubbers (EPR) and the hard version (HEPR) are specified in BS 6899 and BS 6469 respectively. The latter is suitable for use at temperatures up to 90°C.

EPRs have good ozone resistance making them suitable for high voltages. Like natural rubber the material burns and is not particularly oil resistant so insulated cable cores have to be provided with some kind of sheathing.

Butyl rubber was used as a cable insulating medium but its rather limited mechanical strength has caused it to lose favour, although it is still used for some wiring cables. Nevertheless it can be operated in high ambient temperatures and is therefore used sometimes as the entry cable in a luminaire.

Silicone rubber is becoming increasingly popular as the insulating medium for fire resistant cables.

4 Semiconductors and semiconductor devices

Semiconductors

Following on from the early pioneers of electrical science, Faraday's work in the late nineteenth century laid the foundation for the growth of the electrical power industry. This resulted in the development of electromagnetic machines: motors, generators and transformers, together with the necessary switchgear and cables for the transmission and distribution infrastructure to support them. Although greatly improved and refined, much of this equipment and its under-lying principles is recognizably similar to that developed eighty to a hundred years ago. The basic materials of this electrical power industry were copper, iron, and the various insulators described in the two preceding chapters.

In the middle of the twentieth century a new material appeared on the scene. This was to revolutionize electrical technology in the latter half of the century and the early part of this century. It changed the field of electronics beyond recognition and resulted in a blurring of the once clearly definable boundaries of electronics and electrical power. This material is silicon.

True, telephony, radio, radar, and even computers had been developed using thermionic valves in the first half of the twentieth century but, until they were able to benefit from the advantages of the silicon chip, none of these technologies possessed the scope, intricacy and sophistication that we currently take for granted.

It should be noted that while silicon is not the only material possessing semiconductor properties and although few of its properties might be consid-ered ideal, none are so far from ideal as to exclude its use in all but a small fraction of present day devices.

Pure silicon crystals are comparatively good insulators since the crystal lattice does not possess the free electrons normally associated with the trans-mission of an electric current in materials such as copper and silver. This is because the silicon crystal is based on a tetrahedral lattice structure and the silicon atom has four outer electrons, which are locked into the crystal lattice formed by the individual atoms.

If a very small quantity, of the order of 1 p.p.m., of an impurity such as phosphorus or arsenic, having five outer electrons, is added to the crystal this results in a number of atoms within the crystal having one 'surplus' electron. This 'free' electron is available to move within the crystal under the influence of any applied potential difference, thus rendering the material a conductor. Because the conduction is by means of movement of a negative charge as in the conventional current flow, material containing such an impurity is known as *n type*. Alternatively an impurity such as boron or gallium which has three outer electrons can be added to the crystal. These result in an electron 'shortage' at points within the lattice, which are termed *holes*. Holes can also move within the crystal under the influence of a potential difference and, since the presence of a hole results in the crystal having a net positive charge, material thus treated is said to be *p type*.

The intersection of a p-type material with an n-type one gives rise to a p-n junction which is a rectifier, but it is also the basic building block of almost all other semiconductor devices.

The desired impurities may be added to the melt during the growth of the single-crystal material. A long single crystal may be produced up to 200 mm diameter. This will then be sawn many times along the same well-defined crystal plane and each slice is then ground and polished on one of its two faces. Further doping may then be carried out during fabrication of the various devices by the introduction of impurity atoms in a furnace at temperatures of around 1000°C. This may be done by creating a mask of silicon dioxide on the slice surface which is then etched to reveal the silicon in the required pattern. This ensures that the further doping will only be carried out in the required areas. An alternative more costly but more accurate technique is to use ion implantation. This involves firing a beam of suitably charged ions at the surface of the slice at the points where the additional doping is required. Because of the surface damage caused by the ion bombardment the slice must be annealed for a prolonged period following this process.

Early semiconductor equipment used discrete devices, i.e. individual diodes, transistors or thyristors to produce amplifiers, logic circuits, or rectifiers in a similar manner to those which had been designed using valves. But the major benefits which accrue from the use of semiconductors in electronic equipment arise when these are incorporated into *integrated circuits*. It should be noted, however, that for power devices, i.e. those carrying currents in the range of several hundred milliamps to possibly 5 kA, discrete devices are still the norm.

There are many advantages from the use of integrated circuits and there is constant pressure to make these integrated circuits smaller in order to maximize the advantages. One important benefit to be obtained from size reduction is in minimizing the effects of defects in the silicon crystal. In many integrated circuits the silicon chip will carry a large number of identical circuits, for example to perform a logic function in a computer. If the chip is large there will be an increased likelihood that it will contain defects. It must, therefore, have a number of redundant circuits to allow for the defective ones to be discarded. Clearly, if the circuits can be made smaller the chip can be smaller and the likelihood of defects will be reduced. It is also the case, to a certain degree, that the smaller the circuits the lower the voltages which need to be applied to produce the required movement (mobility) of electrons or holes within the material and achieve a given speed of response. Reduction in the applied voltage reduces operating power requirement and, of course, heat generation due to losses. Reduction in the size of the components coupled with reduction of input power, losses, and cooling requirements also has benefits by way of reduction of overall size and cost of equipment.

Thick- and thin-film microcircuits. Thick and thin conducting films are used in microcircuits to connect active components such as transistors and to replace passive components such as resistors and capacitors. Where a mixture of thick and thin films is used to connect discrete active devices, the overall system is known as a *hybrid microcircuit*. Microcircuit construction principles were developed from the techniques used during the second world war to reduce the size and weight of electronic circuitry. These used printed carbon resistors on ceramic substrates.

Thick-film hybrids use screen printing techniques to deposit patterns of conducting pastes onto ceramic substrates. These are suitable for relatively simple circuits duplicated in large quantities as used in, for example, telephone exchanges. Firing at high temperature fuses the pastes to the substrates to form circuit elements of less than 0.025 mm thick. Even so, these remain 'thick' films compared to the thickness of those used in 'thin'-film technology. Other substrate materials such as high purity alumina can be used and are constantly being developed. The requirement is that these must be insulating, flat, non-reactive, and thermally stable.

Multilayer hybrids can be built up from layers of conductor tracks separated by layers of dielectric glaze and interconnected windows (called *vias*) in the glaze.

To attach components with solder, solder cream is printed over conductor pads in the substrate, into which the component feet are placed prior to reheating of the solder. The solder cream contains the solder and flux and is printed through a coarse mesh screen.

All types of components can be attached to hybrids in this way including semiconductor chips packaged in a variety of ways, as well as capacitors ranging in size from 1.25 mm × 1 mm to 6 mm × 5 mm and in value from 1 pF to 1 μF.

The preferred substrates for thin-film microcircuits are glass or 99.6% pure alumina. The surface finish needs to be very smooth to allow the deposition of a uniform conductive metal film. Typically, resistor lines may be no more than 10 μm wide with conductor lines 50 μm in width. Circuit designs must be produced on a magnified scale and reduced photographically. A thin film of photosensitive material is coated onto the surface and the pattern is exposed and developed. This film is known as a photoresist. The metal film can then be etched away in areas not protected by the photoresist to provide the required circuit patterns.

Thin-film circuits produced in a matrix on glass substrates in this way can be diced using a diamond wheel. The individual circuits are then assembled using similar methods to those employed for thick films, although generally when assembled they are placed into hermetically sealed packages for protection.

Semiconductor devices

Transistors. The first semiconductors in commercial use were *transistors*. As long ago as the 1870s a German engineer called Braun discovered that when a wire made contact with certain crystals an electric current would flow freely in one direction only. The first use of Braun's discovery was the 'cat's whisker' used in early radio receivers.

In 1947 experimenters at the Bell Telephone Laboratory in the USA found that when two wires were brought into contact with certain crystals, a small current flowing from one wire to the crystal could control the flow of a very much larger current flowing in the other wire. This was the first transistor amplifier and was developed commercially as the *point-contact transistor*. The crystal used was n-type germanium. Such a device is shown in *Figure 4.1*. Operation is explained by reference to *Figure 4.2*.

If a potential, positive with respect to the base electrode, is applied to the emitter and a negative bias to the collector, it is found that as the current to the emitter is varied by the voltage, a corresponding and greater current

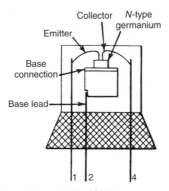

Figure 4.1 Germanium point-contact transistor

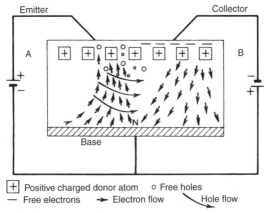

+ Positive charged donor atom	o Free holes
— Free electrons	→ Electron flow
	⟍ Hole flow

Figure 4.2 Operations of the point-contact transistor (RCA Ltd)

change appears in the collector circuit, thus giving rise to an amplification effect. Amplifications up to 100 times (20 dB) can be obtained in this way.

Junction-type transistor. In practice, the point-contact transistor has several limitations and, in consequence, the junction-type transistor was developed. This consists of a suitably prepared thin section of p-type silicon sandwiched between two large crystals of n type. The centre section is known as the base and performs the control function while the outer n-type sections, known as the emitter and collector respectively, correspond to the main current path. This type of transistor is known as the *n-p-n* transistor. It is possible to produce a transistor having n-type material sandwiched between layers of p type, in which case the transistor is said to be of the *p-n-p* type.

Operation. In order to understand the operation of the p-n-p or n-p-n junction transistor it is necessary to consider the conditions existing on both sides of the particular p-n junction. In the n-type material there is a concentration of mobile electrons while on the p-type side there is a concentration of mobile holes. Due to a potential barrier existing at the junction it is not possible for the free electrons to move across to the p-type side or for the holes to move across to the n-type side.

If a d.c. voltage is connected to the p-n junction as shown in *Figure 4.3(a)* an increased bias is exerted on the junction causing the free electrons and the free holes to move away from the junction. When the battery connection is reversed, *Figure 4.3(b)*, the external potential 'overcomes' the internal potential barrier and there is a movement of electrons and holes from the respective materials across the junction and round the outside circuit. This movement constitutes a current flow.

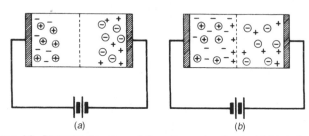

Figure 4.3 Diagrammatic representation of a p-n junction. In (*a*) the battery is connected in such a way as to increase the bias. In (*b*) the bias is removed by reversing the battery, which now assists current flow

Thus it will be seen that the junction acts as a rectifier, with conduction taking place in one direction only. However, it must be appreciated that if the reverse voltage is too large *Figure 4.4* a breakdown of the potential barrier will result and the rectifying qualities of the transistor will be destroyed.

The characteristic curve for a junction transistor is shown in *Figure 4.4*. From the curve it will be noticed that for a small applied voltage in the conducting or forward direction a comparatively large current flows. However, when a large reverse voltage is applied very little current flows until the breakdown point is reached. It is obvious therefore that for correct operation the device must not be used in a circuit where the system voltage exceeds the breakdown voltage.

Theory of action. The p-n-p junction transistor structure is shown schematically in *Figure 4.5*. The letters *e*, *b* and *c* indicate respectively the emitter, base and collector. Two p-n junctions have been formed in the transistor and across each there will be a potential barrier as described earlier. The arrows in the diagram show the direction of the electric fields across the two junctions and are the directions in which positive charges would move under the influence of the electric field.

If the d.c. voltage is applied as shown in *Figure 4.5(a)*, the right-hand p-n junction is biased in the reversed or non-conducting direction, the length of the arrows serving to indicate the strength of the electric fields at the junctions. Under these conditions the collector current is very small, and in a good transistor is only a few microamperes.

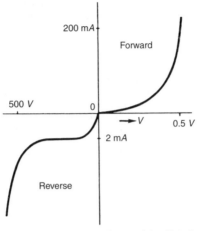

Figure 4.4 Characteristic curve for a junction transistor. Note the different scales for the voltages

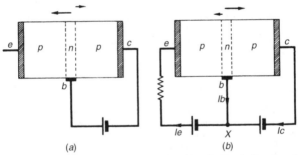

Figure 4.5 A p-n-p junction transistor. The arrow lengths denote the strengths of the fields at the junctions, as well as their direction of operation

When the voltage is applied as shown in *Figure 4.5(b)*, it will be seen that the field across the barrier, in the forward or conducting direction, is reduced, so that it is relatively easy for holes and electrons to move across the barrier. But the material of the transistor is such that there is a large concentration of holes in the p-region and relatively few electrons in the n-region so that the current flowing consists almost entirely of holes and the emitter is said to inject holes into the n-region.

These holes move across the emitter-base junction into the n-region, which has a width of about 0.005 mm, and find themselves near the base-collector junction across which the field is such as to drive the holes towards the collector. Thus the effect of applying a voltage to the emitter is to cause an increase in the collector current by an amount nearly equal to the emitter current. An increase in the emitter voltage causes an increase in emitter and collector current. The amount of emitter current may be several milliamperes and so the collector current due to this is very much larger than the small collector current which flows when the emitter current is zero.

The mode of action of the n-p-n junction transistor is similar to the above except that the applied voltages are reversed and the part played by the holes is taken over by the electrons.

Field effect transistors. Field effect transistors (FET) may be divided into two main categories, namely junction and insulated gate. Basically a junction FET is a slice of silicon whose conductance is controlled by an electric field acting perpendicularly to the current path. This electric field results from a reversed-bias p-n junction and it is because of the importance of this transverse field that the device is so named.

One of the main differences between a junction FET and a conventional transistor is that for the former current is carried by only one type of carrier, the majority carriers. In the case of the latter both majority and minority carriers are involved. Hence the FET is sometimes referred to as a unipolar transistor while the conventional type is called a bipolar transistor.

Another important difference is that the FET has a high input impedance, while ordinary transistors have a low input impedance. Because of this, junction FETs are voltage-operated as opposed to current-operated bipolar transistors.

Insulated gate FETs utilize a sandwich type of construction consisting of a conducting silicon surface in contact with an isolating layer of metal oxide and a layer of semiconductor. Such a device is known as a MOSFET – metal oxide silicon field effect transistor. The metal layer is the control electrode or gate. The metal oxide layer allows the field produced by the gate to control the operation of the semiconductor while preventing any d.c. current flow from the gate to the other electrodes.

Power semiconductors

The above descriptions of transistor devices are given to enable an understanding to be gained of their principles of operation; however as indicated above, discrete semiconductor devices nowadays tend only to be used for power applications. The following descriptions of present day practical discrete devices are therefore related to power semiconductors, of which the most important and widely used are thyristors.

Although power semiconductors utilize the principles already described, their role is generally that of a switch or rectifier rather than as an amplifier or detector as is frequently the case for transistors used in microelectronic circuits.

In their operation as switch or rectifier, power semiconductors are aiming to achieve infinite off-state resistance and zero on-state resistance, an ideal which many modern devices can closely approach. The secondary objectives are, of course, to achieve the above for ever increasing levels of power at an economic cost and with a low level of losses. To achieve these secondary objectives demands a range of different devices each of which strikes a different compromise, generally trading cost against rated voltage, current, losses and operating frequency.

Silicon diodes. Almost all modern power semiconductor devices are made using crystalline silicon. The simplest uses a single junction of p- and n-type material, which through the basic mechanism described above produces rectification. Single-element silicon diodes are available with ratings up to 6 kV and more than 5 kA average current. Typically the largest silicon diodes have a forward volt drop of 1.1 V at 3 kA and a reverse leakage current of less than 100 mA at 5 kV. This low level of forward volt drop is achieved by carefully arranging the silicon doping process to control the disposition of current carriers within the material to provide a mechanism of operation known as *charge modulation*. The very low forward volt drop and reverse leakage current lead to very low power loss, to be dissipated within the device, compared to the power controlled or rectified, and the high efficiency of this type of diode. Normally this type of diode will be mounted in a package which can be double side cooled.

Bipolar power transistors. Almost all bipolar power transistors are n-p-n devices which also rely on an arrangement of silicon dosing to enable charge modulation to occur and thus provide a very low forward voltage drop. The bipolar power transistor is basically a switch and it is operated either in the fully off or in an on state, being driven between these two conditions by means of the bias voltage applied to the base. In the off state the base is held at zero or negative bias to minimize the collector leakage current. Typical ratings at the present time are continuous currents up to 1000 A with peak currents of 1200 A and collector emitter sustaining voltages of 1 kV.

The Darlington bipolar power transistor. This is a single-chip derivation of the n-p-n bipolar power transistor described above which has been specially developed to provide high gain.

Thyristors. The generic term thyristor covers a large and important group of semiconductor devices. It is therefore worthwhile looking in more detail at their basic method of operation.

The basic thyristor is also known as the silicon controlled rectifier (SCR) and is the oldest controllable power semiconductor device. It is still the most widely used in MV applications. Its structure is shown in *Figure 4.6*. As can be seen from the figure, the thyristor is a four layer p-n-p-n junction silicon device which operates as a controllable rectifier. The structure is best visualized as consisting of two transistors, a p-n-p and an n-p-n interconnected to form a regenerative feedback pair as shown in *Figure 4.6*. Current gain around the internal feedback loop G is $h_{fe1} \times h_{fe2}$ where h_{fe1} and h_{fe2} are the common emitter current gains of the individual sections.

If I_{col} is the collector to base leakage current of the n-p-n section and I_{co2} is the collector to base leakage of the p-n-p section, then

for the p-n-p section : $I_{c1} = h_{fe1}(I_{c2} + I_{col}) + I_{col}$

for the n-p-n section : $I_{c2} = h_{fe2}(I_{c1} + I_{co2}) + I_{co2}$

and the total anode-to-cathode current $I_a = (I_{c1} + I_{c2})$

$$\text{from which } I_a = \frac{(1 + h_{fe1})(1 + h_{fe2})(I_{col} + I_{co2})}{1 - (h_{fe1})(h_{fe2})}$$

With a proper bias applied, i.e. positive anode to cathode voltage, the structure is said to be in the forward blocking or high impedance 'off' state. The switch to the low impedance 'on' state is initiated simply by raising the loop gain G to unity. As this occurs the circuit starts to regenerate, each

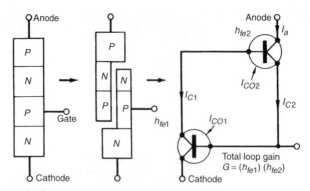

Figure 4.6 Two-transistor analogue of thyristor

transistor driving its partner into saturation. Once in saturation all junctions assume a forward bias, and the total potential drop across the device approximates to that of a single junction. Anode current is then only limited by the external circuit.

To turn off the thyristor in a minimum time it is necessary to apply a reverse voltage and under this condition the holes and electrons in the vicinity of the two end junctions will diffuse in these junctions and result in a reverse current in the external circuit. The voltage across the thyristor will remain at about 0.7 V positive as long as an appreciable reverse current flows. After the holes and electrons in the vicinity of the two end junctions have been removed, the reverse current will cease and the junction will assume a blocking state. The turn-off time is usually of the order of $10-15\,\mu s$. The fundamental difference between the transistor and thyristor is that, with the former, conduction can be stopped at any point in the cycle because the current gain is less than unity. This is not so for the thyristor, conduction only stopping at a current zero.

Other power devices

There is a sizeable family of other two-junction (transistors) and three-junction (thyristors) which have been developed – and continue to be further developed – to meet specific requirements, which usually mean higher voltage or current, lower losses, or faster switching. The following is an outline of some of the more important types.

Thyristors

Gate turn-off thyristor (GTO). This has a similar operating principle to that of a conventional thyristor. It is shown diagrammatically in *Figure 4.7*. It is a three-terminal device reliant on complex gating and snubber circuitry to control switching transients. The turn-off gain is typically 4–5. Currently maximum blocking voltage is of the order of 6 kV with current ratings of 4 kA (average anode current <2 kA). Switching frequency is typically less than 1 kHz.

Integrated gate commutated thyristor (IGCT). This is a development from the GTO to achieve improved switching performance and with the objective

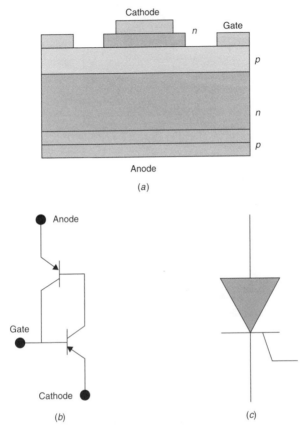

Figure 4.7 Gate turn off thyristor

of simplifying snubber circuitry. A gate drive is mounted in close proximity to the GTO and the switching action is achieved by employing a large gate-current pulse. It has a turn-off gain of unity. Maximum blocking voltage is of the order of 5.5 kV with peak current ratings of 1.8 kA (1.2 kA r.m.s.). At the present time practical IGCTs can operate up to about 500 Hz.

Metal oxide silicon turn-off thyristor (MTO). This device arises out of the objective of achieving the same power handling capability as the previous device but with lower switching losses and a simpler gate drive. See *Figure 4.8*. It is still under development and appears to be promising. The small turn-off gate allows substantial cost saving and improved reliability compared to GTOs and IGCTs. The MTO structure lends itself to double cooling.

Insulated gate bipolar transistor (IGBT). This device has the low conduction loss of a bipolar junction transistor and the switching speed of a MOSFET.

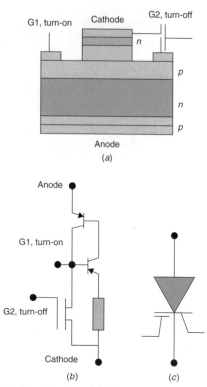

Figure 4.8 Metal oxide silicon turn off thyristor

See *Figure 4.9*. It is voltage driven and suitable for snubberless operation. High voltage IGBTs have a positive temperature coefficient of resistance in the on state and this enables them to be paralleled. A basic chip can thus be rated at 50, 75 or 100 A and higher current ratings achieved by paralleling IGBT chips inside a single package. Compared with other devices the IGBT has a high forward voltage drop but switching losses are low compared with GTOs. Recent IGBT technology has produced reliable low cost devices for up to 3.3 kV, 1200 A r.m.s. Short-circuit protection is possible by a simple gate drive control. Under short-circuit the IGBT current is inherently limited by the device characteristics and can be reduced by a reduction in applied gate emitter voltage before turn-off. Gate drive power requirement is very low; less than 1 W for a 3.3 kV 1.2 kA device operating at 2 kHz. Higher frequency switching means lower harmonic content in the output waveform, lower losses in the driven plant, reduced filtering requirements and lower acoustic noise. To achieve a higher blocking voltage switch, power devices can be connected in series. Heavy investment in IGBTs by almost all semiconductor manufacturers is currently leading to a reduction in their cost.

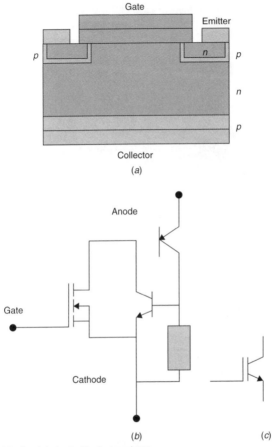

Figure 4.9 Insulated gate bipolar transistor

Cooling. High power thyristors have to dissipate any losses in a small volume and thus artificial means are often provided to take the heat away from the devices. It is usual therefore to mount the thyristors on aluminium alloy castings which act as heatsinks, see *Figure 4.10*. For very high power installations this natural cooling may be supplemented by forced draught means.

Applications of Power Semiconductors

The chief use of diodes is as d.c. supplies for electrolytic process lines, for traction substations and for general industrial purposes. Early uses of thyristors

Figure 4.10 Stack assembly of thyristors showing the heatsink arrangements. It is rated to control up to 120 A on a 415 V a.c. system (International Rectifiers)

included phase-controlled converters producing variable d.c. supplies for motor drives, turbine-generator and hydro-generator excitation, vacuum-arc furnaces, electrochemical processes, battery charging, etc. These are still major applications, but increasingly thyristors and power transistors are being employed in new switching modes as d.c. choppers or as inverters producing a.c. of fixed or variable frequency. Other applications for thyristors are given below.

Excitation of synchronous motors. Most specifications for synchronous motors state that they must be capable of withstanding a gradually applied momentary overload torque without losing synchronism. This was detailed in BS 2613, withdrawn in 1978 and superseded by BS 5000 Part 99. The latter standard does not have a similar section dealing with this matter.

In the past the overload torque requirement has been met by using a larger frame size than necessary to meet full-load torque requirements, the torque being proportional to the product of the a.c. supply voltage and the d.c. field produced by the excitation current.

With thyristor control of excitation current it is possible to use a smaller frame size for a given power rating and arrange to boost the excitation by means of a controller to avoid loss of synchronism under torque overload conditions.

The excitation current of a synchronous motor may be controlled by supplying the motor field winding from a static thyristor bridge, using the motor supply current to control the firing angle, *Figure 4.11*. A pulse generator varies the firing angle of the thyristors in proportion to a d.c. control signal from a diode function generator. Variable elements in the function generator enable a reasonable approximation to be made to any of a wide range of compensating characteristics.

When the motor operates asynchronously, i.e. during starting, a high e.m.f. is induced in the field winding, and the resulting voltage appearing across the bridge must be limited to prevent the destruction of the bridge elements. This may be done by using a shunt resistor connected as shown. Where more exacting requirements have to be met, current feedback can be applied to

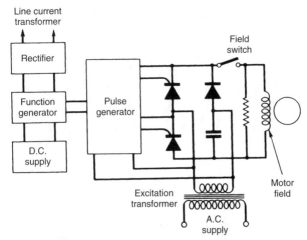

Figure 4.11 Simple compensated excitation circuit

eliminate effects of non-linearity in the pulse generator and rectifier bridge and it will also improve the response of the system to sudden changes of load. Automatic synchronizing is possible without relays by incorporating a slip-frequency sensing circuit to control the gate which supplies the control signal to the pulse generator.

Variable frequency supply. It is possible to use a cycloconverter to control the speed of an induction motor. The cycloconverter is a rectifier device first developed in the 1930s but with the improved control characteristics of thyristors and better circuit techniques, a continuously variable output frequency is possible.

Figure 4.12 illustrates the process of conversion for 15 Hz output from a 50 Hz supply. During the conduction cycles starting at point *a* the output voltage reaches a maximum since there is no firing delay. At point *b* commutation from phase two to phase three is slightly delayed and commutation is further delayed at point *c*.

At time *e* the firing delay is such that the mean output voltage is only just possible. In the diagram the low frequency load power factor is 0.6 lagging so that although the mean load voltage crosses the axis at *X* the current remains positive until *Y*. Consequently the rectifiers which conduct at instants *f*, *g* and *h* are giving positive load current and negative voltage, that is inverting. From *i* the system behaves as a controlled rectifier using the negative group of thyristors, *d*, *e* and *f* until the mean output voltage becomes positive when an inversion period starts again.

D.C. chopper controlling a d.c. motor. The basic circuit is shown in *Figure 4.13*, the supply being applied to the motor by firing *T*1. At standstill the current rises rapidly to a value controlled by the circuit resistance and at a rate controlled by the inductance of the motor.

After a certain interval *T*1 is switched off, the flow of current round the motor and through the diodes *D*1 being maintained by the inductive energy

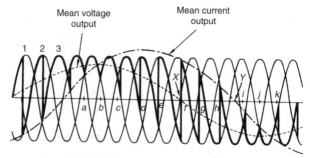

Figure 4.12 Synthesis of 15 Hz from 50 Hz (load power factor 0.6 lagging)

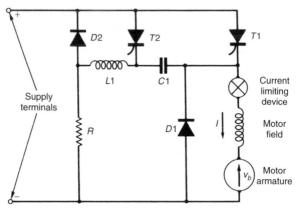

Figure 4.13 Basic circuit

of the system. When this current has decayed to a certain level, $T1$ is again fired. In order to maintain the constant motor current the current limiting device provides a control signal which varies the length of the times $T1$ is on or off.

In *Figure 4.13* the circuit components $T2$, $C1$, L, R and $D2$ are included for the purpose of switching off $T1$.

Control of fluorescent lighting banks. Dimming of fluorescent lighting by means of thyristors is possible, the circuit being shown in *Figure 4.14*. A separate transformer is necessary to maintain constant preheating currents to the electrodes irrespective of the setting of the dimmer. This entails use of a 4-wire distribution system. Control is effected by simply varying the firing angle of the thyristor.

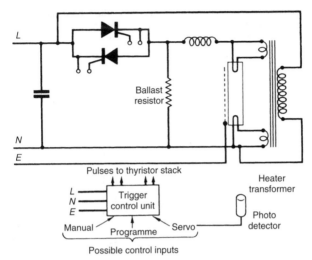

Pulses to thyristor stack

L ——
N —— Trigger
E —— control unit

Manual   Programme   Servo

Heater transformer

Photo detector

Possible control inputs

Figure 4.14 Circuit for control of fluorescent lamp, including dimming facilities

Thermionic Devices

As indicated in the introduction to this chapter, many of the functions which are now performed by semiconductors were, in the early days of electronics, carried out using *thermionic* devices, that is they achieved their operation by controlling the flow of electrons between two electrodes, generally in a vacuum but occasionally in a gas. The control could be simply on/off depending on the polarity of the applied voltage or could be modulating control by the action of a voltage applied to a further electrode, or electrodes. The supply of electrons was produced from a *cathode* coated with a material which freely emits these on heating. The cathode was usually indirectly heated by means of a low voltage heating element. This gave the devices the generic name of thermionic *valves,* presumably because they enabled current flow in a circuit to be regulated in a similar manner to the regulation of fluid flow by a valve in a pipework system. Nowadays few thermionic devices are used. Those which are, are almost exclusively high power devices, thyratrons, magnetrons, klystrons and ignitrons used in telecommunications transmission and radar.

Hydrogen thyratron. The hydrogen thyratron has become the most widely used switching device for medium and high power radar pulse modulators and other applications where accurately timed high power pulses are required. Thyratrons were first used in radar systems but they are now being employed in various circuits concerned with particle accelerators and high power energy diverter systems. The English Electric Valve Co.'s range includes both hydrogen and deuterium-filled triode and tetrode thyratrons with either glass or ceramic/metal envelopes. Multigap ceramic tubes are available for operation up to 160 kV.

Basically all thyratrons consist of a gas-filled envelope containing an anode, a control grid and a thermionic cathode, see *Figure 4.15.* Gas may be mercury vapour, hydrogen or an inert gas. The tube remains in a non-conductive state with a positive voltage on the anode if a sufficiently negative voltage is applied to the grid. Value of the grid voltage depends on the anode voltage and the geometry of the tube. If the grid voltage is made less negative the ability to hold off the positive anode voltage is reduced. This is the basis of operation of the negative grid thyratron.

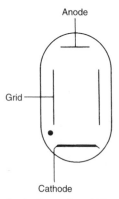

Anode

Grid

Cathode

Figure 4.15 The structure of a simple negative grid thyratron (English Electric Valve Co. Ltd)

The cathode emits electrons which, in the absence of grid control, are accelerated by the anode voltage and collide with the gas atoms present, to produce an ionized column of gas. A very low voltage, typically 50–100 V for hydrogen and deuterium, then exists across the discharge, through which a wide range of currents can be passed, their magnitudes depending on the external circuitry. During conduction the grid is sheathed with ions which effectively prevent any control being exercised. The tube returns to its non-conducting state only when the anode voltage is removed or reversed for a time sufficient to allow the charge density to decay to a low value. After this the grid regains control and when the anode voltage is reapplied the tube will not conduct. Thus the tube acts as an electronic switch, turned on by a positive-going charge of grid voltage and turned off only by removal or reversal of the anode voltage.

In some designs a disc baffle is fixed close to the cathode side of the grid and this usually modifies the tube's characteristics so that the potential of the grid may be taken positive without the tube conducting when the anode voltage is applied. This arrangement is called a positive grid thyratron. Baffles may also be used to protect the tube electrodes from deposition of cathode material which may cause malfunctioning. In practice this means that a positive pulse with respect to the cathode potential must be applied to the grid to initiate a grid cathode discharge.

Mercury-vapour filled thyratrons have been superseded by hydrogen-filled tubes because of the low limiting voltage that could be developed across the tube, approximately 30 V. A higher voltage across the mercury tube results in positive bombardment causing rapid cathode deterioration.

Most thyratrons can be mounted in any position, though with the larger tubes a base-down position is usually more convenient.

When the reduction of firing time variations is important in a thyratron circuit, the triode type can be replaced by a tetrode unit, which has two grids. These grids may be driven in a number of ways. The first grid may be continuously ionized while the second may act as a gate when pulse driven above its negative bias level. Another way is that of pulsing the grids successively with a delay of about one microsecond between the leading edges of the pulses. A third method is to pulse both grids from a single trigger source arranged to drive them separately.

Multi-gap thyratrons overcome the problems of operating thyratrons in series. When triggered, the multi-gap thyratron breaks down gap by gap in the normal gas discharge mode.

Hydrogen thyratrons for pulse modulator series are available from English Electric Valve and the M-O valve with a peak power output of 400 MW and a peak forward voltage of 160 kV, this tube being a 4-gap ceramic/metal tetrode.

Magnetrons. The cavity magnetron is a compact, efficient thermionic valve for generating microwaves, e.g. radar waves. Basically it is a diode, but it belongs to the family of valves known as 'crossed-field devices'.

Essentially it consists of a (usually) concentric heater, anode and cathode assembly and an embracing magnet (*Figure 4.16*). The magnetic field is at right angles to the electric field. High voltage pulses between the hot cathode and anode produce a spoked-wheel cloud of electrons which spirals round the cathode under the influence of the interacting electric and magnetic fields. The electron cloud only exists in a vacuum. The copper anode is so designed that a number of resonant cavities are formed around its inside diameter. As the 'tips' of the electron-cloud spokes brush past each very accurately dimensioned cavity the field interacts with the field existing in the cavity. Energy is absorbed and the oscillations within the cavity build up until the cavity resonates. The energy released is beamed through the output window and along a waveguide or cable to the aerial.

Pulsed-type magnetrons are available in the frequency range from 1 to 80 GHz with peak power outputs from a few hundred watts to several thousand megawatts, and mean power a thousand times less than peak power. Tubes for continuous wave operation are used mainly for heating, having powers in the range of 25 kW–200 W at frequencies of 0.9 GHz and 2.45 GHz. Lower power types are available for beacon operation at frequencies of about 9 GHz.

The cathode of a magnetron must be operated at its correct temperature. If it is too low reduced emission can cause unstable operation which may damage the magnetron. Excessively high temperatures lead to rapid deterioration of the cathode. The combination of the magnetic and electric fields in the interaction space results in a back bombardment of electrons to the cathode. This causes dissipation of a proportion of the anode input power. In order to maintain the cathode at its optimum temperature under these conditions it is generally necessary to reduce the heater voltage.

Care must also be exercised when raising the cathode to its operating temperature, before applying the anode voltage. The cold resistance of the heater is typically less than one fifth of the hot resistance. The surge of current when switching on the heater must be controlled. Similar precautions must be taken when designing the modulator output circuit to prevent pulse energy being dissipated in the magnetron heater.

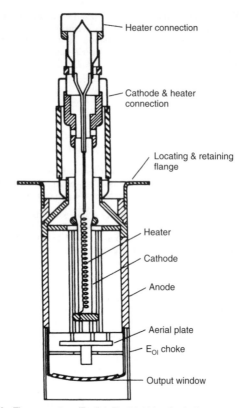

Figure 4.16 The magnetron (English Electric Valve Co. Ltd)

Operation of grid-controlled rectifier on a.c. The most important applications of the grid-controlled rectifier rely on its behaviour on alternating-current supplies.

When an alternating current is impressed on the anode circuit, the arc is extinguished once every half cycle because the valve is fundamentally a rectifier. Hence the grid is given the opportunity of regaining control every half cycle. This means that when the arc is extinguished due to the reversal of the alternating voltage, a negative grid can prevent anode current starting when the anode becomes positive again on the next cycle.

Further, by supplying an alternating voltage of the same frequency but of variable phase to the grid, the average value of the anode current can be controlled. *Figure 4.17* indicates how a grid voltage of variable phase can delay the starting of the arc during the positive half cycle of anode voltage, and thereby permit the mean value of the rectified current to be smoothly controlled from zero to the maximum value, corresponding to a phase change in the grid voltage of 180° to 0°.

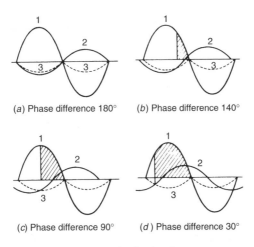

(a) Phase difference 180°   (b) Phase difference 140°

(c) Phase difference 90°   (d) Phase difference 30°

    1   Anode voltage
    2   Grid voltage
    3   Critical grid control voltage

Figure 4.17   Control of anode current by phase variation of grid voltage. The shaded areas show the current conducting periods

An alternative method of grid control having the same effect as phase control is to impose an adjustable d.c. bias voltage upon a constant phase a.c. grid voltage (amplitude control).

Rectification and power control. It will be apparent from the above that by using the grid in conjunction with a phase-shifting device a d.c. load can be supplied from an a.c. source and at the same time the load current can be regulated from zero to maximum. Furthermore, if the phase shifting device is operated by the changes in the external conditions that are to be controlled, completely automatic control can be effected.

Figure 4.18 shows a typical circuit for rectification and power control, using two gas-filled triodes to obtain the advantages of full-wave rectification. The respective control grids are connected to the ends of the transformer secondary, T_2, so that at all times one grid is positive while the other is negative, in the same way as the corresponding anodes. The phase position of the applied grid voltage is readily varied by the phase-shifting device, consisting of a distributed field winding connected to the a.c. supply and a single-phase rotor connected to the primary of T_2. The phase of the induced rotor voltage is determined by the rotor position relative to the field, so that change of rotor position changes the phase of grid voltage in each tube and correspondingly varies the d.c. circuit current. This form of control has been used for many purposes, notably the speed control of motors.

Phase control circuits. The phase-shifting device referred to above is a small induction regulator, constituting a very light controlling element. Alternative methods of applying phase control in terms of voltage are shown in *Figure 4.19*. At (a) the voltage across the inductance-resistance circuit is given

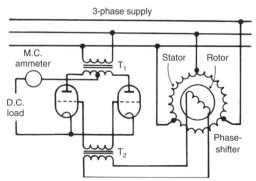

Figure 4.18 Circuit for supplying variable current to d.c. mains using gas triodes

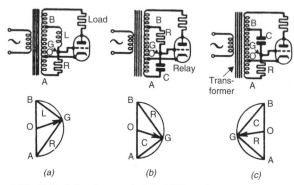

Figure 4.19 Methods for phase-control of grid-bias voltage

by AB in the phasor diagram. The same current flows through the inductance as through the resistance and the voltage drops – AG across the resistance and GB across the inductance – are therefore approximately 90° out of phase with each other. The grid voltage is the voltage of point G relative to point O (which is connected to the cathode), that is, OG in the phasor diagram. By varying R the phase of the voltage OG is varied with respect to the voltage OB, which is the anode voltage; increase of resistance, therefore, produces increase of anode current.

In *Figure 4.19* (*b*) the anode current decreases steadily as the resistance is increased, while in (*c*) the anode current is a maximum or zero depending on whether the resistance is greater or less than a critical value.

Where accuracy of timing is required, the a.c. grid voltage, instead of having a sinusoidal waveform, is arranged to have a peaky waveform, and be superimposed on a steady d.c. bias voltage considerably more negative than the maximum critical value. The phase of the a.c. component is varied so as to cause the positive peak to occur at any point in the positive half cycle of anode voltage at which anode current flow is required to commence.

Control of a.c. loads. The control of alternating-current load currents may be conveniently effected by means of a gas-filled triode used in conjunction with a *saturable reactor*. A single-phase circuit is shown in *Figure 4.20*. The saturable reactor has a laminated core of high permeability iron and carries two a.c. windings in series with the load on its outer limbs. The centre limb carries a d.c. winding supplied with direct current by the gas-filled triode. When the grid voltage phase difference, as controlled by the phase-shift device, is such that the tube is non-conducting, the a.c. winding has a high impedance due to the presence of the high permeability iron core. Under these conditions, the load current is negligible. As the grid-controlled rectifier output current is increased, the iron of the reactor gradually becomes saturated, reducing the permeability of the iron, and thus lowering the impedance of the reactor a.c. winding. At full output of the thyratron, the iron is completely saturated and substantially full-line voltage is impressed upon the load.

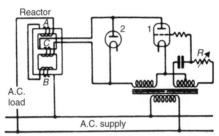

Figure 4.20 Circuit for control of a.c. load current using saturable reactor

During the positive half cycles the thyratron passes grid-controlled impulses of unidirectional current; during the negative half cycles the thyratron can convey no current, but the reactor current is maintained by the discharge of the energy stored in its magnetic field through the rectifier (2) across its terminals. The rectifier may be of the selenium or gas-filled diode type.

Temperature measurement and control. In the control of a heating load where temperatures above $750°C$ are encountered, measurement of the temperature can be carried out by a photoelectric pyrometer. This gives instantaneous response to changes of temperature when viewing the hot body. The small photoelectric currents, which are a measure of the received radiation, are amplified by a vacuum valve amplifier which energizes the control grids of the vapour valves, thus increasing or decreasing the current to the load by saturable reactor control.

For low temperatures, a resistance thermometer connected in a bridge circuit is used with a vacuum valve amplifier for providing the signal voltage on the control grid. In electrically heated boilers an accuracy of control of plus or minus $1°C$ at $300°C$ has been obtained by this means.

Electronic control of machines and processes. Other physical qualities beside that of temperature referred to above may be employed as signals for driving the control grids, provided the quality is converted into terms of voltage. Thus speed, torque, acceleration, pressure, illumination, sound, mechanical movement, and the various electrical qualities are utilized for initiating control in gas-filled or mercury-vapour triode circuits. Sound, for example, can be detected by a carbon, inductive, or capacitor microphone, and R, L, or C used as the variable element in the grid circuit. Mechanical movement can be detected by change of R, L, or C in a carbon pile resistance, slider resistance, variable choke, or capacitor – or by means of a photo-cell. A tachogenerator is employed for registering changes in speed or acceleration, as, for example, in certain types of speed control of d.c. motors.

Ignitrons. The ignitron is a high current rectifier with a mercury pool cathode, usually in a water-cooled steel envelope. In its simplest form it consists of a cylindrical vacuum envelope with a heavy anode supported from the top by a glass insulator, dipping into the mercury pool at the bottom. For some applications tubes may be provided with additional ignitors, auxiliary anodes and internal baffles. Ignitrons have been used in applications calling for high current levels such as resistance welding and high power rectification. There are also types intended for high current single-pulse operation such as discharging capacitor banks; these are used to pulse particle accelerator magnet coils, for electromagnetic forming of metals and similar applications.

The action of an ignitron is similar to a thyratron in that a control signal is needed to start conduction, which then continues until the current falls to zero. When the tube is operating as an a.c. rectifier it conducts during one half cycle of the supply frequency and must be ignited every alternate half cycle for as long as it is required to conduct. *Figure 4.21* shows a cross-section of an English Electric Valve rectifier ignitron.

The ignitor is a small rod of semiconducting material, with a pointed end dipping into the cathode pool. When a suitable current is passed through the ignitor-mercury junction, the ignitor, being positive, a cathode spot is formed on the surface of the mercury and free electrons are emitted. If the anode is sufficiently positive with respect to the cathode at this time, an arc will form between cathode and anode. Once the arc has struck the ignitor has no further control and the tube continues to conduct until the voltage across it falls below the ionization potential of the mercury vapour.

In a three-phase welding control circuit, the ignitron must deionize quickly in order to hold off the high inverse voltage which immediately follows the conduction cycle. This is accomplished by including a baffle (see *Figure 4.21*) which operates at cathode potential. No additional connections are required but the voltage drop across the tube is increased slightly. An auxiliary anode may be provided for power rectification at higher voltages. This is used to strike a small arc in a low voltage circuit separate from the main load. This maintains the cathode spot at low load currents ensuring stable operation under these conditions. The large tubes designed for single-pulse operation are also fitted with an auxiliary anode which may be used to prolong the ignition arc. Little or no baffling is used so as to keep the arc voltage drop as low as possible.

The cathode ray tube. The cathode ray tube consists of a highly evacuated conical-shaped glass container having at its narrow end or throat, a cathode capable of being heated, a control electrode and several anodes. Higher up the throat of the tube are two pairs of deflector plates arranged at right angles to each other, *Figure 4.22*.

The large end or base of the conical container is coated on the inside with a fluorescent compound which emits light when the cathode ray impinges on it.

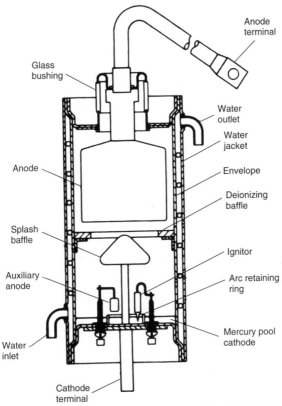

Figure 4.21 Cross-section of a rectifier ignitron (English Electric Valve Co. Ltd)

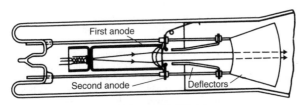

Figure 4.22 Electrode system of electrostatic C.R. tube

The action of the tube is as follows. When the cathode is heated, and the anodes connected to positive high voltages of increasing magnitude, the electrons emitted from the cathode are accelerated and focused into a beam which emerges from a small aperture in the final anode. This beam passes between the first pair of deflector plates, which are located so that when given an electrostatic charge they can attract the electrons, say, to the right or left. The beam then passes between the second pair of deflector plates arranged so that the beam can be attracted up or down. The beam, which consists of a stream of electrons moving at high speed, impinges on the fluorescent screen and causes a spot of light to appear on it.

If now a fluctuating voltage is applied to the first pair of deflector plates, the beam will move from side to side of the screen in synchronism with the voltage applied to the first deflectors. If another voltage is applied between the second deflector plates, the beam, while maintaining its side-to-side motion, will also receive a vertical displacement corresponding to the voltage applied to the second deflectors.

It will be seen that the spot of light on the screen will thus trace out a curve corresponding to the varying potentials applied to the two pairs of deflector plates.

In many applications the horizontal deflectors are given a steadily increasing voltage with instantaneous flyback after the beam or spot has reached the extreme edge of the screen.

The result of this is that the spot travels at a uniform speed from left to right in a given time period, e.g. 1/50 of a second. If a 50 Hz alternating voltage is applied to the vertical deflector plates the vertical deflection of the spot at any instant will correspond to the instantaneous voltage of the supply. Thus the waveform of the supply voltage will be traced out on the fluorescent screen. As owing to the instantaneous flyback the above process is repeated 50 times per second, the waveform of the applied alternating voltage will be seen on the screen as a stationary sine wave with or without harmonics, according to the waveform of the applied voltage.

When the cathode ray tube is applied to television, the horizontal deflector plates are used for causing the spot to sweep across the screen with flyback at the end of each sweep, while the vertical deflectors give the spot a small vertical displacement after each horizontal sweep, with vertical displacement after each horizontal sweep, with vertical flyback when the lower edge of the screen has been 'scanned'.

The incoming television signals are applied to the control electrode and have the effect of increasing or decreasing the intensity of the beam, according to the brightness of the picture spot, which is being transmitted at any instant.

Klystrons. The klystron is a thermionic device in which electrons from a heated cathode are accelerated to full anode potential and formed into a long parallel beam. This first traverses two grids or apertures between which is a high frequency voltage provided by a tuned circuit in the form of a resonant cavity. The resultant small periodic changes in speed eventually lead to conventional current modulation by a process known as bunching. Where this bunching is virtually complete another resonant cavity (catcher) is placed. If its impedance and tuning are correct it will extract from the beam much more power than was used to modulate it. The arrangement then acts as an amplifier. Efficiency is moderate, output power can be very high, and gain can be almost indefinitely increased by putting more resonant cavities between the buncher and catcher. Oscillators up to the highest frequencies can be formed either by returning some of the collected power to the buncher, or by using a single resonant cavity and reflecting the beam back through it.

High performance klystrons require an axial magnetic field to focus and control the magnetic beam during its passage through the drift tubes, see

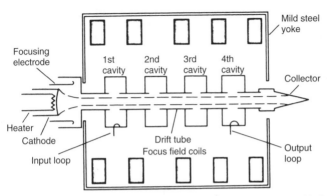

Figure 4.23 Schematic arrangements of a four-cavity amplifier complete with the coils and mild steel yoke used to provide the axial magnetic field (English Electric Valve Co. Ltd)

Figure 4.23. The improved performance generally justifies the complication involved. Damage to a valve will occur if the beam is not focused efficiently and it is essential to protect it against a failure of the focusing system, which would cause a rise in the electron current reaching the drift tubes. If a simpler installation with a reduced performance is acceptable, it may be possible to use a klystron which has no external means for focusing the beam. This eliminates the need for a magnetic field and the power supply required to maintain it. The precautions on the beam focusing are reduced simply to that of screening the electron beam from the influence of any stray magnetic fields present in the vicinity of the valve.

Photoelectric Devices

Photocell relays. The basic component of a photocell relay is an integral light activated switch. It combines a silicon planar photodiode with integrated circuitry on a single substrate to provide a highly sensitive photoelectric device. Operation is such that when light of a selected intensity falls upon it, the device switches on and supplies current to an external load. When the light intensity falls below the critical level the load current is turned off. This critical level can be adjusted within wide limits.

The equipment comprises a projector containing a light emitting diode (LED) and optical system projecting a beam of light either directly or by reflection onto a photocell mounted in a receiver unit. The relay coil is energized when the light beam is made and de-energized when it is broken. Thus the relay contacts can provide a changeover operation which can be used to perform some external control function. The control unit can contain additional circuitry such as time delays or LED failure circuits to meet a wide variety of application requirements. Systems are available for operating over distances from 10–15 mm up to 50 m or more.

Applications include conveyor control, paper breakage alarm, carton sorting and counting, automatic spraying, machinery guarding, door opening, level controls, burglar alarms, edge alignment control and punched card reading.

Photoelectric switch units. Light-sensitive switches are used for the economical control of lighting. They consist of a photocell which monitors the intensity of the light and automatically switches the lighting on or off. Construction of a typical unit is shown in *Figure 4.24* and this will switch a resistive load of 3 A at 250 V a.c. The unit is based on a cadmium-sulphide cell and it incorporates a 2-minute time delay to prevent 'hunting'. Larger units are available with resistive switching capacities up to 10 A.

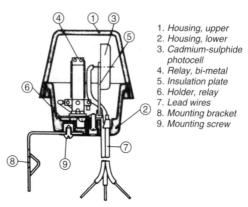

1. *Housing, upper*
2. *Housing, lower*
3. *Cadmium-sulphide photocell*
4. *Relay, bi-metal*
5. *Insulation plate*
6. *Holder, relay*
7. *Lead wires*
8. *Mounting bracket*
9. *Mounting screw*

Figure 4.24 Photain type P B0-2403 photoelectric switch unit

Silicon photoelectric cells. These cells are designed to provide large output current even under low illumination intensities. Currents of several milliamperes are obtainable. Structure of a photoelectric cell is shown in *Figure 4.25* in which it will be seen to consist of a thin p-type layer on n-type silicon. Due to its photovoltaic effect there is no need for a bias power source. A linear output can be obtained by selecting a suitable load resistance for a wide range of illuminance. Like the silicon blue cell, described below, it has no directivity of receiving light, so there is no need to adjust the optical axis as is the case with phototransistors.

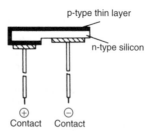

Figure 4.25 Photain photovoltaic cell

Silicon blue cell. The Sharp's silicon blue cell manufactured by Photain Controls was claimed to be the world's first photoelectric diode possessing high sensitivity over the entire visible light spectrum. It is more reliable than the selenium or cadmium-sulphide photocells and has superior time response. No bias power is required, it has a lower noise level than the other two types and it is non-directional.

Applications include illumination meters, exposure meters, optical read-outs of film sound tracks, colorimetry, flame spectrometry, photospectrometry and colour or pattern recognition equipment.

5 Rectifiers and converters

Introduction

Rectification is the conversion from alternating to direct current. Over the last few years the development of inexpensive, rugged, variable speed a.c. drives over a wide range of power ratings has vastly reduced the need for d.c. machines, which were hitherto indispensable for many industrial processes where fine control of speed was necessary. Such applications include rolling mills, electric overhead cranes and traction drives. Direct current nowadays has much more limited use for processes such as electroplating plants, gas production plants, and to supply chargers for standby battery systems. Added to these is the somewhat exotic but limited application for high voltage direct current transmission.

For most of these applications the conversion device will nowadays be a piece of power electronics, consisting of diodes or thyristors, but the principles of operation of rectifier equipments remain much the same regardless of the device which actually performs the rectification. This short chapter will confine itself to describing some of the aspects of rectifiers not covered by the previous chapter but still likely to be encountered in service.

Metal Rectifiers

Three basic types of metal rectifier remain in common use: namely, selenium, germanium and silicon. The last two are generally referred to as semiconductor rectifiers and their theory of operation is described in Chapter 4. One of the major manufacturers of these rectifiers in the UK was formerly the Westinghouse Brake and Signal Co., to whom acknowledgement is made for providing information upon which this chapter is based, although the Westinghouse name is no longer used in connection with the manufacture of metal rectifiers.

Selenium rectifiers. During the past 30 years there has been a continuous development of the selenium rectifier so that the stability of the rectifier is such that it can be operated at relatively high temperatures, i.e. 120°C if required. Alternatively it can be operated for a much longer time than hitherto under normal operating conditions.

Selenium rectifiers have been used widely for all low power requirements where initial cost is important and the ability to withstand substantial and repeated overloads eliminates the need for special protective devices that may be required for silicon or germanium rectifiers. Although the efficiency and performance of selenium rectifiers may be slightly inferior to the other two and the size somewhat greater, these features are often of less importance for outputs below 25 kW.

Applications include electroplating where oil-immersed units provide currents up to 200 kA. At high values of current, water-cooled germanium units are used due to their smaller weight and the limited space which they require.

Electrostatic precipitation for removal of dust particles from gases where d.c. voltages from 30 to 100 kV are required is a common application for selenium rectifiers. Oil-immersed equipment is reliable in operation, robust and provides an efficient means of rectification. Equipments with outputs of 60 kV and currents from 60 mA to 1 amp are in widespread use. These equipments are transductor controlled with very rapid arc extinction.

Cinema arc power supplies can be provided by transductor or choke/capacitor constant current selenium equipments for both high and low intensity arcs and can be operated from either single-phase or three-phase supplies.

Germanium rectifiers. Germanium rectifiers are used extensively for low power and medium power industrial applications where the ambient temperature is not high and particularly at voltages below 100 V where high efficiency at such voltages is of overriding importance. The size of the installation may be in the megawatt range. The germanium is extracted from coal and zinc deposits and refined to a high degree of purity. The resultant grey metallic material is pulled into a single crystal which is specially cut to form small wafers. By heat treatment, an indium button is welded on to and diffused into the germanium wafer. The bond between the germanium and the indium forms the rectifying junction which is mounted in a hermetically sealed housing.

Both germanium and silicon rectifiers are used extensively in equipments providing power for the electrolytic production of chlorine and hydrogen. Outputs in excess of 27 kA at 120 V have been provided from these equipments, germanium offering the slightly better efficiency at d.c. voltages below 100 V, but they may not be economical in countries where the ambient temperature is high.

Industrial d.c. power supplies are frequently provided from germanium and silicon semiconductor equipments although as stated earlier for powers below 25 kW the selenium rectifier still remains the most attractive plant. Both types have been used extensively for large telephone exchanges. Chargers for battery electric vehicles are generally of the germanium rectifier type. Welding is another area where both germanium and silicon rectifiers compete with each other.

Silicon rectifiers. Silicon is a metal obtained from sand and is refined to a high degree of purity, drawn into large monocrystals and cut into wafers. A thin plate of aluminium is bonded to and diffused into the silicon wafer and the junction between the aluminium and the silicon forms the rectifying junction.

Many of the applications for this type of rectifier have already been outlined but this type of rectifier was considered to be the best for railway traction supplies. Variable speed drives is another area that this type of rectifier has been widely employed. The facility of the grid control made it ideal for speed control of d.c. machines up to the largest ratings.

Comparison of three types. The semiconductor germanium and silicon rectifiers are, for a given output, more compact than selenium. This is due in part to the low forward resistance per unit area of rectifiers, and in part to the high voltages which they withstand in the reverse direction. Silicon will withstand a higher reverse voltage than germanium and will also operate at a higher temperature. The forward resistance of germanium is lower than that of silicon.

The d.c. voltage/current characteristics of the three types of rectifier are shown in *Figure 5.1*. It will be seen that with a rectifier operating at normal

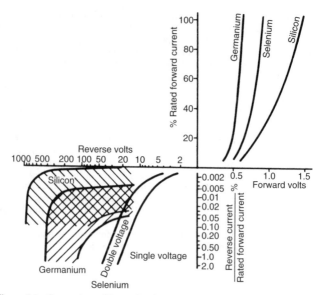

Figure 5.1 Comparison of forward and reverse characteristics of the three types of rectifier

current density, one may expect a low voltage drop with germanium, a higher voltage drop with selenium and a still higher one with silicon. In practice it is not usual for the reverse current to exceed about 0.5% of the forward current and, therefore, the selenium rectifier is capable of being operated at about 32 V, germanium 70–90 V and silicon 100–300 V. The voltage at which the rectifiers are operated is determined by many factors such as duty cycle, circuit connection, temperature, etc.

Temperature characteristics. There is a maximum critical temperature above which each of these three rectifiers cannot safely operate. It is important therefore to ensure that the temperature rise of the rectifier and the ambient temperature together does not exceed the critical value. Forward ageing is the limit for a selenium rectifier and such units are suitable for temperatures up to 70°C with specials being available for use at total temperatures up to 130°C.

Although the maximum temperature at which germanium rectifiers can safely operate from forward and reverse ageing is 90°C, another factor limits the safe working temperature to 70°C. Due to the low thermal mass of this type of rectifier, relatively small overloads can cause rapid temperature increase. It is considered advisable therefore to limit the total operating temperature to 50°C. For similar reasons the total operating temperature of the silicon rectifier is reduced from 200°C to about 160–170°C to give an added factor of safety.

Overloads. Selenium rectifiers are formed on a substantial metal base and operate at a low current density. They can therefore withstand severe current overloads of short duration without damage. As mentioned above the low thermal mass of the other two types does not permit them to withstand current overloads.

Voltage overloads give rise to increased leakage currents and self-heating. Again the silicon and germanium devices are more sensitive to these and care must be taken to protect them from such overloads under normal operating conditions.

Parallel and series connections. With germanium and silicon rectifiers care has to be taken when connecting them up in series or parallel to ensure even load sharing. Performance ratings should be closely matched or ratings should be substantially reduced. It is desirable to have individual protection of diodes by fuses or an equivalent.

Size and efficiency. A germanium rectifier is roughly one third the volume of a selenium rectifier for the same power output. A silicon rectifier is about one third the volume of a germanium device or one tenth of the volume of a selenium rectifier.

The overall efficiency of an installation is determined by a consideration of all the equipment and to some extent by the circuit connection, load, etc. *Table 5.1* lists the efficiencies of the three types of rectifiers.

Table 5.1 Comparison of efficiency of rectifiers (in %)

D.C. voltage	Selenium	Germanium	Silicon
6	85	91	83
12	91	95	90
25	91	97	94
50	91	97	97
100	92	97	97
500	92	97	98
750	92	98	98

Rectifier Equipments

Physical arrangements. In its simplest form a silicon rectifier comprises an assembly of silicon diodes on heatsinks together with fuse and surge voltage protection components. These are all normally housed in a sheet enclosure and supplied by a separately mounted double-wound transformer. Additional items may include control cubicles, switchgear, voltage regulators and connections. A controlled rectifier will be similar, with thyristors in place of the diodes and with appropriate electronic firing gear and controls.

The rectifier assembly is generally contained in a floor-mounted cubicle with the associated transformer. Cooling may be air natural or in some cases air blast. If the atmosphere is corrosive, dusty or damp, closed-circuit air cooling or liquid cooling may be used. Heavy current rectifiers usually employ a water cooled welded-up heatsink/busbar assembly as the best means of achieving the high rating suited to the electrochemical environment.

Basic connections. The choice of single-way (half-wave) or double-way (full-wave bridge) connection depends partly upon the d.c. voltage required.

With single-way the d.c. passes effectively through only one diode at a time instead of through two in series as in the bridge connection. The forward voltage drop is therefore half the value of that for the bridge connection. Hence the losses are also half. However, transformer losses and cost are higher for single-way operation. The voltage above which double-way connection is used is generally determined by a combination of efficiency and cost considerations.

Figures 5.2 to *5.7* show a number of different rectifier configurations. Only one diode per arm is shown although there may, in an actual system, be a number of diodes in a series-parallel arrangement for each arm.

Three-phase bridge (6-pulse). The double-way connection, *Figure 5.2* gives 6-pulse rectification with a 120° conduction angle. The voltage regulation is virtually a straight line over the normal working range. This is the most commonly used connection for industrial power supplies.

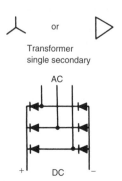

Figure 5.2 6-pulse circuit for 2-wire single bridge d.c. output

Double-star (6-pulse). With this connection the six-phase secondary windings are separated in the two opposed star groups, the neutral points of which are connected through an interphase reactor, *Figure 5.3*. The voltage equalizing effect of the interphase reactor enables the two star groups to share the current, and 120° conduction occurs at all currents above the very small value required to magnetize the interphase reactor. The voltage regulation curve is virtually a linear slope while 120° conduction is taking place but there is a sharp rise of about 15% as the interphase reactor becomes demagnetized. With modern core steels this point is usually well below 0.5% load and the rise is not normally objectionable or harmful. A small shunt load permanently connected or automatically switched in at low loads can be provided if necessary to eliminate this effect. This connection is primarily used for power supplies for low voltage process work of a continuous nature, where overall efficiency is of prime importance.

Multi-circuit connections. For 12-pulse working, two 6-pulse rectifier equipments, two groups of equipments or two rectifier circuits need to be phase shifted 30° from each other. This is usually achieved by having a star/delta relationship between transformers or transformer windings associated with the two 6-pulse rectifiers. Phase displaced rectifiers closely paralleled in this way usually require an intercircuit reactor to equalize the voltage

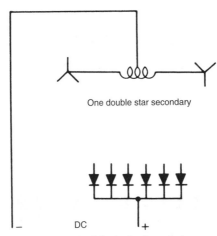

Figure 5.3 Half-wave 6-pulse circuit for 2-wire d.c. output

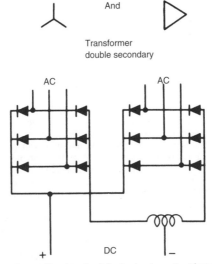

Figure 5.4 Transformer 12-pulse circuit for 2-wire d.c. output from parallel bridges with intercircuit reactor

between the phase-displaced groups (see *Figures 5.4* and *5.5*) but these are sometimes omitted if there is sufficient d.c. circuit reactance in connections or in other reactors. On very large installations, pulse numbers greater than 12 can be obtained by suitably phase shifting a number of 6- or 12-pulse rectifier equipments.

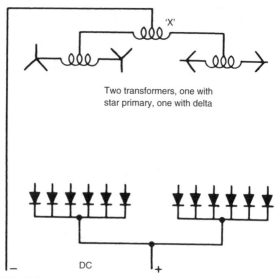

Figure 5.5 Half-wave 12-pulse circuit for 2-wire d.c. output. The reactor *X* can sometimes be omitted

Three-wire circuits. Three-wire d.c. outputs may quite easily be obtained with several rectifier and transformer connections. Choice of a particular connection depends upon the required out-of-balance current, the pulse number of the outer to mid-wire d.c. voltage, and the cost involved. The arrangement shown in *Figure 5.6* is the most economic, comprising a normal three-phase bridge rectifier and a transformer having a star or interstar connected secondary with the neutral point brought out. This connection provides for not more than 20% out-of-balance current in the mid-wire with a simple star winding or 100% if an interstar winding is used. In such circuits the d.c. voltage waveform contains the harmonics of 6-pulse operation across the outers and 3-pulse operation from mid-wire to outers. The r.m.s. value of the harmonic voltage is about 6% of rated d.c. voltage for 6-pulse working and about 25% for 3-pulse working.

The connection shown in *Figure 5.7* consists of two three-phase bridge rectifiers connected in series and fed by a transformer having two secondaries, one star connected and the other delta connected. This connection will handle 100% out-of-balance mid-wire currents with d.c. voltage harmonics outer-to-outer of only about 3% r.m.s. (12-pulse) and the outer to mid-wire of about 6% (6-pulse).

Supply harmonics. It is important that the harmonics drawn from the supply by a rectifier system be kept within the limits laid down by the supply authority. Recommendations normally specify the maximum ratings of rectifiers, having a stated pulse number and assuming no phase control, which can be fed from supplies at various voltages at the point of common coupling for two or more consumers.

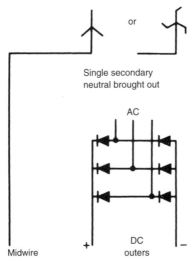

Figure 5.6 Circuit for 3-wire d.c. output with 6-pulse outers and 3-pulse mid-wire. Star secondary for up to 15% mid-wire current and interstar secondary for 100%

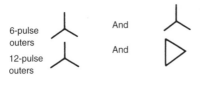

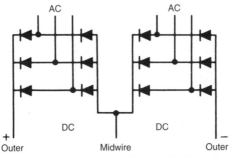

Figure 5.7 Circuit for 3-wire d.c. output with 6- or 12-pulse outers and 6-pulse mid-wire. Suitable for mid-wire current up to 100%

Table 5.2   Maximum permitted rectifier loadings at specified voltage in the UK

Supply system voltage (kV) at point of common coupling	Type of converter	Permissible kVA capacity and corresponding effective pulse number of three-phase installations		
		3-pulse	6-pulse	12-pulse
0.415	Uncontrolled	–	150	300
	Half controlled	–	65	–
	Controlled	–	100	150
6.6 and 11	Uncontrolled	400	1000	3000
	Half controlled	–	500	–
	Controlled	–	800	1500
33	Uncontrolled	1200	3000	7600
	Half controlled	–	1200	–
	Controlled	–	2400	3800
132	Uncontrolled	1800	5200	15000
	Half controlled	–	2200	–
	Controlled	–	4700	7500

Table 5.2 shows the general recommendations for the UK; these requirements are usually more onerous than those in other countries where larger
rectifier loadings may be permissible. It should be remembered that a three-
phase bridge circuit has a pulse number of 6. Pulse numbers less than 6 are
used only on rectifiers of a few kW rating or by special arrangement with
the supply authority. The table is only a guide and prospective rectifier users
should always consult their local supply authorities who are aware of other
consumers' rectifier plants that may already be connected to the system.

Regulation.   The inherent regulation of a rectifier equipment is defined as
the rise in voltage from full load to light load, and is expressed as a percentage
of the rated full load d.c. voltage. Light load is taken as 5% of rated load.

With certain rectifier circuits, the d.c. voltage can rise sharply below 5%
load because of the demagnetization of an interphase transformer. Steps can
be taken to prevent this occurring.

Small rectifier equipments incorporating fuse protection will have an inherent regulation of the order of 5%. Larger equipments designed to be protected
by switchgear, and having a higher rectifier transformer reactance, will have an
inherent regulation of approximately 8%. Lower regulation values than these
can be obtained but they will usually cost more.

Rectifier protection.   Rectifier equipments are protected against overcurrent and surge voltages, and are designed so that internal component failure
cannot cause damage. There are three basic overcurrent schemes: circuit-
breakers, fuses or a combination of both.

When fuses are used on their own they are of the fast acting type, specially
designed to match the characteristics of the diodes. They are connected in
series with the diodes but will also disconnect the diode should it have an
internal fault, allowing the passage of current in the blocking period. When
diodes are connected in parallel they normally have individual fuses. Indicator
fuses, of the striker pin type, are also connected across the diode fuses to
assist in identifying a blown fuse and the faulty diode. For small equipments

containing six or less diode fuses it is often acceptable for these to be the sole means of protection.

Circuit-breakers are used where it is not desirable to lose the complete d.c. supply due to a feeder fault on the d.c. distribution network. This is generally applicable to large rectifier equipments. In such cases fuses are provided for protection of the diodes but d.c. feeder faults are cleared by high speed d.c. circuit-breakers. This leaves the rectifier supply available for healthy circuits.

When d.c. overcurrents are likely to occur often but are not normally of maximum severity, i.e. not short-circuits, the overcurrents may be cleared by a moulded case a.c. circuit-breaker with diode fuses as back-up protection for severe faults. Such protection applies to small equipments, up to 200/300 kW, and only where the a.c. supply voltage is under 1000 V. A slightly higher transformer reactance than normal allows the a.c. breaker to clear most faults.

Surge voltage protection is provided by capacitor/resistor networks within the rectifier equipment. These networks limit transient voltage surges and commutation voltage peaks to levels substantially below the transient rating of the diodes. In most cases the surge absorbing circuits are themselves protected by fuses. On larger equipments, local or remote indication of fuse operation can be provided.

When regeneration can occur it is necessary to protect the d.c. network by some other means than the surge absorbing circuit. Any motor attempting to feed power back into the rectifier terminals raises the d.c. voltage to a level which may cause damage. Protection for this condition can be provided by a loading resistor that can be permanently connected or switched into circuit under regenerative conditions.

Converting Machines

The term converting machines is used to cover those arrangements whereby a.c. is converted to d.c. by machines having rotating parts. Although their use is gradually dying out because of the advent of static converters, nevertheless there are still many in operation today. There are three main types as follows.

Rotary converters. These consist of a wound rotor revolving in the field of a d.c. generator, the rotor being fitted with slip-rings at one end and with a commutator at the other. If while rotating at synchronous speed an a.c. supply is connected to the slip-rings, d.c. can be taken from the commutator. There is only one winding and the power to keep the machine running and to supply the electrical and friction losses is taken from the a.c. side.

A rotary converter will run from the d.c. side when a.c. can be taken from the slip-rings – this arrangement being called an inverted rotary converter. Practically all rotary converters of any size are polyphase; three-phase for small and medium outputs and six-phase for larger outputs.

Ratio of transformation. The d.c. voltage will be $\sqrt{2}$ or 1.41 times the a.c. voltage for a single-phase, and the various ratios for polyphase machines are given in *Table 5.3*.

The relationship is given by a.c. volts between slip-rings

$$= \frac{\text{d.c. volts}}{\sqrt{2}} \cdot \sin\frac{\pi}{m}, \quad \text{where} \quad m = \text{no. of slip-rings}$$

Table 5.3 Transforming ratios

	Single phase	Three phase	Six phase	Twelve phase
Volts between slip-rings as a percentage of d.c. volts	70.7	61.2	35.4	18.3

As d.c. voltages are usually in the region of 220 to 240 V it will be seen that with normal a.c. supply a transformer is needed to give the required voltage for the supply to the a.c. side.

Six-phase machines are the most usual since it is fairly simple to obtain a six-phase supply from the secondary of the transformer, using a delta connection on the h.v. side.

Voltage regulation is obtained either by varying the power factor (by excitation control), thus using the reactance of the transformer, or by using a booster. Voltage control can also be obtained by using an induction regulator.

The efficiency of a rotary converter varies from 90 to 94%; it has a high overload capacity, and as the power factor is under control it can be kept approximately at unity.

Normally, rotary converters are not self-starting from the a.c. side, but a starting winding can be wound on the *stator* to act as an induction motor. Other methods include starting from the d.c. side and the use of an auxiliary starting motor. With these methods careful synchronizing is necessary before switching on to the a.c. supply otherwise serious damage may result.

Motor generators. These consist of two entirely separate machines (from the electrical point of view), and any two machines (e.g. a motor driving a generator) form a motor converter.

Normally they are coupled machines for converting a.c. to d.c. and consist of either an induction motor or a synchronous motor driving a d.c. generator. They also convert in the reverse direction from d.c. to a.c. when the latter is required where there is no supply. Also motor converters are used as frequency changers (i.e. an a.c. motor driving an alternator).

As there are losses in both machines the efficiency is not high and not usually above 90%. One of the advantages is that high voltage can be taken to the a.c. side and a wider control is obtainable as to the voltage on the output side than with rotary converters. Motor generators are sometimes used as standby supplies for computer installations.

Motor converters. These consist of two machines mechanically coupled together but also connected electrically. The motor portion consists of an induction motor with a wound rotor, the rotor being connected to the winding of the rotor of the generator portion. The efficiency at full load varies from 86 to 92% with a power factor of 0.95 and over. They are rather more stable than rotary converters and are self-starting from the a.c. side. They are not quite so efficient at light loads. Both motor generators and motor converters are discussed in some detail under 'High integrity power supplies' on page 139.

6 Computers and programmable controllers

Compared to any other branch of electrical engineering, the field of computing has seen the most radical changes in recent times. Indeed the area has expanded so much that the study of computers and computing is a subject in its own right. Here it is intended to give only the briefest overview of the many systems now available, with the intention that the reader, knowing what systems are capable of, will find the detailed information he requires in a more specific publication.

The reduction of the many facets of computing activity to a single phrase results in a definition which could be 'computers are devices which process information'. In the industrial environment this involves the control of other pieces of equipment, but even this control is based on information signals received, and a set of instructions against which the signals should be interpreted.

Two aspects are fundamental to the way computers carry out this information processing:

1. The data must be represented digitally, this means using *binary* code. In the binary or 'base 2' arithmetic system there are only two states which may be represented by the digits 0 and 1. However, any number can be represented using this notation as long strings of 1s and 0s. By simple substitution, letters and other typographical characters can be represented. Similarly a photograph can be described digitally, with every dot, or *pixel*, being given a number corresponding to a particular shade of colour. To describe a large colour picture in this way is clearly an extremely data intensive exercise, but the point is that all information is reduced to a form in which it can be processed by a machine.
2. The second fundamental aspect of modern computing is that the digital information is processed electronically. Electronic circuits can be built which act on the digital data. Provided the numbers are either 1 or 0 then a circuit to add two numbers, or compare two numbers, is relatively simple to construct. Data comprising of 1s and 0s is easy to store. Although the number of individual bits of information is large, for even, say, a short English word, this does not matter. Computers have enormously large numbers of circuits, and they work incredibly fast. Indeed the power of a computer is a function of the speed at which it operates, and the number of bits of information it works with at any one time.

To a user it appears that pressing buttons on the keyboard causes letters to appear on the screen. In the background, the computer is working on many different fronts to achieve this task.

First there is the *hardware*, the actual physical components which comprise the computer system. The keyboard is a panel of pushbuttons. The screen might be either a cathode ray tube, or a grid of light emitting diodes. For a desktop style computer the main processing functions will be carried out in a single integrated circuit on a silicon chip. Other chips will provide the 'memory'. There will be an electrical power supply, maybe a cooling fan, and one or more long-term memory or data storage devices using magnetic media.

In a field where the pace of development is so rapid, giving examples of the capacity of storage devices, or the speed of processors, serves only to date the text, but at the present time storage capacities are counted in *gigabytes* where the prefix 'giga-', abbreviated by the symbol G, indicates multiplication by the factor 10^9. A *byte* represents a string of individual 1s or 0s which go together to represent a particular character or piece of information. Processor speeds are determined by the operating frequency of the device, which is counted in *gigahertz* where the prefix 'giga-' has the meaning defined above and one hertz represents one operation per second.

The most technically complex of the components is the processor chip itself. By the time the data arrives for processing, it will already have been converted to digital 1s and 0s. The basic control of how data is moved around inside a computer is the function of the *operating system*. This is a set of instructions (a program – or, more precisely, a program *protocol*) which the computer follows to undertake basic tasks, like getting the instructions to do more complex activities. The operating system is controlling the actions of electronic components in the processor chip, so this makes an operating system specific to a particular design of computer. In reality many common standards have been agreed within the industry. There are a few major operating systems, and most hardware designs are broadly similar. The resulting general data processing platforms are then customized to do specific tasks by loading into them particular *software*. This could be word processing, controlling a machine, robot or large production process, or performing calculations and displaying the results.

The processing power available, the software available, and the facility for users to create their own software for their particular needs make computers extremely versatile tools. As processing power increases, users find ways to utilize the extra power to give more and better functionality. The cost of the technology comes down all the time. As a result, it is practical to embed computers into hand-held devices or telephones, and development is still continuing.

Office and Home Computers

The computer familiar to many will be a case of 4 to $6\,dm^3$, a keyboard, and a screen. Although it may have been manufactured by any one of a number of companies, this PC, or personal computer, is based on a generic design by IBM. Indeed, for a long time the design was known as 'IBM compatible' machines. The most prevalent operating system is Microsoft Windows, a product which has achieved widespread use through marketing along with the hardware. The combination of similar hardware design and similar versions of software mean that almost all office and home computers will work in pretty much the same way. This has obvious advantages for the user. Apple computers use different hardware and operating systems, and are frequently defended by their users as being technologically better for the tasks they perform, but they are a minority of the systems in use today.

As a standalone device the computer is a powerful tool. It can be used for calculations, preparing printed material, processing music and many other activities. The true potential of computers is released when they are connected together into networks. Networks allow many computer users to access data stored centrally, in processors with large data storage capacity called *servers*. In an office, peripheral devices like printers can be shared among all the users of the network. The network connection makes it easy for the data on one computer to be sent to another. This could be between two adjacent desks in an office, or to the other side of the world. In the same way that telephones transmit voice signals down wires, through exchanges, and via satellites, computers can pass their digital information from one machine to another.

Early computer networks actually used the telephone system, and most home computers still use this to make their connections, but for large users there are now dedicated data connections. In exactly the same way that a telephone exchange switches a telephone call through to its destination, the computing equivalent, the *router*, directs the flow of digital data through connecting wires and satellite links to its destination. This worldwide connection of computers is known as the *Internet*. There is a consistency of standards in use and an incredible amount of data can be freely accessed using the protocols of the World Wide Web.

There are so many uses that computers can be put to that it would not be possible to list them all. The more important ones are worthy of note, however, since they illustrate the widespread use of such systems.

Word processing. This is the creation and editing of text, laying it out on a page, possibly adding pictures, and then printing the results on paper. The computer controls the printer, it checks spelling and can arrange the formatting automatically. This has totally replaced the typewriter in everything from writing a short paragraph, to typesetting books and newspapers. In its most advanced form, this is known as desktop publishing.

Spreadsheets. Another activity carried out in office applications. Spreadsheets are programs for storing and arranging data, in a format which is very easy to use. They are also good for producing graphs, and doing calculations.

Email. The need to send messages from one computer to another was questioned in the early days, but today it is an extremely widespread use of computer networks. A concern has been expressed that the ability to broadcast information to large numbers of people has reduced the effort users put into the quality of what is sent. Files and other information can be attached to email message covering letters.

Databases. These are used by companies often as a core part of their business. For example, in the case of a mail order business a single database might contain the names and addresses of all customers, lists of everything they have ever purchased, what they have on order at the moment, and their account balance. This could all be tied-in with the purchasing database for ordering stock from suppliers. Programs within the main database can automatically generate dispatch notes and invoices, order new stock, and print out a warehouse pick list.

A utility company may have a database containing every meter reading, for a customer's domestic energy supply details. The same data is used for billing. Banks keep account information in databases, and the share dealing and commodity brokerage systems used by financial institutions also involve large databases. The common factor is that a lot of information is stored and indexed, and various programs can process or change the data in order to effect the desired transactions.

In fact the amount of information stored on computers nowadays is so extensive that it has led to the Data Protection Act, an Act of Parliament which limits how organizations can use electronically stored data, and give the people who are the subject of the data certain rights.

Technical calculations. In the engineering field, it is the calculating power of computers which is of benefit. Mathematical models representing real life situations (*algorithms*) can be created, and tested in all conceivable circumstances. These can range from modelling the loads in an electricity network, to calculating the thickness of the structural members of an aircraft fuselage. Computer aided design goes together with drawing packages which enable the finished output to be produced on paper. Alternatively the design can be taken directly to the computer numerically controlled (CNC) machine tool, and produced without the need for prototypes.

Digital sound and pictures. Since sound and pictures can be converted to numerical patterns, or *digitized*, computers can process music and film. Whereas previously video signals were recorded in analogue form on tape, now the information is stored digitally, so it is easier to edit and refine.

Computers in education and in the home. On a smaller level, computers in schools and in the home can be excellent educational tools. Their use in schools at as early a stage as primary school level also has the benefit of teaching computing skills to the next generation at a very early stage in their education. Links to the World Wide Web enable information to be found from many sources, and learning to access this information is a skill that is taught from an early age. Finally computers can also be used for playing games whose complexity is limited only by the programmers' skills.

Security

The ability of computers to connect to other machines introduces security risks. Many systems are password protected, theoretically allowing only authorized users to access the data. However, computer hackers can find ways round passwords and gain unauthorized access to private information. Most large corporations take the issue of computer security extremely seriously, and invest in the latest technology to defeat hackers. Another threat to computer systems is through viruses. These are programs which can modify or delete data, or have other undesired effects. The programs can run on any computer in the network and copy themselves on to other machines, hence the name. Virus programs are usually introduced into a system in an email message or inadvertently downloaded from the Internet. Software is available to detect and remove them.

Industrial Computing

The widespread use of computers has undoubtedly revolutionized the way we work in offices and relax at home. Industrial computing, robotics, and process control all seem less fundamental changes by comparison, but to the industries they affect they are no less important. Machinery for performing repetitive or strenuous tasks has been around since the industrial revolution, but what is new is the way in which machines can now be controlled. Again, computers are the controlling device, but in this case the information they are processing represents aspects of the operation of a machine.

Previously an automatic machine would have relied on a panel of electro-magnetic relays to sequence its operations properly. The wiring was configured such that a signal on a particular *sensor* would trigger output of a certain *actuator*, provided other criteria were met. Large panels of relays have now been replaced with the programmable logic controller, or PLC.

Programmable controllers. A PLC is a small computer. It has inputs, and outputs, where connections can be made to the external circuits. It has a processor to control the outputs in relation to the signals on the inputs, and it has memory to store the program. What may not be apparent is a screen or keyboard. Often these are removed after programming, since the PLC does not need them to function.

The PLC has many advantages over hard-wired relay panels. In terms of processing power per unit size, it is much smaller. As there are no moving parts, it is orders of magnitude more reliable. If the user wishes to change the logic, then it is easy to plug in the programming keyboard and edit the program; rewiring would have been necessary with a relay panel. The degree of control available is much more sophisticated. PLCs can have digital inputs from switches, or analogue inputs from temperature, pressure or flow sensors. The computing commands available cater for all industrial requirements. The device could perform logic control and interlocking, it could sequence various plant operations, or it could control process flows and conditions using PID (proportional, integral and derivative) algorithms.

Many manufacturers produce a range of PLC components. The entire device may be contained in one single 'brick', suitable for mounting in a small box, with the input and output terminals arranged along the top and bottom of the case. The larger systems are usually built up from a number of components. On to a pack plane would be fitted a power supply module, a processor module, some analogue and some digital input modules, some analogue and some digital output modules. The rating of some output modules may be small, to keep the size down, and so to enable large loads to be controlled, it may be necessary to specify a relay output as opposed to a transistor output. Usually the modular nature of the systems makes this an easy activity.

High integrity control. Where PLCs are used in critical process areas, for example in safety shutdown systems in petrochemical plants, it is frequently required to have *redundancy*, that is more than one device or system is used to provide the control function. Many systems permit this by linking two or more processors together. If one processor fails, then the system will automatically switch over to the other. In triplex redundant systems, three processors are working together, and the output modules operate according to the voted commands based on a 'two-out-of-three' rule. Thus if one processor gives a spurious result this will be ignored. Manufacturers claim that such systems are so reliable that they will never maloperate over their entire service life. Such a guarantee is obviously dependent on careful commissioning of the controller initially.

PLC networks. Another feature of the larger types of PLCs is the ability to network them. In just the same way as the productivity of office computers goes up when they can all share data, when PLCs are connected together, the data they are processing can be monitored. A group of PLCs on a production line can be linked to one operator station. This could be a specific control panel by the same manufacturer, or a standard office computer running special software. From this station it might be possible to make running changes as the line is operating. It might be possible to diagnose faults, or reprogram the controllers when they are off-line. The running totals of production quantities can also be monitored.

System-specific PLCs. The PLC systems described above are made from standard products, and specifying and indeed programming the systems is straightforward. Nothing about the controller is specific to the task to which the customer is going to put it. This leads to ranges of good value, yet reliable systems, suitable for a wide variety of applications. However, there are some areas where a specific control system is called for. An example might be that of a gas turbine controller. A control system like this is designed by the manufacturer of the primary plant, and the two are sold together. The control algorithms might be proprietary information and so once the equipment is set up, the end user or customer is not given access to the source code. For normal operation of the equipment this does not matter, but for servicing, OEM (original equipment manufacturer) support is needed. The manufacturer might put a premium price on this assistance, but this is usually offset by the reduced frequency of servicing which goes with computer controlled systems.

Control of entire plant. The highest level system of all would be that which controls a whole process plant or power station. In such systems the various processors are often distributed about the plant, earning their name of *distributed control systems*. It is a relatively short cable run from any field instrument to the nearest processor cubicle. In large systems there would usually be redundancy of controllers, and redundancy of the network connections between them. A single control room could have a number of operator screens on the same network. Although each processor is substantially autonomous and does not need the network to operate, it uses the links to send plant data back and forth. Thus every aspect of the plant can be both monitored and controlled from a single point. In such systems the computing power is used to:

1. Keep the system running as efficiently as possible. The algorithms might tune themselves in order to maximize the desired output. Very sophisticated control actions can be taking place.
2. Make the operator's life as easy as possible. Information is presented on plant mimic-diagrams, with graphics that represent flows in pipes and the levels of tanks and reaction vessels. Alarm conditions are automatically linked to information about the problem. Fewer people can look after more plant.
3. Store massive amounts of data about the plant. Often, for reliability reasons, the two functions of controlling the plant and recording data are performed on separate processors, but both are connected to the same network. On some plants, several thousand analogue values, and maybe ten thousand binary signals are logged with a 5-second sampling interval for the 30-year life of the plant. This requires colossal amounts of storage capacity, but given the way costs fall, it is economic to do. The data available to plant

engineers for maintenance, fault diagnosis, and performance improvements is invaluable.

Major industrial computing systems such as these can have their networks connected to the more commonplace office networks, usually with some bridging computer running special software. Thus a plant manager can have real-time information on his desktop about the plant conditions.

Microprocessor-based Devices

It is now so easy to design a microprocessor-based computer, and so cheap to manufacture these, that many products now contain them. This adds new features to existing products, so more can be achieved with fewer components. Protection relays for use in electrical systems now use microprocessors. Individual cubicles in a motor control centre may have integrated microprocessor protection and control modules. When designing a large plant, this can reduce on-site wiring, and enable many commissioning tests to be carried out from a single point. However, this introduction of different systems has brought with it new areas which must be carefully thought out. Making one computer system work with another is not always straightforward. Although there are common standards for data communication in industrial systems, it can be complicated to commission a brand new design.

Although the only moving part might be a hard disk which is in a hermetically sealed enclosure, the environment in which a computer is located is important. Server rooms in companies will be air-conditioned. The power supply will be filtered so that high frequency transients or spikes are removed. Since an unexpected power interruption can cause loss of data and corruption of files, uninterruptible power supplies frequently have to be provided. When installed in an industrial environment, a clean, dry, cool location should be selected. Since computers use low signal levels, interference from other items of electrical apparatus can be a problem. Optical-fibre cables can be used for long routes where this is the case.

Computers and computing is a field in itself, there is much jargon to come to terms with as well as new skills to learn. Working with computers and microcontrollers introduces new challenges, but also some exciting opportunities. The technology is relatively cheap, and getting cheaper all the time. Similarly the computing power available is increasing all the time. In the office, new ways of working have resulted, and in industry the automation has brought cost savings and productivity improvements. There are still many more developments to come, especially in the areas of mobile computing, and it is likely that these will seem every bit as revolutionary as what has happened in the last ten years.

7 Electricity generation

Generation may be direct current or alternating current but the versatility of the latter together with the availability of units with high ratings, and the robustness and reliability of alternating current plant, have made it universally dominant. Only alternating current generation is dealt with here.

Synchronous Generator Theory

An alternating current generator consists essentially of a magnetic field system produced by direct current, and an armature having windings linking this field system so as to induce an alternating voltage. The field is almost invariably carried on the *rotor* of the machine which is driven by the power source, possibly a steam or gas turbine in the case of a large unit, or an internal combustion engine for a small generator up to, say, 25 MW. The rotor may be cylindrical or it may have *salient*, that is readily identifiable, poles. The armature, which produces the alternating output voltage is the stationary part of the machine, or *stator*. For all but the very smallest machines the output is generally three-phase, at a frequency of 50 Hz in the UK and Europe, and 60 Hz in North America.

The form of construction of the machine depends greatly on the power source and speed. To generate at a frequency of f Hz when driven at a speed of n r.p.m. the generator must have $2p$ poles, such that

$$2p = \frac{120f}{n}$$

where n is the synchronous speed, that is the speed of the rotating magnetic field which would be produced by currents at frequency f circulating in the three-phase stator winding. It will be seen that a low speed machine will thus have a large number of pole-pairs and must therefore have a large diameter to accommodate these, while a high-speed machine of similar power will be longer with a smaller overall diameter. The highest machine speed corresponds to two poles ($p = 1$) and will be 3000 r.p.m. for 50 Hz or 3600 r.p.m. for 60 Hz.

The electrical output, S, of a generator can be determined from its dimensions, speed of rotation and electrical and magnetic loadings as:

$$S = KD^2 LnAB$$

where K is a constant, D is the diameter of the stator bore, L is the stator active length, A is the electrical loading (stator ampere turns per unit stator bore circumference) and B is the flux density at the stator bore. For any given choice of materials and method of cooling, the electrical and magnetic loadings (A *and* B) are essentially fixed, so generators for different outputs are designed by selecting different diameter and length of stator bore. These are constrained by the ability to build corresponding rotors capable of withstanding the centrifugal forces which increase as the diameter increases, and of avoiding damaging critical speeds which decrease as the length increases. It is also necessary to ensure adequate cooling of both stator and rotor which becomes more difficult as the length increases.

The e.m.f. equation. Consider the three-phase, two-layer winding of *Figure 7.1*. While the field system moves through a distance equal to one half a pole-pitch the induced e.m.f. changes from zero to E_{max}. Regardless of the space distribution of flux in the air-gap its r.m.s. value will be k_1 times the average value, where k_1 is the *form factor* of the flux distribution. Since the waveforms of the induced e.m.f. distribution in any conductor cut by a flux field travelling at constant speed and the flux distribution are identical, this too will have form factor k_1.

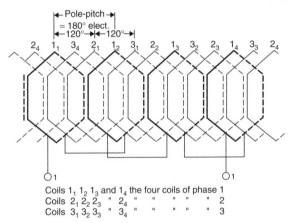

Coils $1_1\ 1_2\ 1_3$ and 1_4 the four coils of phase 1
Coils $2_1\ 2_2\ 2_3$ " 2_4 " " " " " 2
Coils $3_1\ 3_2\ 3_3$ " 3_4 " " " " " 3

Figure 7.1 Three-phase two-layer winding

Now the average e.m.f. induced in this conductor, E_{av}, is given by

$$E_{av} = \frac{\text{flux cut}}{\text{time of cutting}}$$

and the flux which cuts the conductor as the pole system moves one half a pole-pitch is $\Phi/2$, where Φ is the total flux per pole, while the time of cutting is one quarter of the periodic time, namely $1/4f$ sec.

Hence

$$E_{av} \text{ per conductor} = \frac{\Phi/2}{1/4f}$$

$$= 2\Phi f \text{ volts}$$

If there are Z conductors in series per phase, then the average voltage of all conductors summated will be $2\Phi Zf$ and the r.m.s. voltage will be $k_1 E_{av} = 2k_1\Phi Zf$ volts.

However, in establishing this expression the assumption has been made that all the e.m.f.s of the Z conductors can be added arithmetically. In reality the conductors of a phase will be distributed over several slots and it is also likely that the width of the coils is not the full pole-pitch. The terminal e.m.f. must therefore be reduced by factors to take account of the physical disposition of the winding. Hence, the induced e.m.f. per phase is given by

$$E_{\text{ph}} = 2k_1k_2k_3\Phi Zf \text{ volts}$$

where k_2 is a factor to take account of the distribution of the winding over a number of slots and is known as the *distribution factor*, and k_3 is a factor to take account of the fact that coils do not have the full pole-pitch and is known as the *coil-span factor*.

Machine reactances. In order to determine the response of a generator to changes of load, excitation or other factors it is necessary to have a mathematical model. This requires that reactances and resistance values are defined. Because of the lack of symmetry of the machine rotor about its central axis, it is usual to consider these parameters resolved along two axes at 90 electrical degrees, where the angle between two adjacent poles is 180 electrical degrees. These are known as the direct or simply 'd' axis, taken along the centre line of the rotor poles, and the quadrature or 'q' axis being along the interpolar axis. The d axis and q axis values can be calculated from a knowledge of the machine geometry. At low values of magnetic flux when the reluctance is all in the air-gap, the reactances are said to be *unsaturated*. Under short-circuit conditions when very large currents flow, the flux paths in the stator core and rotor body can become saturated and reactance values will be reduced. These are known as *saturated* values.

Leakage reactance. The stator leakage reactance results from flux which passes along the air-gap from tooth to tooth without entering the rotor, or it may arise from the stator endwinding. Since these leakage flux paths are essentially independent of the angular position of the rotor, the direct and quadrature axis values are considered to be the same.

Armature reaction. When load current flows in the stator winding, this produces a magnetomotive force (m.m.f.) which combines with the m.m.f. produced by the rotor winding to modify the flux which would be produced by the rotor winding on its own. This is known as *armature reaction*. On no load, with no stator current, there is no armature reaction. On load, the way in which the excitation flux is modified by the armature reaction depends on the power factor of the load. Consider the single-phase two pole machine of *Figure 7.2*. *Figure 7.2(a)* shows the e.m.f. developed on open-circuit with a polarity directly related to that of the adjacent poles. *Figure 7.2(b)* shows the current distribution with a unity power factor load. Since current and voltage are in phase this is identical to the e.m.f. distribution of *Figure 7.2(a)*. The m.m.f. produced by the stator is at 90 electrical degrees to the rotor m.m.f. so that the resultant flux is shifted through an angle determined by the relative magnitudes of rotor and stator m.m.fs.

In *Figure 7.2(c)* the stator current lags by $90°$, i.e. zero power factor lag. It will be seen that the m.m.f. produced is in phase opposition to the rotor m.m.f. so that the resultant flux is considerably reduced. By similar reasoning it can be shown that when the stator current is at zero power factor lead there will be an increase in the nett flux (*Figure 7.2(d)*).

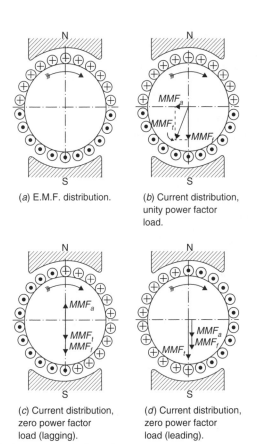

(a) E.M.F. distribution.

(b) Current distribution, unity power factor load.

(c) Current distribution, zero power factor load (lagging).

(d) Current distribution, zero power factor load (leading).

Figure 7.2 Armature reaction in an alternator

Synchronous reactance. The effects described above can be quantified as follows. Suppose that a generator is operating at zero power factor lag. Then the current I_{ph} can be represented lagging the terminal voltage V_{ph} by $90°$ as shown in the phasor diagram of *Figure 7.3*. Now the leakage reactance drop, E_x, in the machine leads the current by $90°$ so it will be in phase with E_{ph}. The

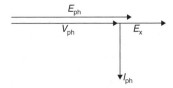

Figure 7.3 Operation at zero power-factor lagging

terminal voltage, V_{ph}, is thus the difference between E_{ph} and E_x. However, it has been shown, above, that the effect of armature reaction is to directly reduce the excitation flux and since the actual induced e.m.f. is the result of this reduced flux and not the open-circuit flux, it is clear that the armature reaction has the same effect as increased leakage reactance. Hence the armature reaction can be ignored and a notional value of reactance assumed which would have the same effect as the combined leakage reactance and armature reaction. This notional reactance is called the *synchronous reactance*.

Transient and subtransient reactances. These are the reactance values which determine the contribution of the generator to an external short-circuit. (They also determine the effect of external faults on the machine windings.) The *subtransient* reactance, denoted by the symbol X″, is the parameter which determines the contribution to the initial peak current, the *transient* reactance, denoted by X′, determines the fault current on a longer timescale.

Generator open- and short-circuit characteristics. Prediction of a generator steady-state performance is based on its open-and short-circuit characteristics shown in *Figure 7.4*. With the stator terminals open-circuited a curve of open-circuit volts, E_o, is plotted against exciting current I_f. At rated speed the abscissa Ob corresponds to the field current for rated volts, bg. Oa is the proportion of the field current required to overcome the reluctance of the air-gap and ab that required to generate the flux within the iron circuit.

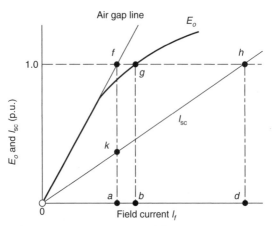

Figure 7.4 Open- and short-circuit characteristics

With the stator winding short-circuited, stator current can be plotted against field current to produce the curve Oh such that field current Od circulates rated stator current I_{sc}. On short-circuit the stator winding represents a load of almost zero power factor lagging in view of the very low winding resistance. Armature reaction is in direct opposition to the field m.m.f. so that the resultant flux is very small and insufficient to cause any saturation. Oh is therefore a straight line. The stator induced e.m.f. E_{sc} which circulates rated stator current I_{sc} against the stator leakage reactance X_1, neglecting the stator winding resistance, is $I_{sc}X_1$.

Since the relationship between I_{sc} and I_f is linear, E_{sc} and X_1 are both proportional to speed so the short-circuit characteristic is, to a first approximation, independent of speed. It is nevertheless usually obtained at rated speed.

Considering the open-circuit and short-circuit characteristics and again neglecting saturation effects and stator resistance, it will be seen that a field current of magnitude Oa produces an e.m.f. equal to af on open-circuit and ak on short-circuit. Hence on short-circuit the stator appears to present a reactance $E_o/I_{sc} = $ af/ak, a constant representing the *unsaturated* direct axis synchronous reactance, X_{du}.

Types of Generator

Generators are generally classified in accordance with speed and construction into the following categories.

Turbogenerators. Are driven by steam or gas turbines and cover power outputs from a few MW up to the largest built, currently around 1300 MW. They are usually two-pole machines having cylindrical rotors in which the field winding is housed in axial slots. The machines may be cooled by circulating air by means of a shaft-mounted fan either drawn directly from external atmosphere or using an enclosed circuit having secondary air/air or air/water heat exchangers. For the largest ratings, say over 200 MW, the rotor and stator core may be cooled by circulating hydrogen, and the stator winding by passing cooling water down the centre of the winding conductors. This water must be *demineralized* in order to have the high resistivity necessary, and the water-pipe connections to the stator windings are normally of PTFE (see Chapter 3) to insulate these from the high voltage between the stator windings and the earthed heat exchangers.

Hydrogenerators. Are driven by water turbines and are generally very much slower speed than steam or gas turbine driven generators. They are normally salient pole with a large number of pole-pairs and operate at 50–1000 r.p.m. depending on available water head and flow rate. Low-speed machines can be very large diameter with correspondingly short axial length. Generally such machines will have a vertical spindle with the generator mounted above the turbine.

Industrial generators. Are widely used for standby power supplies or supplies in remote areas where a public supply is unavailable. Often the prime mover is a diesel engine driving a salient pole machine at up to 1500 r.p.m. Ratings may be up to around 12 MW with the larger machines running at lower speeds.

Induction generators. Are similar in construction to an induction motor and draw their magnetizing current from the power system. They generate an output when driven at a speed slightly greater than synchronous speed and have ratings not normally greater than about 3 MW running at up to 1000 r.p.m.

Generator Construction

Rotors. The rotor shaft is usually a steel forging which is supported horizontally on bearings, ball or roller bearings for small generators and journal bearings on the larger machines. Hydrogenerators often have a vertical shaft which is supported on a thrust bearing designed to carry the weight of the rotor and possibly the turbine runner together with any hydraulic thrust. For a cylindrical rotor generator, slots running the length of the rotor body are machined in the rotor forging to house the copper conductors forming the rotor winding, and high strength wedges to hold the conductors in place against the high centrifugal forces. The conductors normally form concentric coils about the pole centreline which is not slotted for the winding. At the ends of the rotor body, the overhangs of the concentric coils are held in place against the centrifugal forces by high strength end-rings which go over the coils and are supported from the ends of the rotor body. Salient pole generators may have pole bodies forged integrally with the shaft or may have pole bodies, assembled from steel laminations, secured to the shaft by T headed slots. Slower speed and larger diameter generators, which can be over 10 m in diameter, may have a bolted laminated steel rim supported on a fabricated spider from the shaft. Again the pole bodies, assembled from steel laminations, are secured to the rim by T headed slots or similar. These very large rotors can exceed transport limits in terms of weight and/or dimensions so they are split into sections for transport or are built for the first time on site.

Rotor windings. The rotor winding carries direct current in order to provide the magnetic flux. *Figure 7.5* shows a typical salient pole machine in cross-section while *Figure 7.6* shows the cross-section of a cylindrical rotor turbogenerator. Simplified flux distributions are shown in both diagrams. The disposition of the field windings on the salient pole machine is apparent from the diagram. For the cylindrical rotor machine the field winding must be distributed in slots covering a considerable section of the rotor face. These field winding arrangements play a part in determining the shape of the air-gap flux wave, which in turn determines the stator output waveform. Distribution of the field coils in the rotor slots assists in shaping the flux wave. The machine designer is aiming for the stator output waveform to be as near as possible to a true sine wave. For the salient pole machine the pole faces are shaped so as to have increased air-gaps at the pole edges. This assists in producing the required flux pattern which otherwise would be basically rectangular, that is opposite the pole face there would be flux and outside this band there would be no flux.

Stator core. The core is made from thin laminations of special magnetic steel sheet cut to form rings, with the rings segmented for the larger generators. Each lamination is stamped from sheet to give the correct profile including slots for the stator winding and any ventilation holes, and is then insulated to reduce eddy current losses. The rings of laminations are built up to the required core length and then pressed and held under pressure within a stator frame to hold them securely. Any loose laminations may vibrate and damage the stator windings, or the teeth between the winding slots may fracture. Special care is taken with the stator core of a two-pole generator which experiences a compressive force in line with the pole centreline and rotating at synchronous speed. For transport purposes the stator cores of large generators can be split into segments.

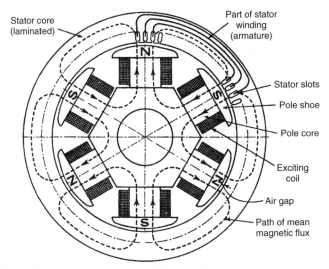

Figure 7.5 General arrangement of rotating field alternator

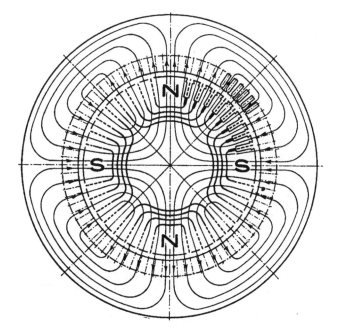

Figure 7.6 Cross-section of cylindrical rotor turbogenerator

Stator casing. This holds the stator core and frame to the foundations, transmits the forces under normal and abnormal operating conditions and contains/directs the cooling medium. Large generators may have hydrogen, at perhaps up to 3 or 4 atmospheres, in the casing to give improved cooling of stator and rotor. The casing must then be a gas tight construction with special seals around the rotor shaft. These take the form of journal or thrust bearings which hold oil at a higher pressure than the gas so that there is a small oil flow to the gas side.

Stator windings. *Figure 7.7* shows the disposition of a simple single-phase winding. It has only one conductor per pole and in addition to making very poor use of the available periphery of the stator, this would have a very small output e.m.f. *Figure 7.8* shows ways in which turns can be accommodated and distributed in slots around the stator. Of course, in reality, each slot may contain a number of conductors so the winding will be made up of a number of coils containing several turns each. In *Figure 7.8(a)* coils have two turns each distributed over two slots per pole. It can be seen from the diagram that for this winding the coils' ends or *overhang* occupy only half the stator periphery. This is inefficient use of space and would lead to a more costly machine. By utilizing coils whose width is slightly less than a pole pitch as shown in *Figure 7.8(b)* the overhangs are evenly distributed around the whole periphery. The windings shown in *Figure 7.8* are *single-layer* windings because one slot holds the conductors of one coil only. Most modern

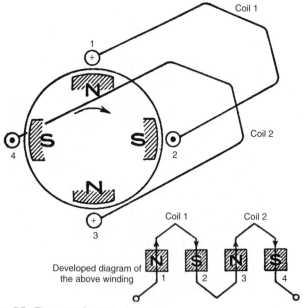

Figure 7.7 Elementary four-pole alternator

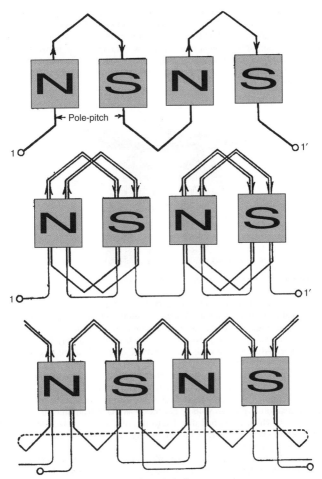

Figure 7.8 Concentrated and distributed windings

generators have *two-layer* windings in which each slot holds the conductors of two coils. These are arranged so that the lower half slot contains one coil-side of which the other side occupies the upper half in its return slot.

Three-phase windings. The windings described above are for a single-phase machine. In a three-phase machine each phase-group of coils occupies an arc of 60° electrical under each pole. In order to achieve the required 120° electrical displacement between the stator induced e.m.fs., phases a and b will occupy the outer positions and phase c, connected reverse polarity, will occupy the middle.

Practical considerations. The physical make-up of the coils is dependent on the size of machine. Small machines operating at modest voltages utilize *mush coils*. These are wound from round wire, enamel insulated, laid at random in open or semi-closed slots (*Figure 7.9*). The main insulation between coil and core is a slot liner of polyamide or aramid paper (Nomex). Two-layer windings have either a separate liner for the top coil-side, or simply a separator between top and bottom halves of the slot.

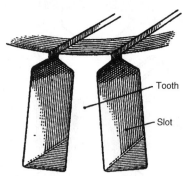

Figure 7.9 Semi-enclosed slots in generator stator periphery

Larger machines have preformed coils wound from rectangular conductor. Each conductor may be subdivided into a number of subconductors to reduce eddy current loss. These may be enamel insulated, or enamel plus a thin covering of braided glass or polyester-glass. Subconductors may be transposed along the coil-side so that each links a similar level of leakage flux (produced by the load current) thus balancing out the leakage flux induced e.m.fs. Coils may be wound as flat loops which are then formed into shape, having the correct length for the stator slots and the appropriate slot pitch. Interturn insulation is usually applied by overwrapping after this forming stage. The main insulation to earth on large high voltage generators is made from mica bonded with epoxy resin which is applied directly to the subconductor stack and pressed and cured to give an integral structure having excellent mechanical, electrical and thermal properties.

Special attention is paid on very large generators, under normal operation as well as under fault conditions, to the forces on the conductors in the slots and in the overhang.

Testing

Apart from site built generators it is usual for all generators to undergo some proving tests before leaving the factory. Full load testing is impracticable except on the smallest generators. Testing can include:

Winding resistance
Insulation resistance

Winding high voltage
Open-circuit excitation
Open-circuit losses
Short-circuit excitation
Short-circuit losses
Zero power factor excitation
Temperature rise
Sudden three-phase short-circuit
Overspeed

Some of these tests, referred to as *type tests*, may only be carried out on one generator of a particular design.

Generator Protection and Synchronization

All generators have some form of electrical protection which is aimed at limiting the damage to the generator from an internal fault or external fault that gives rise to damaging conditions. Also the protection aims to limit the disturbance to the external electrical system from a generator fault. The extent of protection applied is determined by the value, i.e. size and rating, of the generator and disturbances to the external electrical system. Protection can include:

Generator differential
Overcurrent
Stator winding earth fault
Rotor winding earth fault
Negative phase sequence current
Loss of excitation
Pole slipping
Reverse power
Rate of change of frequency
Over/undervoltage
Over/underfrequency
Overfluxing
Overexcitation

Microprocessor protection offers several of these functions within one equipment so that the user may select those to be used. Large generators may have a two-channel arrangement which duplicates some protection functions and gives alternative cover for other fault types.

Automatic synchronizing equipment which controls prime mover speed to match frequency and voltage phase angle, and controls voltage to match voltage magnitude before automatically closing the generator breaker, is used widely. Synchronizing is achievable within fine limits.

Connection to Electrical Network

At voltages up to about 11 kV, the rated generator stator voltage is normally selected to match the voltage of the system to which the generator is connected.

However, even at these voltages, a small output at a high voltage or a large output at a low voltage may make the generator design uneconomic. In these cases, and where a generator is to be connected to a higher voltage transmission or distribution system, it is usual for the generator output to go via a generator transformer which matches the voltages. In these cases the generator designer is able to optimize the selection of rated voltage and current. For example, for the 500 MW and 660 MW generators in the UK, the rated generator stator voltages are 22 kV and 23 kV, giving rated stator currents of about 15 kA and 20 kA. These rated currents are too large for cable connections so phase-isolated busbars comprising a hollow aluminium conductor inside a circular aluminium earthed outer enclosure are used. These virtually eliminate the risk of phase to phase faults at the generator voltage with the attendant large currents and forces. The use of a generator transformer gives the option of control of voltage at the electrical system interface by changing generator voltage or by keeping fixed generator voltage and using an on-load tapchanger on the transformer. The latter arrangement is used in the UK and gives the required voltage droop characteristic with reactive power by virtue of the transformer reactance (see section on excitation control) and permits power station auxiliaries to be supplied via a unit transformer connected to the fixed voltage generator terminals.

Operation of Generators

The operating characteristics of an a.c. generator are slightly different when running singly compared to when it is running connected to a large system. The latter arrangement is, of course, by far the most common method of operating an alternating current generator.

Operation singly. When a generator runs alone to supply power to an isolated system, the power demanded by the system has to be supplied by the prime mover. If the demand is increased the generator will initially tend to slow down but action of the engine governor will increase the throttle setting to maintain the speed and frequency constant. The system power factor will likewise be determined by the load, but if the load conditions change such as to change the system power factor the generator output voltage will initially change. This will be maintained constant by the automatic voltage regulator having control over the field. By convention, a lagging power factor load is considered to require that the machine *export* reactive volt-amperes (VArs) and a leading power factor load requires the machine to *import* VArs. A lagging power factor load will cause the machine output voltage to fall compared with generation at unity power factor, so that increased excitation (field current) will be required to maintain the system voltage. A leading power factor load will require reduced excitation compared with that at unity power factor to maintain a constant output voltage.

Operation on a large system. When it is connected to a large system the action of the individual machine controls cannot have any significant effect on the system as a whole. Hence opening the throttle of the prime mover to increase power input to the machine cannot affect system frequency but the power exported by the machine will be increased, while it continues to run at synchronous speed. Similarly, increasing the machine excitation cannot increase the voltage on the system but this simply increases the VArs exported by the machine to the system, that is the machine power factor moves further lagging. In the same way reducing the machine excitation causes the power factor to move further leading. Reducing the excitation too far will cause the power factor to move too far to the leading condition so that the machine could become unstable and pole-slip. The field may be seen as the means of transferring the input power delivered by the prime mover to the rotor to the output system connected to the stator. If this field is allowed to become too low it is no longer capable of transmitting the output power demanded by the system. Generally, in order to avoid any risk of pole-slipping, leading power factor operation is limited to about 0.7 at zero load to 0.9 at full load. (See description of automatic voltage regulators later in this chapter.) In the case of a generator connected to the system via a generator transformer, the change in reactive power and consequently the change in excitation is achieved by operation of the on-load tapchanger on the transformer.

Excitation Systems

The generator excitation system provides the source of supply to the rotor field coils. Control of the field current produced by the excitation system is exercised by the *automatic voltage regulator* or a.v.r. which is described later.

Control of the field current must ensure that the machine runs at the desired voltage or, when the generator is operating connected to a large system, that it imports or exports the required level of reactive kVArs or MVArs.

The combination of excitation system and a.v.r. must:

- Control the machine voltage accurately in response to slow changes in power or reactive var demand.
- Limit the magnitude of voltage excursions in response to sudden changes in load.
- Maintain steady-state stability.
- Ensure transient stability in response to system faults.

It is the final three requirements which largely determine the type of excitation system used.

For many years the standard method of providing the excitation current was to use a d.c. generator coupled to the shaft of the synchronous machine. For larger high speed generators coupling was via a gearbox to reduce the exciter speed to typically 1000 or 750 r.p.m. in order to simplify construction and avoid commutation problems with the d.c. generator. The exciter output was fed to the machine rotor via slip-rings. Generally the exciter itself was separately excited from a directly coupled pilot exciter. Control of the generator excitation was by controlling the field current of the main exciter.

A.C. excitation. The advent of solid-state rectifiers made possible the elimination of the commutator on the d.c. exciter and the substitution of the exciter by an a.c. generator operating at any convenient frequency between 50 and 250 Hz and providing an output through slip-rings to be rectified in locally mounted diode cubicles. Rectifier output is then fed to the machine rotor via slip-rings as for the d.c. excitation system.

The main exciter field is supplied by a pilot exciter which is usually a permanent magnet generator. The main exciter is generally three-phase with the diodes arranged in a bridge network. Bridge arms normally have a number of diodes in parallel and each diode is individually fused to remove it from the electrical circuit in the event that it faults to the short-circuit condition. Cooling may be by natural or forced air ventilation, or in the case of very large machines the rectifiers may be water-cooled.

Brushless excitation. The next advance in the a.c. excitation system described above, with rectifying diodes in cubicles adjacent to the exciter, was to mount these on the exciter shaft. The main exciter has its output armature winding on the rotor with its field on the stator. Rotor output, which can be at a frequency between 150 and 250 Hz, is then connected directly to the shaft mounted diodes. Output from these can then be taken directly to the rotor of the main generator. The pilot exciter, which might operate at up to 400 Hz, will very likely be a permanent magnet generator or, on small generators, the supply to the exciter field can be taken from the generator terminals.

Thyristor excitation. Thyristor excitation represents the ultimate development of the theme. Control of the thyristor rectifiers enables direct control of the generator field current to be obtained providing a very much faster response than can be obtained by controlling the exciter field current, which thus enables the fastest response to system transients to be obtained. This can be particularly important for installations where rapid response to faults or disturbances is critical for system stability. The disadvantage of thyristors is that a means has not yet been developed commercially for shaft mounting these, the reason being the difficulties of reliably providing control signals to them from stationary equipment. Hence it is necessary to provide sliprings and brushes to connect to the machine rotor.

Excitation power may be provided by fairly conventional directly coupled main and pilot exciters or it may be direct from the main generator terminals via a step-down transformer. The step-down transformer ratio is normally selected so as to provide full output during the occurrence of a system fault which reduces the machine terminal voltage. This has the disadvantage that the field winding insulation level must be such as to withstand the higher voltage existing when system conditions are healthy. Elimination of the time delay associated with the exciter and pilot exciter means that transformer-fed excitation systems have the fastest response of all.

Automatic Voltage Regulators

As indicated above, the purpose of the a.v.r. for a generator running singly is to maintain the steady-state voltage within specified limits and contain the extent of voltage excursions when sudden changes of load occur. In addition, for a group of generators running in parallel some further controls may be required to ensure satisfactory sharing of reactive load. In the case of a machine running in parallel with a large interconnected system, control of steady-state and transient stability is an important additional requirement. The demands of this may be such as to render manual control inadequate and require that automatic control be provided.

Basic principles. A signal proportional to generator terminal voltage obtained from the rectified output of a voltage transformer is compared to a stabilized reference voltage obtained within the regulator. Any difference, or error signal, is amplified and used to control the excitation supply, raising or lowering the input to the main field winding or exciter field, as appropriate, to reduce the error signal to zero or an acceptable value. Adjustment of the set voltage is obtained either by adjustment of the reference voltage or by adjusting the proportion of machine voltage compared to the reference voltage. This basic scheme is shown diagrammatically in *Figure 7.10*. The stabilizing loop is included to prevent hunting.

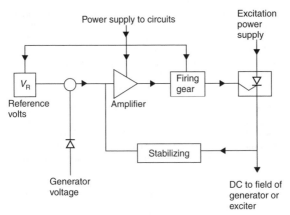

Figure 7.10 Basic circuit of an a.v.r.

Control range. Generators are normally required to deliver from zero to rated output over a voltage range of $\pm 5\%$, at power factors from 0.8–0.9 lag to 0.9 lead. The a.v.r. setting controls must provide the appropriate range of excitation as well as a voltage down to about 85% of nominal at no-load. Accuracy of control is normally between $\pm 2.5\%$ and $\pm 1\%$ of set value over the load range.

Manual control. Manual control is generally provided as a back-up to the automatic control although, as stated above, many large machines cannot be satisfactorily operated except in automatic and the manual setting may only be used for the initial commissioning. Most regulators, when operating in automatic, are provided with 'manual follow-up' which ensures that the manual setting matches the automatic setting so that if failure of the automatic system causes a 'trip to manual' this will not result in any change of set-point. A balance meter may be provided so that if changeover is to be made manually, a check can be made before effecting the change.

Parallel operation. The a.v.r. can be arranged to provide a fall in generator terminal voltage (voltage-droop) for increasing reactive load to ensure satisfactory reactive load sharing when generators are operating in parallel. The droop is normally between 2.5 and 4% voltage fall over the reactive range at full load.

Excitation limits. Fault conditions on the system which result in a fall in system voltage will cause the a.v.r. to drive the excitation to ceiling value to try to restore normal voltage. A timer is provided to return the excitation to normal after a few seconds to prevent overheating of the excitation system if the fault persists. This must be co-ordinated with the design of electrical system protection to reduce the risk of system voltage collapse.

A reactive power limiter, or VAr-limiter is provided to prevent the excitation falling so low that the generator will become unstable. The reactive power at which this operates is automatically adjusted to follow the limit of stability as generator voltage and output power are varied.

A more restricted reactive power limit applies when the excitation is under manual control. Whenever the generator is operating between this restricted limit and the automatic control stability limit, it is desirable for the 'manual' setting control to be restrained at a setting no lower than the 'manual' limit so that should there be a trip to manual control, the generator excitation will be increased to a safe level. Under this condition, there will be a standing error on the balance indicator, requiring operator correction.

Overfluxing protection. It is operationally desirable that the a.v.r. should remain in service when the machine is shut down and subsequently run up. To prevent overfluxing of the machine and any associated generator or unit transformer, the a.v.r. overfluxing prevention, which is sometimes known as V/f protection, will operate to ensure that below about 95% rated speed excitation will decrease at least in proportion to frequency.

Dual-channel a.v.rs. Large generators may have dual channel a.v.rs which effectively have two sets of control equipment. Each channel is able to perform the control function independently and may either operate in parallel or in main and standby mode. If one channel fails the other performs the duty alone and an alarm is initiated.

Digital a.v.rs. Even with the sophistication built into the a.v.rs described above, there remains a risk of reliability problems due to the failure of some contacts of the many electromechanical relays incorporated. It was in an effort to improve on this aspect that consideration was first given to replacing the relay logic by a digital system.

This was initially done by the use of a series of programmable logic controllers, which could be used with analogue inputs as well as digital ones. It soon became apparent that the whole 'front end' electronics could be made to operate in a digital mode. The characteristics required of the a.v.r. could be digitally programmed into the processor and these could be tested by simulation in the factory before installation on site. Problems encountered in analogue amplifiers such as drift can be eliminated as can noise and/or contact problems in setting rheostats. A simplified block diagram, *Figure 7.11*, shows the basic operating principles. Almost all present day excitation control systems use this basic operating philosophy.

129

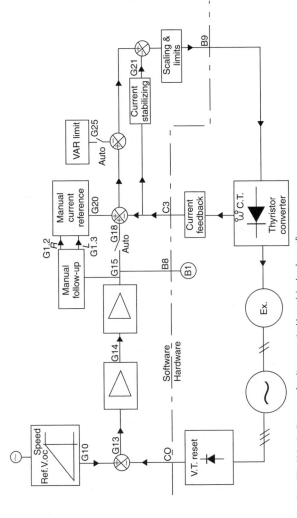

Figure 7.11 Simplified block diagram of voltage control loop (single channel)

Power Generation for Public Electricity Supply

Any description of a.c. generators cannot be complete without some examination of the means whereby the generator is driven, since the act of designing generating plant involves a matching and integration of the prime mover and the generator. Mention has already been made, for example, of the basic differences between turbogenerators and hydrogenerators which arise because of the difference in speed of rotation. The majority of generators used for large-scale power generation are turbogenerators, and until recently these were driven by steam turbines. Since the early 1990s, for reasons which will be explained, there has been an enormous increase in the use of gas turbines so that the two types now co-exist as the main, and diverse, sources of supply of electrical energy throughout the world.

The subject of diversity in electricity supply will be discussed in the next chapter. It is appropriate here to look in more detail at the various methods of power generation.

Steam turbines. There are a number of ways in which steam can be produced for power generation. The principal ones are by burning fossil fuels, coal or oil, or from nuclear power.

Coal- and oil-fired stations. In the UK a large programme of power station construction in the 1960s and early 1970s resulted in a significant proportion of power generation from large coal- and oil-fired stations. Larger unit size led to higher thermal efficiencies so that the first of this new generation of stations had units of 500 MW compared to those of around 200 MW built hitherto. In order to provide a mix of fuels, those in the coal producing areas were coal-fired while oil-fired units were located conveniently for the use of imported oil. Later stations standardized on the use of 660 MW units. The escalation of oil prices in the 1970s meant that the oil-fired units rapidly became uneconomic so that their use was soon greatly curtailed. At an oil-fired station on the Thames at Isle of Grain, originally intended as a 5 × 660 MW station, only the first three units were completed and commissioned.

For these large units steam is delivered to the turbine stop valve at pressures exceeding 160 bar at 566°C. These units are capable of achieving thermal efficiencies of just about 40%. At the time of privatization of the Central Electricity Generating Board (CEGB), the public authority responsible for electricity generation and transmission in England and Wales, in 1989, greater pressures, up to 310 bar, were being contemplated with unit sizes of 900 MW and with temperatures up to 590°C. These higher temperatures associated with the higher pressures require special austenitic steels and many technical problems had still to be resolved. These higher temperatures and pressures were expected to enable percentage efficiencies into the lower 40s to be achieved. Privatization, and the move to gas turbines as prime movers, described below, prevented this development taking place.

Gas turbines. Gas turbines were initially introduced at the 500 and 660 MW unit fossil fuelled stations as peak-lopping generators, as an independent supply to power station auxiliaries in the event of prolonged system operation at a low frequency, and to permit start-up in the event of a transmission system failure. These used only slightly modified aero engines as their source of power and because of the high temperature of their exhaust gases they had very poor thermal efficiency. The output of the aero engine meant that most of these units have electrical ratings of only 17 MW, although some of the later units are 25 MW. Their big advantage is that they can be started, run up and synchronized to the system and loaded to full load in a matter of minutes compared to the hours necessary to perform the equivalent operation on a steam turbine. Hence their duty of 'peak-lopping,' that is being used to support the supply system at times of rapid increase in demand. They also had a capability for 'black-start'. That is they could be started and run up without any input from the grid. Once run up and generating power, the output could then be used to start up a large fossil-fuelled unit. Hence, in any emergency situation which led to the collapse of a part of the grid system, this could be recovered using the gas turbine generators. These benefits also resulted in the CEGB building a small number of pure gas turbine stations based on purpose built units having electrical outputs of the order of 70 MW.

Combined cycle gas turbines (CCGT). The CCGT aims to overcome the poor thermal efficiency of the OCGT (open cycle gas turbine) described above by recovering the heat from the exhaust gas. This is done by passing this through a heat recovery boiler to raise steam which is then used to drive a conventional steam turbine. A CCGT unit will thus consist of one or more gas turbines and associated generators with a heat recovery boiler taking all the exhaust gases to generate the steam to drive a single large steam turbine. The overall thermal efficiency of this 'combined-cycle' arrangement was around 55% for the units commissioned in the 1990s, with the next generation aiming at achieving the 'holy grail' of 60%. This step increase in thermal efficiency compared with large fossil-fuelled steam turbines has resulted in approximately 32% of the UK installed capacity being CCGT in the year 2000 compared with virtually nil some ten years earlier.

In the UK the 'dash for gas', as it has been termed, has been so marked, and the rise in use of the CCGT through much of the world has been so extensive that many commentators have attempted to establish the reasons for this. These are not simply the increased thermal efficiency, since this benefit had been recognized for a number of years.

Industrial gas turbines, that is those designed specifically for non-aero applications, had been developed almost concurrently with those for aircraft. However, the unit size was modest for power generation purposes and the economics of installing heat recovery boilers for such small units were not attractive. It was not until a number of 'power blackouts' occurred in the USA in the 1970s that the problems of re-establishing a major network following shutdown were driven home, and the role that units with rapid start-up capability were able to perform in assisting this re-establishment was recognized in all quarters. This led to further development of the larger gas turbines. If, say, two 70 MW units are installed at a common location it becomes more practicable to provide a heat recovery boiler to handle the combined exhaust gases, leading to the first commercially attractive CCGTs. Once the attractions of the process came to be recognized there were more incentives in developing and transferring aero technology to improve the industrial gas turbines. In addition, more and larger gas fields have been established, and in Europe at around the time of privatization of the UK electricity supply industry, the EU directive of 1975 which stated that natural gas was too precious a resource to be used for power generation was rescinded. These factors occurred coincident with recognition of the 'greenhouse effect' and the pressures to reduce emissions of CO_2 and other gases such as sulphur dioxide and oxides of nitrogen present in the exhausts from fossil-fuelled stations. A CCGT produces very much smaller quantities of these.

Another significant factor which gave added momentum to the surge in the building of CCGT stations was the time taken to complete them. Whereas a coal-fired unit might take seven to eight years from commencement to producing power output, a CCGT can achieve this in as short a time as 18 months. Furthermore some output can be obtained within about 12 months by operating the gas turbine generator(s) alone before the heat recovery boiler is completed. This means that a financial backer can start to get some return on investment much more quickly by building a CCGT.

Nuclear power. Britain implemented a substantial nuclear power programme between the mid-1950s and the early 1990s so that by the year 2000 commercial nuclear power stations produced a significant proportion of the country's base load electricity.

UK commercial nuclear power stations fall into three basic types. Magnox and advanced gas-cooled reactors (AGRs) both use CO_2 as the heat transfer medium (or primary coolant) extracting heat from the nuclear reaction and passing this through heat exchangers to generate steam for the turbines. Pressurized water reactors (PWRs), of which Sizewell B was the only one built in the UK by 2000, as the name suggests, use water as a primary coolant. Again this coolant is taken to heat exchangers where this is used to heat secondary water to raise steam for the turbogenerators.

Under the first commercial programme, nine magnox stations with a total design capacity of 4800 MW were commissioned between 1962 and 1971. Of these only Dungeness A – 445 MW, Sizewell A – 430 MW, Oldbury – 440 MW and Wylfa A – 1050 MW remained in commission by 2002.

The second phase of the nuclear programme was based on the AGR. Electrical outputs of the AGR stations are: Hinkley B – 1220 MW, Dungeness B – 1110 MW, Hartlepool – 1210 MW and Heysham – 2400 MW in England, and in Scotland, Torness – 1250 MW and Hunterston B – 1190 MW.

Britain's first PWR station, Sizewell B, with a rating of 1188 MW, was commissioned in 1995. This was estimated to cost £2.03 billion (1991 price).

In May 1988 the CEGB published a case for another PWR station at Hinkley Point C for submission to a public inquiry. Permission was granted with the funding permission being delayed pending the 1994 Government Review. Permission to build two more PWR stations at Wylfa B and Sizewell C has been withdrawn from Nuclear Electric (now British Energy), who took over the responsibilities of these stations from the CEGB when the industry was privatized.

AEA Technology, formerly the UKAEA, operated the 250 MW fast breeder reactor (FBR) at Dounreay. This was closed down in 1993. British Nuclear Fuels Ltd (BNFL) was responsible for the magnox stations at Calder Hall (200 MW) and Chapel Cross (196 MW), both of which have now been shut down. In the UK there is now a total of about 10.6 GW of available nuclear capacity in operation.

Hydroelectric generation and pumped-storage schemes. Hydroelectric generation is mostly confined to the more mountainous and wetter regions in the west and north of the UK. Generation in Scotland amounts to some 1200 MW whereas in England and Wales it is some 100 MW.

Pumped-storage schemes consist of two reservoirs, one at high level and the other at low level, so that during the day hydroelectric generation takes place, allowing the lower reservoir to be filled from the high level reservoir. During the night the water from the lower level is pumped back to the higher level. Reversible pump turbines can be used for this purpose. This system of generation was developed to operate in conjunction with the high merit plant of large coal-fired and nuclear power stations, which acted as base load units. When the demand is low during the night, the output from these low energy cost stations is utilized to transfer water from the lower to the high level reservoir, so that at peak demand the pumped-storage generation is able to displace some of the high energy cost generation that would otherwise be needed.

Dinorwig and Ffestiniog, both in North Wales, have a combined output of 2088 MW and a storage capacity of 10 600 MWh. The initial concept was that they would be used to provide a rapid source of power to cope with sudden increases in demand for electricity, thus providing system security and stability. Dinorwig has the capability to produce around 1320 MW in approximately 12 s. With privatization the operating philosophy has become subject to commercial considerations such that only 600 MW of the installed capacity is now normally available for system operational support. In Scotland, the 400 MW installation at Cruachan, where the operating head is 365 m comprises four vertical reversible Francis pump/turbines and motor/generators having ratings of 100 MW each when generating and 110 MW when pumping. The Foyers pumped storage scheme at Great Glen has a generating capacity of 300 MW and absorbs 305 MW when in the pumping mode.

Renewable sources of energy. Because of the world's dwindling energy resources and the environmental impact of increased fossil fuel use many countries are examining the possibilities of developing renewable sources of energy. Research is being carried out into a number of areas including the use of wind, landfill, geothermal, tidal, wave, solar, waste, tyres, sewage gas and chicken litter. In the UK the Government has provided support for the development of these new sources of energy by introducing, in 1990, a Non-Fossil Fuel Obligation (NFFO) to encourage the electricity companies to purchase some of their electricity from sources which do not use fossil fuels. The NFFO was funded by the Fossil Fuel Levy and provided a guaranteed price for electricity over a defined term, for qualifying sites.

NFFO only applied to England and Wales; in Scotland a similar scheme known as the Scottish Renewables Obligation (SRO) applied and another NFFO-type scheme applied in Northern Ireland. Five rounds (or 'tranches') of NFFO contracts were awarded, and three of the SRO scheme. Initially the unit price paid for electricity was high, with a short term, but as the schemes developed the unit price fell and the contract term was extended to 15 years.

The NFFO/SRO schemes were replaced in 2001 by a system of tradable Renewable Obligation Certificates.

Wind energy. Wind energy is the best developed of the true renewable energy sources. Growth in installed capacity of wind turbines has been very fast, particularly in Europe and the USA, with an estimated installed capacity by the end of 2000 of 12 800 MW in Europe, about 2700 MW in the USA and a few hundred MW in other countries of the world. Although the UK is estimated to have about 40% of Europe's wind resources, installed capacity at the end of 2000 was only 406 MW, about 3% of Europe's total. This is due to problems with obtaining planning consent (an NFFO or SRO contract does not guarantee planning permission), and the fact that most of the windy areas are located where the grid and distribution infrastructure is weak, placing unacceptably high costs on wind turbine developments where distribution system reinforcement is needed. Other countries have different support mechanisms and planning systems, and these have been more successful than in the UK at encouraging wind energy development. Germany, for example, had an installed capacity at the end of year 2000 of over 6000 MW.

Virtually all wind turbines installed today are horizontal axis machines with two or three blades. Turbines can be fixed speed or variable-speed, and the fixed-speed machines may be either pitch regulated or stall regulated. Variable-speed machines, by their nature, need to use pitch regulation.

The electrical generator in fixed speed wind turbines is invariably an induction generator, using a more or less standard induction machine (see earlier in this chapter) driven through a gearbox at a speed slightly above synchronous speed. Induction machines are used because they are economic, robust, readily available from a variety of manufacturers, need no separate excitation system and require very little maintenance. All these are important factors for a generator installed in a constricted location at the top of a steel tower, which may be 60 or 80 metres high. Some designs use a two-speed generator, having two sets of windings, to increase the efficiency of energy capture at a wide range of wind speeds.

Variable-speed wind turbines can use several different configurations; two widely used machine designs include, respectively, an induction generator driven through a gearbox, and a large diameter salient-pole alternator (see earlier in the chapter) driven directly. The use of variable speed turbines has several advantages, including reduced drive train torque above rated wind speed, reduced demand on pitch actuators and increased energy capture at lower wind speeds. Variable-speed operation inevitably produces variable frequency output and requires the use of static conversion equipment to interface the generated variable frequency output to the local grid.

Generation for outputs up to 1 or 2 MW is at voltages in the region of 480 to 690 V, and this has to be transferred to ground level. As the rotor and nacelle assembly of the wind turbine rotates to face into the wind, this has to be done via a loop of cable to take up the rotation. A control system keeps track of cable twist and initiates nacelle rotation to untwist the cable when a predetermined number of rotations (in either direction) have been completed.

At ground level are installed the conversion equipment (for variable-speed turbines), switchgear, control equipment and a transformer to match the output to the voltage of the local distribution network. The transformer is generally an oil-filled unit with off-circuit taps, although dry type transformers are sometimes used. The planning authorities often prefer the transformer to be mounted inside the turbine tower base for visual amenity reasons, rather than having the transformer pad-mounted adjacent to the tower base. In this case, the space restrictions within the tower base, and particularly the access restrictions, mean that a special transformer design is required, and these dimensional constraints have caused problems with some transformer designs.

Where static conversion equipment is required for a variable-speed machine, this may be a full-power system, in which the whole output of the generator is converted by the equipment to grid frequency, or, for variable-speed induction generators, it may use a system in which a wound rotor induction generator is used, with only the rotor power passing through the conversion equipment. In both cases the control of the turbine blade pitching, the generator rotor excitation and the firing of the various parts of the static conversion equipment will be done in accordance with sophisticated algorithms which are, of course, the well-guarded secrets of the manufacturers. There are two advantages to the use of static conversion equipment as compared with direct connection of induction generators. One is that it is usually possible to control the power factor of the supply to the distribution system, avoiding the problems with reactive power consumption associated with the use of directly connected induction generators. The second is that such systems generally contribute significantly less additional fault current to the distribution network than with fixed speed turbines using induction generators. A disadvantage of static conversion equipment is the possible generation of harmonics, which can cause 'harmonic pollution' of the distribution system if not suppressed.

Reverting to consideration of fixed-speed turbines using induction generators, these have two major characteristics which need to be considered in relation to their connection to the local distribution network. The first of these is their reactive power consumption. This can cause distribution system voltage control problems, and particularly can cause voltage dips on starting and during operation in gusty conditions. This is seen as flicker by other consumers in the area, and if severe needs to be corrected. The voltage dip at start-up is usually dealt with by installing a soft-start device to limit the current drawn at connection to only slightly more than normal full-load current. Other problems associated with reactive power consumption are dealt with by installing power-factor correction capacitors to each turbine. Designs vary but commonly capacitors are fitted to compensate for the no-load reactive power consumption, and sometimes further capacitors are switched in at increasing power output levels as well. The number of capacitors installed must never overcompensate for the reactive power consumption of the induction generator, as this could under certain conditions give rise to self-excitation.

The second problem with induction generators is the high contribution (relative to their output) to making peak fault levels (see Chapter 15). This needs to be studied at initial design, in conjunction with the distribution company's own design engineers, to ensure that switchgear ratings on the local distribution network are not exceeded. In some cases the fault level contribution may impose a limit to the number of turbines which can be connected at a particular location.

Because wind turbines are often located in mountainous areas, electrical system earthing can be a problem, as rocky areas often have a high ground resistance. Special measures are often necessary, including buried ring earth electrodes around substations and bare earth conductors buried in trenches along with the site cabling. Many wind turbine sites are 'hot' sites and special attention needs to be paid to equipotential bonding.

Lightning is a significant source of damage to wind turbine blades, and turbine blades and towers need to be designed to cope with lightning strikes. In general the steel structure of the nacelle and tower is used to conduct lightning safely to ground, and special contact brushes are placed around the rotating joint between tower and nacelle to ensure good contact. At the tower base one or two buried ring electrodes of bare copper cable are installed to provide potential grading – these are of course integrated into the earthing system design as well.

Until the year 2000 the vast majority of wind turbine developments around the world have been on land. However, there have been one or two pioneering offshore developments, and this is set to expand rapidly over the next few years. Offshore development is attractive because:

- The wind speed is generally higher offshore than on land.
- The wind is less variable (less gusty) offshore.
- The overall resource is greater.
- There are fewer perceived planning issues than on land.

However, development is proceeding cautiously in UK waters for various reasons:

- The number of sites with shallow water (for turbine foundations) in locations with a good grid connection on the nearest shore is limited.
- The costs of installing structures offshore are very high.
- Electrical connection costs are high.
- Maintenance access to the turbines can be severely restricted by the weather.
- There is risk of shipping damage to seabed cables (burial is possible but more expensive).

Initial approval was given by the Crown Estates (which controls the seabed around the UK) in the spring of 2001 for 13 offshore wind developments with the potential for generation capacity of about 1500 MW. This will be the next major development area for the UK's wind generation industry.

Landfill schemes. Landfill sites generate predominantly a mixture of methane and carbon dioxide and they have been exploited since the 1970s but only recently have they become key issues as sources of energy for generation purposes. The UK is the second largest exploiting nation (after the USA), some 50% of the gases being used for electricity generation and the rest being employed by industry to fire kilns and boilers, etc. There are about 5000 landfill sites in the UK, of which only about 450 are producing enough methane for power generation, according to a survey carried out by the Energy Technology Support Unit (ETSU) of Harwell, a body set up in the 1980s by the then Department of Energy now the Department of Trade and Industry (DTI).

During the 1990s ten companies utilized 22 landfill sites, the largest having a capacity of 2.25 MW at Chorley, Lancashire, commissioned in mid-1991 by Biffa Waste Services. In 1991 another scheme was proposed which will build up to about 4 MW over several years. ARC ordered and installed the generating sets for a 1 MW site at Huntingdon, Cambridgeshire, and two further projects totalling about 1 MW are being completed. United Utilities and Land Fill Gas have several installations in the north of England totalling around 15 MW.

Geothermal. Geothermal energy derives from the heat flowing outwards from the earth's interior, and in some places from the decay of long-lived radioactive isotopes of uranium, thorium and potassium. It can be tapped from hot dry rocks (HDR) and aquifers, with most effort being concentrated on HDR. The former Department of Energy funded hot water aquifer research for 10 years to 1986, but results overall have shown that there is little prospect of them being exploited economically as a heat source on their own. That programme was curtailed, although financing of the Southampton City Centre borehole continued. This is a joint private sector/local authority district heating scheme.

HDR research has been under way since 1976 by a team from the Cambourne School of Mines at Rosemanowes in Cornwall in three phases. Phases I and II were completed (depths of 300 m and 2500 m) and the main feature of Phase III was the development, with industrial companies, of a conceptual design for a 6000 m deep commercial-size prototype HDR system. Work in connection with Phase III has been redirected to the European HDR programme which involves the construction of a pilot plant at Rosemanoires, Alsace, France, or Bad Urach, Germany.

Tidal energy. Tidal energy in Britain does not rank alongside wind energy in its potential as a renewable energy source for electricity generation. In the 1990s a review aimed at reaching the point where decisions could be made whether to go ahead with the construction of a scheme. The largest part of the programme was devoted to the Severn Barrage, jointly funded by the then Department of Energy and the Severn Tidal Power Group primarily concentrating on the estuary's sediments, particularly in the Bridgwater Bay area and the barrage's potential output. Present power estimates for a 16 km barrage between Cardiff and Weston-super-Mare are 8640 MW from 216 bulb-turbines. The scheme remains for the future and its costs would require treatment similar to the Channel Tunnel project. Other areas being studied include the Mersey Estuary (700 MW), the Conway Estuary, North Wales (35 MW) and the Wyre Estuary, Lancashire (50 MW).

Wave energy. The DTI has scaled down its programme on shoreline wave energy but work is continuing on small shoreline devices. Support is being given to an experimental installation, based on the Queen's University, Belfast, concept, utilizing a Wells turbine driving a 75 kW marine wound rotor induction generator on the island of Islay in the Inner Hebrides. The equipment is located in a natural rock gully north of Portnahaven and consists essentially of a concrete box, internally 10 m long, 4 m wide and 9 m high built over the gully. The wave entering and receding in the gully forces the column of water in the chamber to oscillate, driving the air above through the Wells turbine coupled to the generator. The turbine efficiently converts the pneumatic power of the air flow into rotational motion at a speed compatible with the 4-pole generator. Average power levels of about 20 kW/m are experienced. Present energy costs are estimated to be slightly less than 8 p/kWh assuming a payback period of 30 years and a discount rate of 5%. In mid-January 2002 the Department of Trade and Industry announced plans for a cluster of wave power stations in the Western Isles intended to advance the development of shore-based wave technology. The intention is that these will be technically innovative and will act both as demonstration plants and also as commercial generators.

Islay has the only commercial wave power station in the world, with a capacity of 500 kilowatts, which was developed and installed by Wavegen. Support for the development by the same company of an offshore device, which is likely to be tested in Orkney, is ongoing and momentum will also be maintained on the shoreline and near shore systems to encourage the refinement of that technology. Offshore farms could be used in future to produce power for seawater desalination plants which in turn could produce hydrogen gas which could be used as a clean fuel in a CHP plant to meet all local energy needs.

Solar energy. Present day emphasis is on passive solar building designs, utilizing the form and fabric of the building to capture solar radiation and so reduce the need for artificial light, heat and cooling. Active solar heating employs collectors to capture and store the sun's heat primarily for space and water heating. Photovoltaics is a method to convert the sun's energy into electricity by means of semiconductor devices. Costs at present are high but are expected to fall with time; however, to compete with other renewable sources, costs would need to fall by a factor of 10 to 20. Active solar heating is unlikely to become cost-effective in the UK except for special applications and so the DTI halted its R&D work in this field.

Other renewable energy sources. The gases produced from incinerating municipal waste can be used for electricity generation. The introduction of a tax on the use of landfill sites has meant a rapid increase in the provision by several municipal and other authorities of power generating incinerators. The waste is sorted and screened for ferrous and other metals, then burnt in special boilers to generate steam for the steam turbines. The aggregate total of this type of generation is continually increasing. SELCHP burns waste from three London Boroughs generating up to 30 MW. Hampshire County Council is interested in a 30 MW scheme. A scheme in its early stages in South East London is being proposed by Cory Environmental and has a possible capacity of 96 MW.

There are a number of installations utilizing sewage gases for generation purposes and some utilities are using sewage gases to generate electricity and provide heat in CHP schemes. Increasingly CHP schemes are finding favour especially where surplus power can be exported under an NETA contract or utilized locally, e.g. on a large trading estate.

Elm Energy & Recycling UK constructed a plant, which came on stream in 1993, producing 22 MW through a tyre burning scheme. Designed to initially consume 22 million tyres a year, there were a number of teething problems and the plant gave rise to protests from local residents resulting in it being taken out of service, although clearly this example illustrates how novel ideas, if properly developed, can be exploited.

Chicken litter is another source of energy being utilized by Fibropower at Eye in Suffolk and Fibrogen at Glandford, Humberside, each project being capable of producing 13 MW.

Industrial Generation

Small generating plants are available for base-load operation, as standby power supplies or as temporary mobile power sources. They may be geared steam or gas turbine or diesel driven sets although some of the larger ratings may be associated with the production of process steam in a combined heat and power installation.

Base-load diesel or gas turbine generating sets are used as sources of prime power for remote areas at considerable distances from the national supply. Pumping stations for water and oil pipelines, oil fields, lighthouses, sewage schemes and defence installations are examples.

Some islands such as the Channel Islands and Isle of Man and overseas countries use diesel generating sets for their national supplies, but these are considerably larger, of the order of 20 MW or more.

The most important application for small generating plants is undoubtedly as a source of standby power, emphasized particularly during 1971 and 1972 when the UK experienced widespread power cuts.

Where loss of supply can endanger life, e.g. in hospital operating theatres or nuclear plants, the installation of alternative power sources is a necessity. In the former case power is usually provided by an automatically started diesel generating set, while in the latter case either a large diesel generator or a gas turbine generator would be used. Industry too can suffer badly from the effect of a supply interruption. There is a consequential loss of production, and in many cases, damage to process plant. Where continuous processes are involved, manufacturers install standby power plant either by provision of an alternative mains supply and automatic switching, a battery backed uninterruptible power supply (UPS) or a diesel engine driven generator.

Construction. A number of companies in the UK specialize in manufacturing a standard range of generating sets, and while these can be tailored to a specific application, the advantage of using the standard range is the availability of spares and the relative ease with which they can be repaired quickly should they break down. The basic generating set comprises an engine coupled to a brushless a.c. generator, complete with an overhung exciter, radiator and engine cooling fan mounted upon a bed-plate, with a control panel, voltage regulator and an electronic speed governor.

Water-cooled diesel engines are the most commonly used because of their heavy-duty build, low running costs and rapid starting qualities. They can also include turbocharging which provides higher power ratings and higher efficiency. Every diesel engine should include radiator or air cooling, protective guards over moving parts, a governor to regulate the engine speed and thus the frequency of the generator, and fuel, oil and air filters which can be replaced during servicing.

Most sets have an electric starting system which usually comprises starting controls, battery and starter. Some are started by a key switch on the panel, others are fully automatic on mains failure. Larger diesel engines (above about 2 MW), however, are started by compressed air. A typical fully automatic diesel generator is shown in *Figure 7.12*.

Each generating set should include a fuel tank and pipes, engine exhaust silencer, foundation bolts to secure the baseplate or anti-vibration mountings, engine tools and operating manuals. Fully automatic sets should also include a separate contactor panel which provides the automatic changeover of the load from the mains to the generating set and vice versa.

The generator is typically of the screen protected, drip-proof construction. Where the generator is used in a standalone situation the neutral of the generator and all non-conducting metal must be effectively earthed.

Gas turbine engines are lighter, cleaner and virtually vibration free making small units ideal for rooftop installations, but they are more expensive than diesel engines.

High Integrity Power Supplies

The increasing sophistication of information technology (IT) equipment and automated industrial systems and their increased performance levels, e.g. the speed of data processing, the real-time interconnection of telecommunications systems, continuous and automated operation, etc., means that they are more vulnerable and dependent on their electrical power supply.

Figure 7.12 Typical installation of a fully automatic Dale generating set. This one is powered by a Perkins 6.3544, rated at 50 kVA. It shows the control panel with cable to the load, starter battery and leads, air-inlet ducting to the radiator, exhaust outlet pipe and silencer, air filters, steel guards around moving parts and anti-vibration mountings

This electrical energy is distributed in a waveform making up a single- and three-phase sinusoidal system characterized by its:

- frequency
- amplitude
- shape (wave distortion)
- system symmetry.

While at the power station output, the voltage is virtually perfect, the same cannot be said by the time it reaches the user, where several different types of disturbances can be observed:

- transients
- sags/brownouts
- frequency variations
- outage – blackouts...

The source of these disturbances is related to power transmission, distribution and to both the atmospheric (electrical storms, snow, frost, wind, etc.) and the industrial environment (machine anomalies, consumers' harmonic current pollution, network incidents, etc.). Therefore, in spite of constant improvements to distribution networks and to the quality of the electricity supply, it can still cause problems to sensitive equipment.

However, the increasing sophistication of many computing and telecommunication processes make the consequences of these disturbances increasingly serious.

For example, any file server and its attached hard disk should be connected to a UPS (see below):

• The file server directory for most network systems is held in RAM (Random Access Memory) for ease of access. A split second power cut can erase it completely.
• The UNIX environment calls for all system files to be permanently opened in RAM. If power is lost, even momentarily, the entire operating system may have to be reinstalled, together with application software. Server protection is only the first line of defence, workstations need UPS protection against power cuts too.

Less visible and therefore all the more harmful are the effects, which can be seen in terms of premature ageing of equipment, as well as deterioration of its reliability and dependability.

Solutions to Power Problems

To eliminate all four problems, previously mentioned, the product developed was the Uninterruptible Power Supply (UPS). More than 25 years after they first appeared, Uninterruptible Power Systems now represent more than 95% of back-up power interfaces sold and over 98% for sensitive IT and electronic applications.

The UPS is an interface between the mains and the critical load. UPSs supply the load with continuous high quality electrical power regardless of the status of the mains, whether present or not. UPSs deliver a dependable supply voltage free from all mains disturbances, within tolerances compatible with the requirements of sensitive electronic equipment.

Static power supplies. Static power supplies are generally made up of three main subassemblies:

1. Rectifier-charger to convert the alternating current into direct current and charge the battery.
2. Battery, generally sealed lead-acid type enabling energy to be stored and instantly recovered as required.
3. A static inverter to convert direct voltage into an alternating voltage that is perfectly regulated and conditioned in terms of voltage and frequency.

These are the basic building blocks; in addition other features include a static bypass for synchronized transfer to the reserve mains in case of overload or fault. This is normally supplemented by an external mechanical maintenance bypass enabling the UPS to be completely isolated. Various signalling will be available for remote information and interrogation.

Rotary power supplies. Rotary power supplies have moved on from the original systems which consist of d.c. motor, flywheel and alternator. Rotary

UPSs have become hybrid systems containing static technology together with electrical machine technology and are made up of four subassemblies:

1. Rectifier-charger, as for the static solution.
2. Battery, as for the static solution.
3. A simpler version of the static inverter used to drive an a.c. motor.
4. The motor generator comes in various forms depending on the manufacturer.

This system does not have a static bypass for overload and fault conditions, relying on its subtransient current from the generator to clear faults.

For the online solution, this system is not as reliable or efficient as static, but arguments often put forward in favour of rotary are high short-circuit current, galvanic isolations and low internal impedance, providing good tolerance to non-linear loads. Due to the ratio of physical size to kVA rating and the high noise level (70 to 90 dBA) of rotary systems, they tend to be installed in plant rooms.

The perfect UPS. The perfect UPS would have the following characteristics:

- Perfect sine wave voltage output (with any distorted load current waveform).
- 100% availability of conditioned power (no use of main bypass).
- Perfect sine wave input current (with poor input voltage supply, negligible input harmonics and unity power factor).
- High efficiency, greater than 97%.
- Small in size (minimum footprint for highest power).
- Quiet (nearly silent).
- Low initial cost and inexpensive to maintain.

With the advances in technology, especially semiconductors, static UPSs are moving closer towards these parameters.

Application of UPS. The UPS has become an integral part in any high integrity power supply requirement whether this is 250 VA for a personal computer or 4000 kVA for a data centre. The types of UPS, passive standby operation, line interactive and series online, suit different applications.

It will be seen from *Figure 7.13* that in the normal mode of operation, the load is supplied by the a.c. input primary power via the UPS switch. Other devices may be included to provide some conditioning. The output frequency is dependent upon the a.c. input frequency. When the a.c. input supply voltage is out of UPS preset tolerances, the UPS enters stored energy mode of operation, the inverter is activated and the load transferred to the inverter directly or via the UPS switch (which could be electronic or electromechanical).

The battery/inverter combination maintains continuity of load power for the duration of the stored energy time or until the a.c. input supply voltage returns within the UPS preset tolerances, when the load is transferred back, whichever is sooner. With the a.c. input supply voltage within tolerance, the battery charger starts its duty.

This type of UPS is normally used with standalone personal computers, as it is a low power, low cost item.

For the system shown in *Figure 7.14* in the normal mode of operation, the load is supplied with conditioned power via a parallel connection of the a.c.

• **UPS passive stand-by operation**

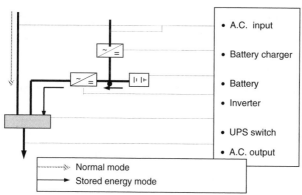

Figure 7.13

• **UPS line interactive operation with bypass**

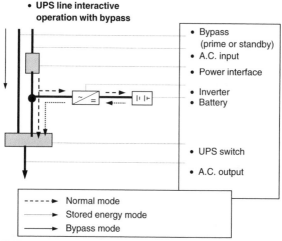

Figure 7.14

input and the UPS inverter. The inverter is working to provide output voltage conditioning and/or battery charging. The output frequency is dependent upon the a.c. input frequency. When the a.c. input supply voltage is out of the UPS preset tolerances, the inverter and battery maintain continuity of the load power.

When in the stored energy mode of operation, the switch disconnects the a.c. input supply to prevent back feed from the inverter.

The unit runs in stored energy mode for the duration of the battery capability or until the a.c. input supply returns within the UPS design tolerances, whichever is sooner.

This type of UPS is generally used in low to medium power applications, groups of PCs and possibly by local area networks.

Both the passive standby and line interactive have their limitation based on the a.c. input mains. When this is out of tolerance for long periods or numerous short periods, the stored energy is depleted thus removing the conditioned supply. To avoid this situation, the input to the UPS rectifier has to have a wide a.c. input mains tolerance, this leads to the most common type of UPS used.

The On Line Double Conversion

Description of the system. The system is configured in the true online mode and ensures that sensitive and critical loads are protected against all forms of mains power supply disturbances, providing a stable and closely regulated source of a.c. power at all times. The standard system comprises: rectifier/battery charger, inverter, static switch, battery and manual by-pass, shown in block diagram form in *Figure 7.15*.

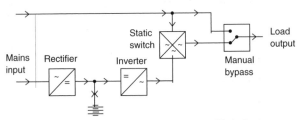

Figure 7.15 Static uninterruptible power supply system, block diagram

The principle of operation is as follows:

Normal	The load is continuously supplied by the inverter. The rectifier/charger derives its power from the incoming electrical supply and supplies d.c. power to the inverter while simultaneously float charging the battery.
Emergency	Upon failure of the incoming a.c. supply, or in the event of it deviating beyond specified tolerances, the load continues to receive its supply from the inverter, which derives its power from the battery for a predetermined period of time dependent upon capacity and magnitude of connected load.
Recharge	Upon restoration of the a.c. input supply, the rectifier/charger will again supply the inverter and simultaneously recharge the battery.

Static switch	To provide even greater security to the load, an alternative source of supply is made available to which the load is automatically connected in the event of a sustained overload or fault occurring in the inverter. The transfer is achieved without interruption through the static switch.
Manual bypass	This is also included so that the load may be manually connected to the alternative supply. This facility provides isolation of the inverter and static switch for maintenance purposes. The transfer is achieved without interruption to the load.

The photograph, *Figure 7.16*, shows a typical installation of this type of UPS.

Figure 7.16 Siel UPS-100 kva modules (TCM Tamini Limited). (Photo on right shows the subassemblies)

Description of subassemblies. The UPS subassemblies are as follows.

Rectifier. Configuration of the rectifier/battery charger is shown in *Figure 7.17*. The function of the main components are as follows.

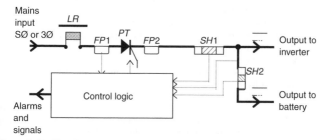

Figure 7.17 Rectifier/battery charger, functional diagram

'*LR*'	The inductance that reduces the voltage spikes and distortion reflected in the mains supply generated by operation of the '*PT*' rectifier bridge.

| '*PT*' | The 6-pulse, internally controlled rectifier bridge consisting of six thyristors, which rectify the three-phase alternating input voltage into direct voltage/current. Utilizing a phase controlled rectifier allows for large input voltage deviations up to 30% without use of the battery. |
| '*FP1*' and '*FP2*' | Super rapid fuses protecting the '*PT*' rectifier bridge thyristors against accidental overloads and short-circuits. |

The 'control logic' manages the overall operation of the rectifier in order to guarantee direct output voltage stability for any change in the load and input voltage within specified limits. It also controls the maximum output current to the inverter bridge (through *SH1*) and the battery-charging current (through *SH2*), implementing the automatic recharging cycle defined by the I-U characteristic of the DIN 41773 standard. The 'control logic' also has the task of displaying the operational status of the rectifier by governing both local and remote alarms and signals.

Inverter. The configuration of the inverter is given in *Figure 7.18.* The functions of the main components are as follows.

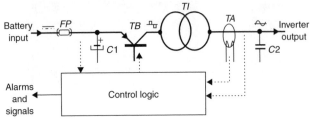

Figure 7.18 Inverter, functional diagram

'*FP*'	The super-rapid protection fuse against accidental overloads or short-circuits on the d.c. power line.
'*C1*'	The d.c. filter which blocks the pulsating current reflected by the inverter commutator bridge.
'*TB*'	The transistorized inverter commutator bridge (IGBT) which converts direct current from the rectifier or battery into alternating current.
'*TI*'	The output transformer, which adapts the a.c. power from the *TB* bridge into a.c. power within the limits for supplying the load.
'*C2*'	The output filter, which minimizes harmonics in the a.c. output voltage.

The 'control logic' manages the operation of the various components of the inverter in order to guarantee alternating output voltage stabilized for any change in the load and input voltage within the acceptable limits.

The 'control logic' also has the task of displaying the operational status of the inverter by governing both local and remote alarms and signals.

Static switch. The configuration of the static switch is given in *Figure 7.19.* The functions of the main components are as follows.

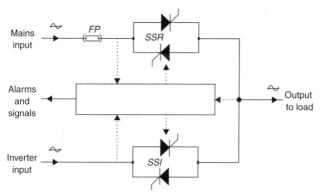

Figure 7.19 Static switch, functional diagram

'*SSI*'   A pair of thyristors connected in positive and negative conduction parallel mode, which switch the inverter voltage. The maximum current passing through these thyristors is limited by the inverter.

'*SSR*'   A pair of thyristors connected in positive and negative conduction parallel mode, which switch the mains power.

The 'control logic' manages the various components of the static switch. More particularly, it keeps the inverter output voltage under continuous control, synchronized with the mains input voltage. It also controls the pair of thyristors on the inverter side, *SSI*, and the pair of thyristors on the mains side, *SSR*.

In normal operating conditions, the load is powered by the inverter when the '*SSI*' thyristors are triggered.

When an overload occurs on the output or the inverter fails, the load is transferred to the mains source, the '*SSR*' thyristors are triggered and the '*SSI*' thyristors have trigger signal removed.

Normal operating conditions are restored automatically or by resetting the inverter in the event of a failure.

Both inverter-mains and mains-inverter switching take place automatically in zero time without causing any disturbance whatever on the load.

These switching operations can also be performed manually.

The 'control logic' also has the task of displaying the operational status of the rectifier by governing both local and remote alarms and signals.

General Requirements for UPS

Certification. The manufacturer should be certified to ISO 9001.

The UPS should be CE marked in accordance with the Safety and EMC Directives 73/23, 93/68 and 89/336, 91/31, 93/68 and designed and manufactured in accordance with the following international standards:

- IEC 6950   Office Equipment Safety
- BS EN 250091-1
- BS EN 250091-2   }   Electromagnetic Emissions, Conducted and Radiated
- BS EN 250091-3

Components. All active electronic devices should be solid state and not exceed recommended operating parameters for maximum reliability.

Neutral connection and grounding. The UPS output neutral should be electrically isolated from the UPS chassis. The UPS output a.c. neutral should be connected to the commercial a.c. source neutral within the UPS.

EMI and surge suppression. Electromagnetic effects should be minimized to ensure that computer systems or other similar electronic loads neither adversely affect nor are affected by the UPS. The UPS should be designed to meet the requirement of EN 250091-2.

Rectifier/Battery Charger

General/input. Incoming commercial a.c. power is converted to a regulated d.c. output by the rectifier/battery charger. The rectifier battery/charger generally comprises a 6- or 12-pulse fully controlled thyristor bridge with constant voltage/constant current characteristics. Each phase of the input should be individually fused with fast acting fuses to prevent cascading failures.

The rectifier battery charger must be compatible for use with the type of battery selected, either valve regulated lead-acid (VRLA), vented stationary lead-acid, nickel-cadmium.

Voltage regulation. The rectifier/battery charger output voltage should not deviate by more than $\pm 1\%$ under each of the following conditions:

- No load to 100% load variation.
- Primary input voltage and frequency variations within the supply limits.

To ensure optimum battery charge and maximize battery life, the float voltage should be automatically adjusted dependent on battery ambient temperature.

The rectifier must be capable of providing its rated power to the inverter without drawing power from the batteries until the input a.c. voltage decreases by more than 25% from nominal.

Walk-in/soft start. The rectifier/battery charger should have a timed walk-in circuit that ensures that the unit gradually assumes the load over a period of 30 seconds after the input voltage is applied.

A facility to ensure that any standby generator will be gradually introduced on to the UPS input and battery recharge inhibited until the mains supply is restored is also desirable.

Power factor. The rectifier/battery charger should have an input power factor equal to or higher than 0.9 lagging with nominal load, nominal input voltage, in automatic float charge state.

Filtering. The rectifier/battery charger should limit the ripple voltage to less than 1%.

Total harmonic distortion. The maximum current THD injected on the mains should be limited to local authority regulations with a 12-pulse rectifier and filter option, the level can be reduced to 5% on larger units.

Capacity. The rectifier/battery charger must have sufficient capacity to support the inverter at the nominal power while simultaneously maintaining the battery in a fully charged float condition. After partial or complete discharge of the battery, the rectifier/charger must automatically power the inverter and recharge the battery.

Automatic battery testing. The status of the battery is generally tested automatically at periodic intervals. The test comprises a weak discharge to confirm that the battery and its associated links, cables and connections are in satisfactory condition. The battery test must be capable of being performed without any risk to the load even if the battery itself should prove defective. A battery fault detected by the automatic test is generally alarmed to alert the user to take action.

IGBT Inverter

The inverter incorporates IGBT power devices and utilizes a PWM principle to generate sinusoidal a.c. output voltage. The inverter must operate within specified parameters over the normal rectifier/battery charger output and battery voltage range.

Voltage regulation. The inverter output voltage must achieve the following typical performance.

Steady state. The steady state output voltage should not deviate by more than $\pm 1\%$ for input voltage and load variations within the specified limits.

Voltage transient response. The transient voltage should not exceed $\pm 5\%$ when subjected to application or removal of 100% load.

Transient recovery. The output voltage should return to $\pm 2\%$ of nominal within 20 milliseconds after a step load change of 100%.

Frequency regulation. The inverter output frequency must be controlled to achieve the following typical performance.

Steady state. The steady-state output frequency when synchronized to the reserve supply is generally selectable over a band from not more than $\pm 1\%$ to not more than $\pm 4\%$.

Frequency control. The output frequency of the inverter is generally controlled by a quartz oscillator. This is capable of operating as a free running unit or as a slave for synchronized operation with a separate a.c. source. The accuracy of the frequency control when free running must be within $\pm 0.05\%$ of nominal.

Total harmonic distortion. The inverter must be provided with harmonic neutralization and filtering to limit the total harmonic distortion of the output voltage to less than 2% with a linear load.

Output power transformer. A dry type isolating transformer is generally provided for the inverter a.c. output.

Overload. The inverter is generally required to be capable of supplying an overload of 125% for up to 15 minutes and 150% for 10 seconds.

Inverter shutdown. On sensing an internal failure, the inverter electronic control will generally instantaneously remove the inverter from the critical load, transfer to reserve provided it is within limits and then shut itself down.

Static Switch

The electronic static switch is a naturally commutated, high-speed, solid-state transfer device rated for continuous duty operation. Each phase of the input is individually fused with fast acting fuses to prevent cascading failures.

The following transfer and retransfer operations must be provided by the electronic static switch:

- Uninterrupted transfer to reserve supply automatically initiated by the following:
  - output overload
  - d.c. voltage out of limits
  - inverter failure
  - overtemperature
- Manual transfer/retransfer to/from reserve supply initiated from the control panel.
- Uninterrupted automatic retransfer from reserve supply initiated whenever the inverter is capable of assuming the load.
- Uninterrupted automatic retransfer must be inhibited under the following conditions:
  - manual transfer to bypass via the maintenance switch
  - failure of the inverter static switch
  - UPS output overload (until overload removed)
  - all no break transfers and retransfers must be inhibited by the following conditions:
    ◇ inverter output voltage or reserve supply out of limits
    ◇ frequency synchronization out of limits.

Overload. The electronic static switch must be capable of supporting overloads, usually:

- 150% for 30 minutes.
- 1000% for 100 milliseconds.

Manual maintenance bypass. Bypass switching must be provided to allow the critical load to be fed from the reserve supply while providing isolation of the UPS and static switch to ensure operator safety during maintenance.

Monitoring and Controls

The UPS will normally incorporate the necessary controls, instruments and indicators to allow the operator to monitor the system status and performance, as well as take any appropriate action.

Mimic panel. These controls and indications are normally provided on a mimic panel having light emitting diodes (LEDs) to indicate the condition of the subassemblies (see *Figure 7.20*).

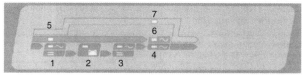

Figure 7.20 Key: Normal operation: LEDs 1, 3, 4 and 5: green, LEDs 2, 6 and 7: off; pre-discharge alarm battery: LED 2: red, LEDs 3 and 4: green, LEDs 1, 5, 6 and 7: off; load supplied by reserve LEDs 1, 3, 5 green, LEDs 2, 4 and 7: off, LED 6: green

Display. An illuminated liquid crystal display (LCD) enables the operating parameters, and all the measurements, alarms of the UPS to be monitored (*Figure 7.21*). Moreover, it is possible to read the operating status of each subassembly with an interval of every 5 seconds. When an alarm occurs, an audible alarm is initiated and the specific alarm message will be displayed. The audible alarm can be muted by pressing the appropriate pushbutton.

During normal operation, the user can control the LCD by pressing the appropriate pushbuttons shown in *Figure 7.21*. The display can be scrolled forward and backwards for information messages.

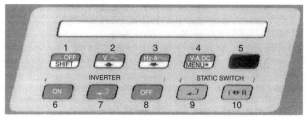

Figure 7.21

Further measurements can be obtained by pressing multiple pushbuttons to obtain the following information:

UPS output voltages
Rectifier input voltages
Reserve input voltages
UPS output currents
Rectifier input currents
UPS output apparent power, active power and power factor
UPS output frequency

Reserve supply frequency
Battery voltage, remaining back-up time (% to the end discharge)
Battery temperature and current

Power history. All the alarms can be stored within the UPS and made available to the user. This facility permits the system status to be analysed after a failure to assist in determining the cause of failure. Additional information is available via the display relating to the number of alarms (total), type, duration and time.

Fibre optic connections. These provide the best way to send data and information, even for long distances, under top security conditions within a hostile electrical environment.

Various interfaces are available for connection to computers. These interfaces allow monitoring of the UPS status with retrieval of all data. The UPS can be controlled, if so desired, by the user.

Standard connections are available for the remote status panel or building management system.

The UPS may be housed in a free-standing cabinet with removable panels to IP20. Forced air ventilation/cooling is provided to ensure that all components are operated within specification, with air entry at the bottom and exit from the rear top.

The UPS must be capable of withstanding any combination of the environmental conditions that may be encountered, generally ambient temperature (operating – excluding batteries) $0°$ to $40°C$ and relative humidity up to 90% (non-condensing) at $25°C$.

Parallel Configurations

Uninterruptible power systems of medium and high power ratings should have the capability of being connected in parallel for multi-module configurations between units of the same rating.

Parallel power. To make up a power supply greater than that available from a single unit; however many units are put in parallel, if one fails all of the units will go to bypass.

Parallel redundant. Where the load is shared between modules, e.g. load 100 kVA, two modules rated at 100 kVA give 100% redundancy or three modules rated at 50 kVA give 50% redundancy. Therefore increasing the reliability of the power supply.

There are two alternative paralleling methods: decentralized and centralized; the main difference being that with centralized, the configured system will require an additional single reserve static switch.

This will result in the situation of being subject to a 'single common point of failure' where in the event of a failure of this section, the whole system will have no reserve facility. For this reason, the majority of manufacturers utilize the 'decentralized' configuration.

The decentralized parallel configuration offers the following benefits:

- increased and superior reliability when compared with the centralized option
- maximum flexibility

- no necessity to foresee future load expansions
- limited initial investment.

Features, performances and monitoring remain the same as those of the single module. The maximum number of UPSs able to be connected in parallel is normally six.

The decentralized parallel philosophy enables UPSs, which include all the subassemblies of a single module, to be connected in parallel, see *Figure 7.22*.

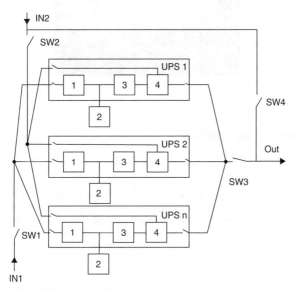

Figure 7.22 UPSs in decentralized parallel arrangement

The subassemblies of *Figure 7.22* are as follows:

IN1   Mains Input
IN2   Reserve Input
OUT   Output to Load
SW 1/2/3/4   Switching Protecting Devices

1 Rectifier/Battery Charger
3 IGBT Inverter
4 Static Switch

These are the standard subassemblies of the UPS

2 Battery

This is always external to the UPS cabinet.

Figure 7.23 shows a three-module parallel redundant system as installed for service.

Figure 7.23 A Siel 3-module parallel redundant system rated 250 kVA per module (TCM Tamini Limited)

Typical Installation

The complete UPS installation shown in *Figure 17.24* includes the LV switchboard distributing non-essential a.c. power to all the building equipment and to the UPS, which provides the essential (high integrity) a.c. power to the critical load, computers, etc. In the event of a complete blackout or extended period outside of the UPS limits, the generator will automatically start. The generator's a.c. power output is checked by monitoring within the LV panel and when it is within the UPS preset requirement, the switches change over and the generator supplies the non-essential power.

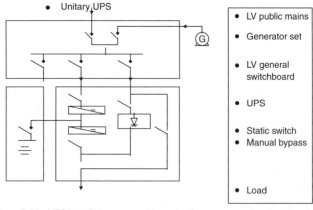

Figure 7.24 UPS installation

The generator sets normally start and are online within minutes, but if the generator fails to start automatically, this will require manual intervention, which takes extra time. The stored energy source (battery) must have sufficient energy to cover this eventuality, the battery is normally rated to support the output load for a period of 15 minutes or greater depending on manual start time.

When the a.c. mains power returns for a satisfactory period, the switches change over to the public mains although there is disruption to the non-essential supplies. The UPS ensures that there are no disruptions or changes to the essential (high integrity) a.c. power to the critical load. The generator is left running on no load for a period before shutdown.

Diesel No Break Systems

These systems are an alternative to the previous diesel, UPS and battery system and are used for very high power loads up to 20 MVA.

System description. The basic components shown in *Figure 7.25* consist of:

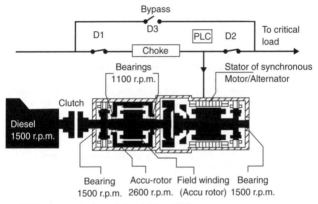

Figure 7.25 Accu-rotor based diesel driven UPS

1. A diesel engine with an electronic speed governor.
2. An electromagnetic clutch.
3. A special brushless synchronous machine consisting of:
   - one four pole motor/alternator
   - one multipole asynchronous brake motor (with its stator mounted on the same shaft as the alternator rotor) and its external squirrel cage rotor (accu-rotor, see below)
   - three exciters and two rotating rectifiers
4. A control and power panel with:
   - PLC
   - mains sensors
   - choke
   - eight electronic control cards

Normal mode operation. Under normal operation motorized breakers D1 and D2 are closed (D3 open) and mains power is fed through the choke to the load. The synchronous machine runs as a synchronous motor.

In the normal mode the following main functions are carried out:

(a) filtration upstream to downstream
(b) filtration downstream to upstream
(c) voltage regulation
(d) power factor improvement
(e) ability to clear fuses without bypass.

Mounted on the same shaft as the synchronous motor is the field winding of an asynchronous motor (rotating at 1125 r.p.m. less slip). A squirrel cage rotor called the accu-rotor follows this field winding with slip at ±1100 r.p.m.

The summation of 1100 r.p.m. plus the 1500 r.p.m. of the shaft means that the accu-rotor rotates at 2600 r.p.m. while its bearings only rotate at 1100 r.p.m. (because these bearings are mounted on the 1500 r.p.m. shaft).

Uniquely the No-Break KS® accu-rotor can store energy for machines up to 2000 kVA and yet only rotate on bearings with a low rpm of 1100 without a disproportionate increase in accu-rotor mass.

The performance in more detail:

Micro-cuts elimination. All micro-cuts, up to 300 milliseconds, are eliminated by the system working at full load without starting the engine. When the load is lower elimination of longer micro-cuts is possible.

Mains voltage regulation. The voltage downstream of the choke is always regulated at the chosen nominal value ±1% even if the upstream mains voltage changes. This is achieved by the electronic voltage regulator, which changes the alternator's excitation current resulting in a variation of reactive current. If the upstream mains voltage variation is higher than 15%, then breaker D1 is opened and the engine starts.

Synchronous compensator (Power factor improvement). The stato-alternator is slightly overexcited in order to produce reactive power, which increases the power factor to 0.99 on the mains side.

Harmonics filtration and short-circuits. The combination of the choke with the very low internal reactance of the stato-alternator (5%) constitutes a perfect filter for all disturbances coming from upstream or downstream, e.g:

• lightning and user induced voltage spikes (reduction up 99%)
• harmonics on mains
• harmonics on load
• low voltage variations in case of upstream and downstream short-circuits.

Peak lopping. By adding watt sensing equipment, it is possible to monitor mains power consumed by the user and to start the engine of the diesel no-break set to operate as a power supply. This eliminates penalties incurred from the supply authority should load power exceed the value imposed by the agreed supply tariff.

Efficiency. All these operations are made with a very high overall efficiency (93% up to 96.4%) and hence ongoing operating costs are kept to an absolute minimum.

Testing. The system is self-testing. A microprocessor initiates diesel start weekly and a full system load test monthly with printout and modem diagnostics.

Mains voltage regulation. The diesel no-break system requires only one inductance located upstream of the stato-alternator. Because the stato-alternator has a very low internal reactance it does not require a further downstream inductance.

This low internal reactance allows the stato-alternator to provide a very high short-circuit current with low voltage variation.

The upstream inductance has, in combination with the stato-alternator, three functions:

1. Filtration of all harmonics.
2. In the case of an upstream short-circuit, limits the fault current to four times full load current.
3. Regulation of mains voltage.
   - the mains voltage downstream of the inductance is lower
   - the stato-alternator feeds reactive current and raises the downstream voltage to the nominal values adjusted with the potentiometer
   - if the upstream mains voltage changes the stato-alternator adjusts automatically the reactive current to maintain the downstream voltage at the same nominal value.

The reactive current decreases when the mains voltage increases and likewise the reactive current increases when the mains voltage decreases.

Solar Energy

Although it has been appreciated by the scientists for a very long time that solar energy represented a vast source of supply, very little has been done in the past to tap it. Within the last few years, however, interest has quickened and today there are many practical devices in operation.

The main component is the solar cell. This is a thin disc of pure silicon containing a minute quantity of boron (or similar substance) to give the silicon a negative potential. A layer of p-type material (see Chapter 4) a few microns thick is diffused on to the upper surface of the disc and the ends lapped over the perimeter.

This portion is then enclosed in a containing case with a glass face, the top surface of the disc being filled with silicon grease to prevent loss by reflection.

As will be appreciated from the description, the arrangement broadly speaking is similar to that found in a transistor where electrons flowing from the n-plate to the 'holes' of the p-plate constitute a current. In the solar cell, power is produced by this process at the barrier junction.

Solar energy has been the primary source of power for spacecraft since 1958. The basic components of a solar cell are shown in *Figure 7.26*. A slice of single crystal silicon, typically 20 mm × 200 mm and 300 μm thick, forms the energy conversion component. In the n on p configuration a shallow junction is formed by diffusing phosphorus into the boron-doped crystal. Metal contacts

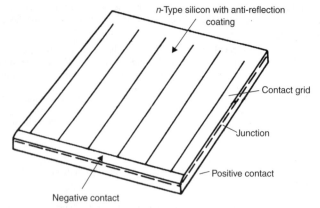

Figure 7.26 Solar cell as used for spacecraft

are plated or evaporated on the front and back of the cell and the active surface is coated with a silicon oxide or titanium oxide anti-reflective layer.

A solar cell of this nature operating at 250°C in normal sunlight above atmosphere has a short-circuit current of 140–150 mA and an open-circuit voltage between 530 mV and 580 mV; the latter is independent of area. Maximum power is between 55 mW and 65 mW and is obtained between 400 mV and 500 mV. Output falls as the cell is turned away from the sun approximately as the cosine of the angle of incidence. A rise in temperature causes a sharp fall in conversion efficiency which is about 11% maximum. An 80°C rise will halve the output.

As bare silicon is a poor emitter, a cell is covered with glass or fused silica. This cover, together with a highly emissive back surface limit the steady-state temperature of a sun-oriented array to about 60°C. In space applications the cover also provides some protection against radiation and micrometeorites.

Many other solar cell materials have been studied for space applications but the only serious contenders are gallium arsenide and polycrystalline cadmium sulphide. Both have failings compared with silicon. However, cadmium sulphide cells are used in photoelectric devices, see Chapter 4.

Although work has been carried out attempting to harness solar energy for large-scale power generation no real success can be reported to provide this power at an economic price. Focusing of the sun's rays to produce concentrated heat energy is the main method adopted, one large installation being in Switzerland. There has been talk of building up an installation of this nature in space and beaming down the energy, but this is not practical at present.

8 Transmission and distribution

Two-wire d.c. Referring to *Figure 8.1*, the volt drop in each conductor $= IR$, therefore the total volt drop $= 2IR$. The voltage drop will therefore be given by $E - V = 2IR$. The power loss in each conductor $= I^2 R$. Therefore total power loss $= 2I^2 R$.

$$\text{Efficiency} = \frac{\text{output}}{\text{input}} = \frac{VI}{EI} = \frac{EI - 2I^2 R}{EI}$$

$$= \frac{VI}{VI + 2IR}$$

$$\text{Voltage regulation} = \frac{E - V}{V} = \frac{2IR}{V}$$

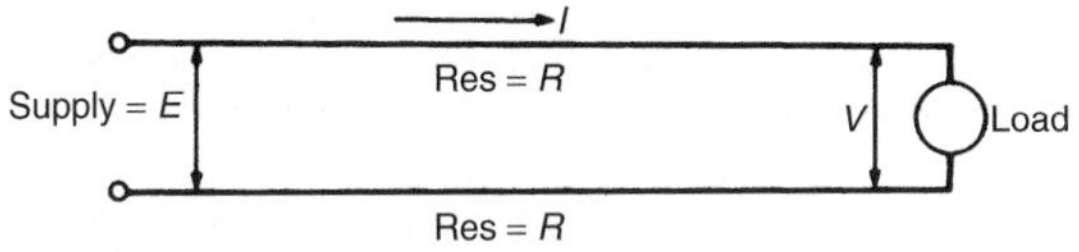

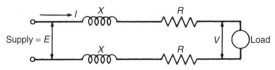

Figure 8.1   D.C. 2-wire supply

Single-phase a.c. Referring to the diagram in *Figure 8.2*, the constants are shown as X and R, where X is the reactance of the conductor and R is its resistance (capacitance is neglected here).

Figure 8.2   A.C. single-phase supply

Taking the power factor of the load as $\cos\phi$, the relation between the volts at the receiving end V, and the sending end E, will be given by

$$E = \sqrt{(V\cos\phi + 2IR)^2 + (V\sin\phi + 2IX)^2}$$

The volt drop and regulation can be found from the values of E and V.

An approximate value for the volt drop per conductor is given by $IR\cos\phi + IX\sin\phi$. So the total volt drop will be $2(IR\cos\phi + IX\sin\phi)$.

The power loss per conductor is $I^2 R$ giving a total power loss for the line of $2I^2 R$. The power factor $(\cos\phi_s)$ at the supply end is found from

$$\tan\phi_s = \frac{V\sin\phi + 2IX}{V\cos\phi + 2IR}$$

159

and the efficiency will be found by

$$\frac{V I \cos \phi}{V I \cos \phi + 2I^2 R}$$

Three-phase a.c. Neglecting capacitance, the line constants will be as shown in *Figure 8.3* and the following details refer to a balanced delta connected load.

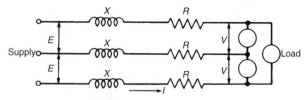

Figure 8.3 A.C. three-phase supply

Reactance and resistance drops per conductor will be IX and IR. But for three-phase, reactance and resistance drops per phase will be $\sqrt{3}IX$ and $\sqrt{3}IR$. The relation between V and E will then be given by

$$E = \sqrt{\left(V \cos \phi_1 + \sqrt{3}IR\right)^2 + \left(V \sin \phi + \sqrt{3}XI\right)^2}$$

The power factor at the supply end can then be obtained from

$$\tan \phi = \frac{V \sin \phi_1 + \sqrt{3}IX}{V \cos \phi_1 + \sqrt{3}IR}$$

The loss in each line will be $I^2 R$, the total loss in this case being $3I^2 R$. The efficiency can be found from

$$\frac{\sqrt{3}V I \cos \phi_1}{\sqrt{3}V I \cos \phi_1 + 3I^2 R}$$

The voltage regulation of the line will be found from $\dfrac{E - V}{V}$.

The vector diagram for a three-phase circuit is shown in *Figure 8.4*, and this can be used for single-phase by omitting the $\sqrt{3}$ before the IR and IX.

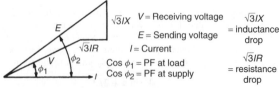

Figure 8.4 Vector diagram for three-phase line

Kelvin's law. In any transmission line it can be shown that the maximum economy is obtained when the annual capital cost of the line equals the cost of the energy loss in transmission during the year. This is known as Kelvin's law and is used as a guide for determining the size which should be used for a transmission line. The result obtained by applying Kelvin's law must be considered also from the point of view of volt drop, current-carrying capacity and mechanical construction.

The capital cost of a line is the cost (usually taken over a year) for the interest on the capital expended, plus depreciation and maintenance. Usually a figure of between 10 and 20% of the capital cost is taken to cover these items. The energy loss in the line during the year can only be estimated and the following equation can be used:

$$\frac{eBa}{100} = mI^2 Rp \times 8760 \times 10^{-5}$$

where e = interest and depreciation in percentage per annum
B = cost per km of line per square millimetre of cross-section a in £
m = number of conductors
I = r.m.s. value of the current taken over a year
R = the resistance of one conductor per km
p = the cost in pence for energy per unit
a = section of line in square millimetres

From the above equation the ideal section for any transmission line can be obtained and the nearest standard size larger should first be considered.

British Regulations for Overhead Lines

The Electricity (Overhead Lines) Regulations 1970 set out the basic criteria governing the design of overhead lines. They were partially superseded by the Electricity Supply Regulations 1988 but more will be said of this later. They permit the exclusion of ice as a loading on the conductor, but increase a wind pressure to 760 N/m on the projected area of bare conductor and the factor of safety to 2.5. This applies to h.v. light construction lines with a conductor size less than 35 mm² copper equivalent and voltage exceeding 650 V but not greater than 33 kV. The minimum temperature assumed is −5.6°C.

For heavy construction lines, i.e. for conductors larger than 35 mm² copper equivalent at voltages exceeding 650 V and for all lines not exceeding 650 V, the factor of safety is 2.0 assuming the same minimum temperature as for light lines but with a wind force of 380 N/m on the projected area of augmented mass of conductor; the augmenting diameter for lines exceeding 650 V is 19 mm and for lines not exceeding 650 V is 9.5 mm.

The factor of safety for supports for h.v. light construction lines is 2.5 with no wind pressure acting on the supports. For h.v. heavy construction lines and all lines not exceeding 650 V the factor of safety for supports is 2.5 with a wind pressure of 380 N/m² acting on the supports, the steelworks and insulators, etc.

The wind pressure on the lee side members of lattice steel or other compound structures including A and H shall be taken as one half the wind pressure on the windward side members. The factor of safety is calculated on the crippling load of struts and upon the elastic limit of tension members.

Mechanical strength of overhead lines. Referring to *Figure 8.5*, the tension in an overhead line conductor will be found from the formula:

$$\text{Tension in conductor} = T = \frac{wL^2}{8s}$$

where w = total equivalent weight of conductor in newtons per metre
L = span length in metres
s = sag in metres
T = tension in newtons

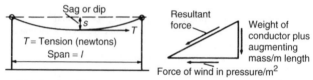

Figure 8.5   Overhead lines

The conductor must then be designed to withstand this stress or tension allowing for the necessary factor of safety (this varies from 2.0 to 2.5). The cross-sectional area of the conductor can be derived from the formula, solving for a:

$$T = a \times f_t \times \frac{1}{\text{factor of safety}} \text{ newtons}$$

and when temperatures are defined, T_1 can be calculated from the following formula:

$$Ea\alpha(t_2 - t_1) + \frac{w_1^2 L^2 Ea}{24T_1^2} - T_1 = \frac{w_2^2 L^2 Ea}{24T_2^2} - T_2$$

where E = modulus of elasticity of conductor in MN/m^2
a = total cross-sectional area of conductor in mm^2
α = linear coefficient of expansion, per °C
t_1 = initial temperature in °C
t_2 = final temperature in °C
w = weight in N/m
w_1 = initial weight in N/m
w_2 = final weight in N/m
L = span length in m
T = tension in N
T_1 = initial tension in N
T_2 = final tension in N
s = sag of conductor in m

d = conductor diameter in mm
r = radial thickness of ice in mm
p = wind pressure in N/m^2
f_t = breaking tension in conductor in N/mm^2

Allowance for augmented mass. The Electricity (Overhead Lines) Regulations 1970 make allowance for ice on the line which is known as the augmented mass. It is derived from:

$$\text{Mass of ice} = w_I = w_i \times r \times (d + r) \text{ kilograms/metre}$$

where w_i = weight of ice in kg
r = radial thickness of ice in mm
d = diameter of conductor in mm

Allowance for wind. Wind pressure is expressed in N/m^2. For overhead lines covered by the Electricity (Overhead Lines) Regulations 1970, Schedule 2, Part 1, for line conductors not exceeding 35 mm^2 of copper equivalent cross-sectional area, where the voltage of the system exceeds 650 V but not 33 kV the wind force is taken to be 760 N/m^2. Schedule 2, Part 1 of the 1970 Regulations applies to line conductors other than those covered in Part 1, and the wind force in these cases is taken as 380 N/m^2.

Allowance for effective weight. In many calculations account has to be made for the allowances for wind and augmented mass acting on the line conductor. The effective weight, w, is the resultant of consideration being taken of the weight of the conductor and wind loading acting on the conductor.

$$w = \sqrt{(\text{weight of conductor} + \text{ice})^2 + (\text{wind load})^2}$$

$$= \sqrt{[w_c + w_i \times r \times (d + r)]^2 + [p \times (d + 2r)]^2}$$

In many cases the cross-sectional area (a) of the conductor is fixed, and it is necessary to find the amount of sag (s) for a stated span length (L) for a specific value of T.

$$T = \frac{wL^2}{8s} \quad \therefore \quad s = \frac{wL^2}{8T}$$

For light construction h.v. lines the wind pressure is taken as 760 N/m^2 acting on the bare conductor.

$$\text{Resultant } w = \sqrt{(w_e + w_i)^2 + w_w^2}$$

where w_e = weight of conductor per metre in grams
w_i = augmenting mass in grams
w_w = pressure of wind in gram/metre^2 on augmented diameter.

Efficiency of Transmission and Distribution Systems

The normal method of comparing the efficiency of any transmission or distribution system is to compare the weight of copper required to transmit a certain load at a given voltage with the same loss in transmission. For this purpose d.c. 2-wire is often taken as a standard and the other systems compared with it as regards the total weight of copper necessary.

Referring to *Figure 8.6*, the d.c. 2-wire system is taken as 100% and the weight of copper required is indicated for each different system. It is important to note that the calculations are made on the basis of the same maximum voltage to earth. In the case of the a.c. systems this means that $V/\sqrt{2}$ (i.e. the

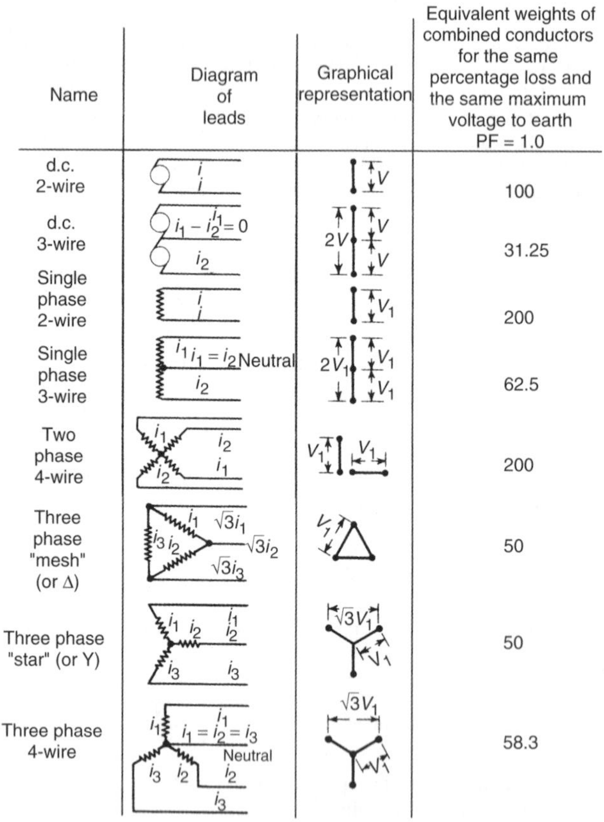

Name	Diagram of leads	Graphical representation	Equivalent weights of combined conductors for the same percentage loss and the same maximum voltage to earth PF = 1.0
d.c. 2-wire			100
d.c. 3-wire			31.25
Single phase 2-wire			200
Single phase 3-wire			62.5
Two phase 4-wire			200
Three phase "mesh" (or Δ)			50
Three phase "star" (or Y)			50
Three phase 4-wire			58.3

Figure 8.6 Comparison of systems. Neutral is taken as half size. In a.c. circuits V is r.m.s. value. Efficiencies based on same power transmitted

r.m.s. value) has to be used in the calculations. For the three-phase a.c. system V_1 is equal to $V/\sqrt{2}$ where V is numerically equal to the d.c. voltage.

In these comparisons the power factor of an a.c. load is taken as unity and it is assumed that in a 2-, 3- and 4-wire system the loads are balanced. It will be seen that except for the d.c. 3-wire system the three-phase 3-wire system scores in that a less total weight of copper is required than for any of the other systems illustrated.

Size of neutral. In the 3- and 4-wire systems employing a neutral, the size of the neutral conductor can be either equal to the 'outers' or half the size of the 'outers'. For the calculations in *Figure 8.6* the neutral has been taken as half-size, and for the case where the neutral is full size allowance must be made for the increase in weight of copper. If a full-size neutral is used for a three-phase 4-wire system the total weight of copper is increased by one seventh, making the comparative figure 67% compared with d.c. 2-wire.

Table 8.1   Minimum ground clearances

Voltage (kV)	At positions accessible to vehicular traffic (m)	At positions not accessible to vehicular traffic (m)
Not exceeding 33	5.8	5.2
Exceeding 33 but not 66	6.0	6.0
Exceeding 66 but not 132	6.7	6.7
Exceeding 132 but not 275	7.0	7.0
Exceeding 275 but not 440	7.3	7.3

Minimum ground clearances. The Electricity Supply Regulations 1988 cover the safety requirements for overhead lines including minimum heights, position, insulation and protection thereof and precautions to be taken against unauthorized access. With regard to minimum heights, these are set out in Schedule 2 of the Regulations, which is reproduced here in *Table 8.1*. The main difference in this schedule compared with the 1970 Overhead Line Regulations is that an upper limit of not exceeding 440 000 volts is added to the highest voltage category where previously there was no upper limit. There is no stipulated maximum working temperature for conductors, it being left to the supply authority to determine its own. The regulations call for a likely minimum ground clearance of a conductor 'at its likely maximum temperature (whether or not in use)'.

9 Cables

Underground Cables

Until the 1970s the type of cable used predominantly for underground power distribution for public supply in the UK at voltages up to 33 kV was the impregnated paper insulated lead-sheathed cable. The standard constructions and requirements for this type of cable are specified in BS 6480.

Later developments, such as use of aluminium sheathing, increased adoption of PME systems by the electricity supply industry and, especially, the worldwide trend towards the greater use of cables with extruded insulations of synthetic materials, have resulted in a substantial reduction in the use of this type of cable for new installations, but it still constitutes a major proportion of the cable already installed.

Solid-type cables. The type of cable described in BS 6480 has traditionally been known as 'solid type', to distinguish it from the gas-pressure and oil-filled types of paper-insulated cables used for voltages above 33 kV and as alternatives to solid cables at 33 kV. The term is perhaps becoming outdated in view of the growing use of the cables with extruded insulations, which might be regarded as more solid than impregnated paper.

The cable conductors are generally of stranded copper or stranded aluminium, although there has been some use of solid aluminium conductors for the 600/1000 V rated cables. A stranded conductor consists of a number of wires assembled together in helical layers around a central wire or group of wires, providing flexibility for drumming, undrumming and handling generally.

The insulation is applied as paper tapes in layers up to the required thickness, determined by the voltage rating. The paper is impregnated with an insulating compound, usually by the process known as 'mass impregnation', which is carried out after the paper tapes have been applied. The cable is dried and evacuated in a sealed vessel to which the hot impregnant is then admitted. An alternative, less used, method involves pre-impregnation of the papers before they are applied to the conductors.

The type of impregnant once used, consisting usually of mineral oil thickened with resin, has now been largely replaced in the UK by non-draining compounds, which contain a proportion of high melting point micro-crystalline wax. Under some conditions of use the fluid oil-resin compounds gave rise to problems due to migration. The non-draining compounds, while fluid at the temperatures employed for impregnation, are, over the normal operating temperature range of the cables, solids of a sufficiently plastic consistency to provide satisfactory bending performance. After its introduction for the 600/1000 V rated cables towards the end of the 1940s the mass-impregnated non-draining cable, abbreviated to MIND cable, was developed gradually for increasing voltages until in the UK it has become virtually the standard paper-insulated solid-type cable over the voltage range.

The impregnated paper insulated lead-sheathed solid-type cable in its simplest form as a single-core cable is illustrated in *Figure 9.1(a)*. A lead alloy sheath is extruded over the insulated core and this is protected by an extruded

(a)

Figure 9.1a Single core 600/1000 V 300 mm^2 lead alloy sheathed, PVC oversheath cable.

1. Circular stranded conductor
2. Impregnated paper insulation
3. Sheath, lead alloy
4. PVC oversheath

(Figures 9.1a to g are reproduced by courtesy of Pirelli Ltd)

PVC oversheath. The metal sheath is an essential component to exclude water, which, in quantity, would destroy the insulating properties of the impregnated paper. The PVC oversheath is the standard form of protection for single-core cables because they are frequently used for interconnectors installed at least in part inside buildings, such as at substations, where a clean finish which will not readily propagate fire is preferable.

Single-core cables have limited use, and (except for very high currents, demanding large conductors, and for fairly short interconnectors) multicore constructions are normally used. The cable of 600/1000 V rating which embraces the 230/400 V standard voltage for domestic supplies has four conductors, three for the phase currents and a neutral. The typical design is illustrated in *Figure 9.1(b)*. The conductors, except for the smallest sizes, have a shaped cross-section, so that when the cores (the insulated conductors) are laid up together they form a compact circular cable with minimum spaces in the centre and at the sides at the rounded corners of the cores, to be filled with paper strings or jute yarns. It also reduces the cable diameter and therefore the amount of lead sheath and armouring.

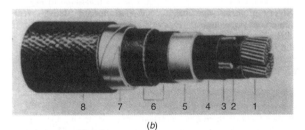

(b)

Figure 9.1b Four-core lead sheathed cable, steel tape armour and served. Suitable for 600/1000 V 3-phase 4-wire systems.

1. Shaped stranded conductor
2. Impregnated paper insulation
3. Filler
4. Impregnated paper belt
5. Sheath, lead or lead alloy
6. Bedding
7. Steel tape armour
8. Serving

Further insulating papers are applied over the laid-up cores, these constituting the 'belt' insulation. The lead sheath, applied after the cable has been impregnated, is protected by paper tapes overlaid with fibrous materials, generally hessian and/or cotton, all impregnated with bitumen. This provides a bedding for the steel tape armour, which in turn is covered with bitumen and further layers of bituminized fibrous material.

Steel tape armour was conventionally used for 600/1000 V cables for the electricity supply industry and galvanized steel wire armour for higher voltage cables, partly to provide identification. For 600/1000 V cables, however, wire armour is used where longitudinal strength is required (e.g. for cables pulled into long lengths of ducting), as well as protection against impact and abrasion. Bituminized fibrous materials have proved generally adequate to protect lead-sheathed cables from corrosion, the metal itself not being susceptible to corrosion in most underground conditions, but in aggressive environments extruded coverings, usually PVC, may be used.

For higher voltages the cables are generally three-core. The construction for voltages up to 11 kV is similar in principle to that of the 600/1000 V cable. The insulation thickness is greater, of course, and the manufacturing processes differ in detail to provide for the higher operating electrical stresses.

For voltages above 11 kV, the belted construction gives place to the 'screened' cable as the standard. At 11 kV both types are provided for in the standards, but the belted type has the greater usage.

In the belted cable half of the thickness of insulation required between conductors is applied to each core and, with the cores laid up, the balance of the required thickness to earth is applied as the belt. In the screened cable the whole of the required insulation to earth is applied to each core, on to which is then lapped a thin metal tape or a laminate of paper and aluminium foil, known as 'metallized paper'. The laid-up screened cores are bound by a tape which includes a few copper wires in the weft, so that when the lead sheath has been applied the screens will be in electrical contact with it. In the screened cable the fillers, which are electrically weak compared with the lapped insulation, are excluded from the electric field and the direction of the field in the cores is radial across the paper thicknesses. Screened cables are sometimes described as 'radial-field cables'.

A typical three-core screened 11 kV cable is illustrated in *Figure 9.1(c)*. The carbon paper screen applied to each conductor is a standard feature for cables rated at 11 kV and above in BS 6480. It is to reduce the electrical stress at the conductor surface by smoothing out the profile and to exclude from the field the small spaces between the wires of the outer layer, which can otherwise be sites for discharge.

A carbon paper screen is also applied over the insulation of single-core 11 kV cables and over the belt of 11 kV belted cables to eliminate discharge in spaces between the outside of the insulation and the inside of the lead sheath in places where the latter does not make close contact.

Another form of three-core screened cable for 22 and 33 kV is the 'SL' type. This has circular cores each separately lead sheathed. The sheathed cores are bound together and the assembly armoured and served. This design virtually eliminates the possibility of breakdown between phases and the cores can be terminated individually. It tends to be more costly than three-core cable under a common lead sheath and has not had much use in the UK.

Consac cable. In the UK the four-core paper-insulated lead-sheathed cable has been replaced to a large extent by CNE cables on public supply

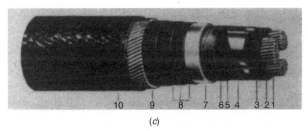

(c)

Figure 9.1c Three-core 11 kV 150 mm² lead sheathed screened cable; single wire armoured and served.

1. Shaped stranded conductor
2. Carbon paper screen
3. Impregnated paper insulation
4. Screen of metal tape intercalated with paper tape
5. Filler
6. Copper woven fabric tape
7. Sheath, lead or lead alloy
8. Bedding
9. Galvanized steel wire armour
10. Serving

networks. The adoption of protective multiple earthing (PME) by the electricity distribution companies has been a major factor in this change. In the main distribution cables it is no longer necessary to keep the neutral conductor separated from earth and in CNE (Combined Neutral and Earth) cables there is effectively a saving of one conductor by combining one of the functions performed by the sheath of the lead-sheathed cable, provision of the earth return path, with the function of the neutral conductor.

The CONSAC cable, formerly the subject of BS 5593 (now withdrawn), has three shaped solid aluminium phase conductors, impregnated paper insulation and an extruded aluminium sheath of dimensions adequate to give a conductance at least equal to that required for the phase conductors. The aluminium sheath is the combined neutral and earth conductor as well as the barrier to water. This type of cable is illustrated in *Figure 9.1(d)*. The bituminized fibrous materials generally used to protect lead sheaths are not adequate for aluminium, which is particularly susceptible to corrosion in

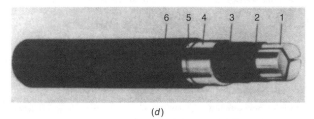

(d)

Figure 9.1d Typical Consac cable for use in low voltage PME systems.

1. Solid aluminium conductors
2. Paper core insulation
3. Paper belt insulation
4. Extruded smooth aluminium sheath
5. Thin layer of bitumen containing a corrosion inhibitor
6. Extruded PVC or polythene oversheath

underground conditions. The CONSAC cable has an extruded PVC over-sheath applied over a layer of bitumen which seals the interface between the aluminium and the PVC.

This is a very economic form of LV distribution cable, but less convenient for the tee-jointing of service cables than the type with extruded insulation and waveform concentric wire neutral which has largely replaced it and which is described later.

11 kV aluminium-sheathed cables. The availability of presses for extru-sion of aluminium sheaths has led also to the adoption of 11 kV aluminium-sheathed cables by the electricity distribution companies in the interests of economy. Although the 11 kV lead-sheathed cable has not been completely replaced, the aluminium-sheathed type constitutes the bulk now purchased. Smooth and corrugated sheaths have been used, protected by PVC oversheaths applied on a layer of bitumen. There are benefits and disadvantages in each type of sheath, but the greater flexibility of the corrugated sheath has been the major factor in causing it to be generally preferred by the users.

The corrugated aluminium-sheathed 11 kV cable is illustrated in *Figure 9.1(e)*. Up to the sheath it is similar to a cable for lead sheathing, but partial filling is needed of the spaces under the sheath resulting from the corrugation. Both belted and screened designs are supplied, the belted having the greater usage.

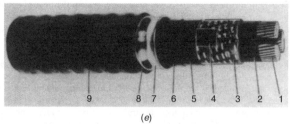

(e)

Figure 9.1e 11 kV belted type cable with corrugated aluminium sheath.

1. Shaped stranded conductor
2. Carbon paper screen
3. Impregnated paper insulation
4. Filler
5. Impregnated paper belt
6. Carbon paper screen
7. Corrugated aluminium sheath
8. Bitumen containing corrosion inhibitor
9. Extruded PVC oversheath

PVC-insulated cables. Well before 1970 PVC-insulated wire armoured cables to BS 6346 had become the established type for industrial installa-tions and power stations for voltages up to 3.3 kV. PVC compounds can be formulated to give a range of flexibility and hardness, but, being thermoplas-tic materials, they soften at elevated temperatures. In applications where the maximum loads to be carried are precisely known and fairly close protection against overload can be provided by fuses or other devices, to ensure that damaging temperatures are not reached, this is not a significant disadvantage. For the l.v. public supply system, however, there is preference for cable which will tolerate overloads of greater magnitude and duration and PVC-insulated cables have only limited uses in that area.

A particular advantage of the extruded types of insulation, such as PVC, is that they are much less affected by moisture than is paper, and terminating is simplified on this account. There are no metal sheaths to be plumbed and there is no need to enclose the ends in water-resisting compound.

The conductors for this type of cable are stranded copper or solid aluminium. Stranded aluminium conductors are quite feasible and are sometimes supplied when requested by overseas users, but BS 6346 does not include them on the basis that, if the more economic aluminium conductors are to be used, then the fullest economy is achieved by the use of the solid form, which is particularly suitable for PVC insulation.

As with paper-insulated cables, the conductors for multicore cables, except for the small sizes, are of shaped cross-section. The PVC compound is extruded on to the conductor by a technique allowing a uniform thickness to be applied to the shaped profile. The insulant is extruded in a hot plastic state and cooled by passage through a water trough. The requisite number of cores are assembled together with a spiral lay and with non-hygroscopic fillers, when required, to give a reasonably circular laid-up cross-section. Either PVC tapes or an extruded layer of PVC compound is applied over the assembled cores to serve as a bedding for the armour. PVC tapes are the cheaper alternative for the cables with shaped conductors and are more often used for cables to be installed in air, unless it is required that the terminating gland should seal onto the bedding, when the extruded form is more suitable. Extruded beddings are also preferable for cables to be buried direct in the ground. The armour may be galvanized steel wires or aluminium strips and, in addition to giving mechanical protection, provides earth continuity in the way that the metal sheath does on the paper-insulated cable.

XLPE-insulated cables. XLPE is the recognized abbreviation for cross-linked polyethylene. This and other cross-linked synthetic materials, of which EPR (ethylene propylene rubber) is a notable example, are being increasingly used as cable insulants for a wide range of voltages.

Polyethylene has good electrical properties and in particular a low dielectric loss factor, which gives it potential for use at much higher voltages than PVC. Polyethylene has been and still is used as a cable insulant, but, as a thermoplastic material, its applications are limited by thermal constraints. Cross-linking is the effect produced in the vulcanization of rubber and for materials like XLPE the cross-linking process is often described as 'vulcanization' or 'curing'. Small amounts of chemical additives to the polymer enable the molecular chains to be cross-linked into a lattice formation by appropriate treatment after extrusion.

The effect of the cross-linking is to inhibit the movement of molecules with respect to each other under the stimulation of heat and this gives the improved stability at elevated temperatures compared with the thermoplastic materials. This permits higher operating temperatures, both for normal loading and under short-circuit conditions, so that an XLPE cable has a higher current rating than its equivalent PVC counterpart. The effects of ageing, accelerated by increased temperature, also have to be taken into account, but in this respect also XLPE has favourable characteristics.

BS 5467 specifies construction and requirements for XLPE and EPR-insulated wire-armoured cables for voltages up to 3.3 kV. The construction is basically similar to that of PVC cables to BS 6346, except for the difference in insulant. Because of the increased toughness of XLPE the thicknesses of insulation are slightly reduced compared with PVC. The standard also covers cables with HEPR (hard ethylene propylene rubber) insulation, but XLPE is the material most commonly used.

From 3.8 kV up to 33 kV, XLPE and EPR insulated cables are covered by BS 6622 which specifies construction, dimensions and requirements. The polymeric forms of cable insulation are more susceptible to electrical discharge than impregnated paper and at the higher voltages, where the electrical stresses are high enough to promote discharge, it is important to minimize gaseous spaces within the insulation or at its inner and outer surfaces. To this end XLPE cables for 6.6 kV and above have semiconducting screens over the conductor and over each insulated core. The conductor screen is a thin layer extruded in the same operation as the insulation and cross-linked with it so that the two components are closely bonded. The screen over the core may be a similar extruded layer or a layer of semiconducting paint with a semiconducting tape applied over it. Single-core and three-core designs are employed, and there is scope for constructional variation depending on the conditions of use, subject to the cores being surrounded individually or as a three-core assembly by a metallic layer, which may be an armour, sheath or copper wires or tapes. A typical armoured construction which has been supplied in substantial quantities is shown in *Figure 9.1(f)*.

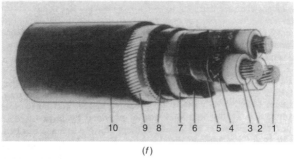

(*f*)

Figure 9.1f   XLPE cable.

1. Circular stranded conductor	6. PVC filler
2. Semiconductor XLPE screen	7. Binder
3. XLPE insulation	8. Extruded PVC sheath
4. Semiconducting tape screen	9. Galvanized steel wire armour
5. Copper tape screen	10. Extruded PVC oversheath

In the UK this type of cable, mainly in single-core form, is favoured for power station cabling, where lightness and convenience of terminating are major considerations. Three-core designs are also used for site supplies.

For underground distribution at 11 kV, the XLPE cable does not compete economically with the paper-insulated aluminium-sheathed cable, but work is in progress on standardizing and assessing XLPE cable design, including trial installations, in preparation for any change in the situation. Overseas, where circumstances are different, XLPE cable is the type in major demand. With manufacturing facilities increasingly orientated to this market, XLPE-insulated cables constitute a large proportion of UK production.

Aluminium waveform cable. The aluminium waveform cable, often described as 'Waveconal', is a type of CNE 600/I000 V distribution cable for public supply utilizing XLPE as the insulation. This now constitutes the largest part of the cables purchased by the electricity supply industry for this purpose. Like the CONSAC cable it has three solid-shaped aluminium phase conductors, but the insulation is of XLPE, with HEPR as an alternative. The laid-up cores are bound with an open-lay tape and covered with an unvulcanized rubber compound into which are partly embedded aluminium wires applied in a waveform constituting the combined neutral and earth conductor. A further layer of the rubber compound is applied over the aluminium wires so that it is pressed between the gaps to amalgamate with the underlying layer. The concentric wires are thus effectively sandwiched in the rubber compound to protect them from corrosion and prevent water spreading between them in the event of the PVC oversheath extruded overall becoming damaged. The cable is illustrated in *Figure 9.1(g)*.

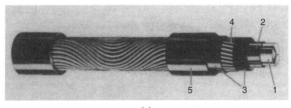

(g)

Figure 9.1g   Typical Waveconal cable as used in low voltage p.m.e. systems.

1. Solid aluminium conductors
2. XLPE core insulation
3. Rubber anti-corrosion bedding
4. Aluminium wires
5. Extruded PVC oversheath

The waveform application of the concentric conductor allows the wires to be lifted from the underlying cable without cutting to give access to the phase conductors to make service joints. This is particularly convenient for maintaining the neutral/earth continuity during live jointing, which can be carried out at the 230 V phase-to-earth voltage when adding services.

Pressurized cables. The electrical strength of solid-type cable is limited by the possibility of drainage of fluid impregnants and by the effects of thermal expansion and contraction which result in the formation of small voids within the insulation. Insulation thicknesses have to be great enough to ensure that electrical stress will not cause severe and destructive ionization within these voids. Consequently, although solid type cables are occasionally used for voltages as high as 66 kV, they are generally uneconomic or impracticable above 33 kV.

For 33 kV and higher voltages, pressurized cables have been developed wherein ionization does not occur even when the electrical stress on the insulation is three or four times as great as the maximum permissible in solid-type insulation.

Oil-filled cables. The earliest form of pressurized cable was the so-called low pressure oil-filled cable, which remains the predominant type for supertension service, being employed at voltages up to 525 kV.

Void formation is prevented by the use of a very low viscosity impregnating oil and the provision of external oil-feed tanks whereby the insulation is always maintained in a fully impregnated state. Channels are incorporated in the cables to permit oil flow resulting from changes in cable temperature.

Oil-filled cables are available in 3-core form at voltages up to 150 kV. Single-core cables are employed for higher voltages than this and are also used for terminating 3-core cables; single-core cables are also used in the 33 kV to 132 kV range when conductor sizes greater than about 630 mm² are required.

The oil-filled cable is designed to operate with an oil pressure within the range 30 kN/m² to 525 kN/m² for normal installations. To withstand the internal oil pressure, the lead alloy-sheathed cable requires to be reinforced with metal tapes; no reinforcement is required for the corrugated aluminium sheath design of cable. Oil feed tanks are provided to sustain the oil pressure within the design range. Several types of oil-filled high voltage cables are shown in *Figures 9.2 (a)* to *(d)*.

Internal gas pressure cables. Whereas in the types of pressurized cable described above the occurrence of voids is prevented by the oil pressure, in the internal gas pressure cable ionization within voids is suppressed by the introduction of nitrogen which permeates the insulation at 1400 kN/m² and increases their breakdown strength to such an extent that their presence no longer imposes any severe restrictions on operating stress.

The paper insulation may be mass-impregnated (as in a solid-type cable) or alternatively the insulation may be built up by use of pre-impregnated paper tapes.

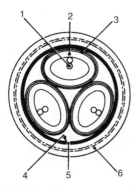

Corrugation of
aluminium sheath

(a)

Figure 9.2a Construction of oil-filled cables. Ductless, fillerless oval-conductor construction used for 3-core 33 kV cables.

1. Conductor
2. Insulation paper
3. Core screen metallized paper
   and cotton woven fabric tape

4. Laid up cores
5. Aluminium sheath
6. Extruded outer corrosion
   protection

By courtesy of Pirelli Cables Ltd

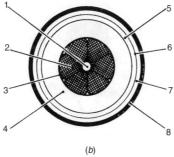

(b)

Figure 9.2b Make up of 400 kV cable.

1. Oil duct
2. Six-segment copper conductor
3. Screening tapes
4. Oil-impregnated paper insulation
5. Screening tapes

6. Lead alloy sheath
7. Tin-bronze reinforcing tapes
8. Cotton binding tapes and
   PVC or PE sheath

By courtesy of Pirelli Cables Ltd

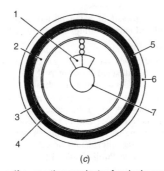

(c)

Figure 9.2c Hollow self-supporting conductor for single-core cables.

1. Conductor, including screen
2. Insulation paper
3. Core screen
4. Copper woven fabric tape

5. Aluminium sheath
6. Polythene or PVC outer sheath
7. Oil duct

By courtesy of Pirelli Cables Ltd

Operating electrical stresses. The insulation of solid-type cables is of such thickness that the maximum electrical stress seldom exceeds about 4.5 MV/m. Pressurized cables, depending on the cable type and system voltage, etc., may operate with electrical stresses of the order of 15 MV/m. For 33 kV service the requirements of impulse strength limit the design stress to about 8.5 MV/m, but even with this modest stress the savings in insulation thickness and cable diameter are sufficient to make pressurized cables economic.

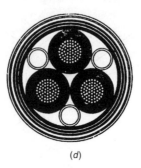

(d)

Figure 9.2d 3-core ducted type served cable. By courtesy of Pirelli Cables Ltd

Current-carrying capacities. The current-carrying capacity of a cable is controlled by the necessity for dissipation of the heat generated by the power losses in conductors, insulation and sheath and the maximum temperature at which its insulation can safely be operated (65–90°C depending on voltage, type, etc.) and by the manner in which it is installed. Generally, the current rating of a cable buried in the ground is rather less than when the cable is installed in air, although in some cases the reverse is true. When cables are buried in unfilled ducts instead of being buried direct in the ground, heat dissipation is hindered and current ratings are decreased.

Aluminium sheaths. Smooth profile aluminium sheaths are widely used nowadays on pressurized cables, particularly gas pressure cables, because they are sufficiently strong to withstand high internal gas pressure without reinforcement. Corrugated seamless aluminium sheath is also available and is particularly appropriate for use on low pressure oil-filled cables. The corrugation design enables sheath thickness to be reduced with resulting cost savings and improved bending performance and in conjunction with modern protections has led to an increasing use of aluminium-sheathed oil-filled cables.

Acknowledgement for the information on cables and for the illustrations used in this section is made to Pirelli Cables Ltd.

Underground Cable Constants

Insulation resistance. The insulation resistance is not directly proportional to the radial thickness of insulation and can be found from

$$R = \frac{\rho}{2\pi} \log_e \frac{R}{r}$$

where ρ is the specific resistance of the insulating material. This is more conveniently expressed in ohms or megohms per km. This is given by

$$R = 1.43\rho \log_{10} \frac{R}{r} \times 10^{-12} \text{ megohms per km}$$

Capacitance. The capacitance of a single-core cable is given by

$$C = \frac{0.024k}{\log_{10} \dfrac{R}{r}} \text{ microfarads per km}$$

where k is the permittivity.

Permittivity of impregnated paper insulation is usually about 3.5; that of polythene 2.3; rubber and PVC 5 to 8.

Voltage gradient. The question of voltage or potential gradient in insulated cables is important, especially in the design of oil-filled and gas-pressure cables. Most supertension cable designs are based on selecting an appropriate maximum voltage gradient having regard to the characteristics of the insulation and to the voltages, particularly impulse voltages, to which the cable will be subjected. The general shape of the voltage gradient curve is shown in *Figure 9.3*, the maximum gradient occurs at the surface of the conductor and its value is given by:

$$g_{\max} = \frac{E}{r \log_e \dfrac{R}{r}} \text{ V/cm}$$

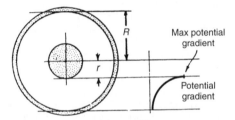

Figure 9.3 Potential gradient in single-core cable

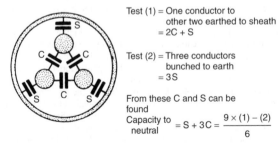

Test (1) = One conductor to
                 other two earthed to sheath
                 = 2C + S

Test (2) = Three conductors
                 bunched to earth
                 = 3S

From these C and S can be found

Capacity to neutral $= S + 3C = \dfrac{9 \times (1) - (2)}{6}$

Figure 9.4 Capacitance in 3-core cables

Economical cable design involves use of the maximum voltage gradient or design stress that can be used without risk of electrical failure. When the values of g_{max} and E are fixed it will be found that a minimum value of R is obtained when

$$\log_e \frac{R}{r} = 1$$

i.e. when $R = 2.718r$. If the smallest possible cable diameter is desired the diameter over the insulation should be 2.718 times the conductor diameter and the radius of the conductor should be such as to satisfy the equation

$$g_{max} = \frac{E}{r} \text{V/cm}$$

For system voltages up to about 132 kV, this approach to cable design is seldom practicable because the value of r so obtained would be too small to allow the use of a reasonable conductor cross-section. At system voltages such as 275 or 400 kV the $R = 2.718r$ relationship is frequently employed in cable design; in the case of oil-filled cables with small copper sections, the required value of r may be achieved by making the internal oil duct of larger diameter than would otherwise be necessary.

Wiring Cables

There are a number of l.v. cables used in the wiring of domestic, commercial and industrial dwellings. In addition the supply to house services and street lighting schemes may be by cables other than PVC- or paper-insulated solid-type cables. These are described in the following paragraphs.

Service cables. The service cables which carry supplies to houses and other premises with small loads are usually single-phase, connected by service joints to one phase and the neutral of the main l.v. distribution cables laid in the streets or pavements. For larger loads three-phase service cables may be used.

One type of service cable is a small version of the Waveconal cable. For single-phase supply it has only one circular phase-conductor insulated with XLPE with the waveform neutral wires embedded in their rubber protection laid around it, the whole being protected by a PVC sheath. For three-phase and neutral supplies the cable is very similar to *Figure 9.1(g)* except that the three phase conductors are circular.

Alternative types are insulated with PVC, one of the few uses of PVC insulation for distribution cables for public supply. The concentric conductors of these are copper wires applied helically without the rubber protection against corrosion required for aluminium wires. The phase-conductors are usually copper also, but may be of solid aluminium.

One type is a CNE cable; in this the wires of the concentric conductor, which serves as neutral and earth, are not individually covered. In the other type, known as 'split concentric', the wires to be used as the neutral conductor are individually covered with a thin layer of PVC while the wires of the earth continuity conductor are bare. The two types are illustrated in *Figure 9.5*. The single-phase split concentric type is covered by BS 4553.

Combined neutral and earth service cables are similar to split cables in design except that all the outer conductors are bare copper wires as shown in *Figure 9.5* (right).

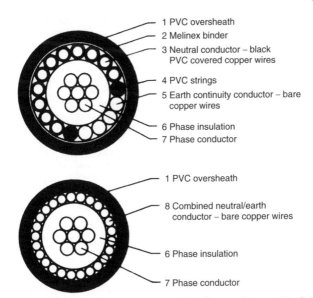

1 PVC oversheath
2 Melinex binder
3 Neutral conductor – black
   PVC covered copper wires
4 PVC strings
5 Earth continuity conductor – bare
   copper wires
6 Phase insulation
7 Phase conductor

1 PVC oversheath
8 Combined neutral/earth
   conductor – bare copper wires
6 Phase insulation
7 Phase conductor

Figure 9.5 Two types of concentric service cable. Above: split concentric. Below: combined neutral and earth (CNE) concentric

Both designs have a black extruded PVC oversheath and are available in sizes ranging from $4\,mm^2$ to $50\,mm^2$.

Split concentric cables have a stranded copper or solid aluminium central conductor, insulated with red PVC and a concentric layer of bare and insulated conductors as indicated above.

Copper conductor cables. Small wiring cables, suitable for lighting and power services in buildings, are generally copper conductor with PVC insulation for conduit and trunking use. For two core and cpc cables a PVC sheath is provided over the insulated cores. The cpc is usually bare copper. Sizes range from $1\,mm^2$ up to about $16\,mm^2$ for domestic and commercial use.

When metric cables were first introduced into the UK, $1\,mm^2$, $1.5\,mm^2$ and $2.5\,mm^2$ sizes had solid copper conductors for both single and multicore cables. The stiffness of the two-core and cpc cable at $2.5\,mm^2$ has resulted in these sizes being made also available in stranded form.

Due to the fluctuating and sometimes high price of copper experiments have been carried out over a period of years to see whether it was possible to use aluminium as a conductor material. To date termination problems, resulting in overheating of accessories, have precluded this material as a satisfactory conductor for house wiring cables.

Mineral-insulated cable. This type of cable is used extensively for general lighting and power circuits, fire alarms and emergency supplies in most types of buildings and industrial installations. A range of cables is available which is BASEC approved to BS 6207 Part I. Mineral-insulated cables are recommended for use in hazardous areas and a full range of BASEEFA approved terminations is available. The cable will neither burn nor support combustion and will not emit smoke or toxic gas. It will continue to operate even when fire occurs in its vicinity thus maintaining essential services such as fire alarms and emergency lighting.

Mineral-insulated cable consists of copper conductors embedded in densely compacted magnesium oxide insulation and contained within a copper sheath which acts also as an excellent circuit protective conductor.

These cables have higher current ratings, size for size, than organic-insulated cables and because they are constructed from inorganic materials, they do not deteriorate with age. They are available in a range of single-core conductor sizes from $6\,mm^2$ to $240\,mm^2$ and in 2, 3, 4 core sizes from $1.5\,mm^2$ to $25\,mm^2$. There are also some conductor sizes available as 7, 12 and 19 core cables. There are two voltage ratings available: 600 V and 1000 V.

As an option, the sheath may have an overall covering to provide protection in environments corrosive to copper. This outer covering may be of a halogen-free material with extremely low smoke emission and flame propagation characteristics.

When the cable is terminated, it is necessary to fit a seal to prevent the magnesium oxide insulant absorbing moisture. A full range of seals, glands and tools is available from the cable manufacturer. The cable is ideally suitable for use on TN-C systems since the copper sheath provides an excellent combined neutral and earth (PEN) conductor.

Elastomer-insulated cables. Where high ambient temperatures are encountered elastomer-insulated cable and flexible cable are available for power and lighting circuits up to 1000 V.

10 Transformers and tapchangers

Transformers

A power transformer normally consists of a pair of windings, primary and secondary, linked by a magnetic circuit or core. When an alternating voltage is applied to one of these windings, generally by definition the primary, a current will flow which sets up an alternating m.m.f. and hence an alternating flux in the core. This alternating flux in linking both windings induces an e.m.f. in each of them. In the primary winding this is the 'back-e.m.f.' and, if the transformer were perfect, this would oppose the primary applied voltage to the extent that no current would flow. In reality, the current which flows is the transformer magnetizing current. In the secondary winding the induced e.m.f. is the secondary open-circuit voltage. If a load is connected to the secondary winding which permits the flow of secondary current, then this current creates a demagnetizing m.m.f. thus destroying the balance between primary applied voltage and back-e.m.f. To restore the balance an increased primary current must be drawn from the supply to provide an exactly equivalent m.m.f., so that equilibrium is once more established when this additional primary current creates an ampere-turns balance with those of the secondary. Since there is no difference between the voltage induced in a single turn whether it is part of the primary or the secondary winding, then the total voltage induced in each of the windings by the common flux must be proportional to the number of turns. Thus the well-known relationship is established that:

$$E_1/E_2 = N_1/N_2$$

and, in view of the need for ampere-turns balance:

$$I_1N_1 = I_2N_2$$

where E, I and N are the induced voltages, the currents and number of turns respectively in the windings identified by the appropriate subscripts. Hence, the voltage is transformed in proportion to the number of turns in the respective windings and the currents are in inverse proportion (and the relationship holds true for both instantaneous and r.m.s. quantities).

Relationship between voltage and flux. For the practical transformer it can be shown that the voltage induced per turn is

$$E/N = K\Phi_m f$$

where K is a constant, Φ_m is the maximum value of total flux in webers linking that turn and f is the supply frequency in hertz.

If the voltage is sinusoidal, which, of course, is always assumed, K is 4.44 and the expression becomes

$$E = 4.44 f \Phi N$$

For design calculations the designer is more interested in volts per turn and flux density in the core rather than total flux, so the expression can be rewritten in terms of these quantities thus:

$$E/N = 4.44 B_m A f \times 10^{-6}$$

where E/N = volts per turn, which is the same in both windings
B_m = maximum value of flux density in the core, tesla
A = nett cross-sectional area of the core, mm^2
f = frequency of supply, Hz

Leakage reactance. Mention has already been made of the fact that the transformation between primary and secondary is not perfect. First, not all of the flux produced by the primary winding links the secondary, so the transformer can be said to possess leakage reactance. Early transformer designers saw leakage reactance as a shortcoming of their transformers to be minimized to as great an extent as possible. With the growth in size and complexity of power stations and transmission and distribution systems, leakage reactance – or, in practical terms since transformer windings also have resistance, impedance – gradually came to be recognized as a valuable aid in the limitation of fault currents in networks. The normal method of expressing transformer impedance is as a percentage voltage drop in the transformer at full-load current and this reflects the way in which it is seen by system designers. For example, an impedance of 10% means that the voltage drop at full-load current is 10% of the open-circuit voltage, or, alternatively, neglecting any other impedance in the system, at ten times full-load current, the voltage drop in the transformer is equal to the total system voltage. Expressed in symbols this is:

$$V_z = \text{per cent } Z = \frac{I_{FL} Z}{E} \times 100$$

where Z is $\sqrt{R^2 + X^2}$, R and X being the transformer resistance and leakage reactance respectively and I_{FL} and E are the full-load current and open-circuit voltage of either primary or secondary windings. Of course, R and X may themselves also be expressed in terms of percentage voltage drops. The 'natural' value for percentage impedance tends to increase as the rating of the transformer increases with a typical value for a medium sized power transformer being about 9 or 10%. Occasionally some transformers are deliberately designed to have impedances as high as 22.5%.

Equivalent circuit of a transformer. Considering the effect of a transformer on an electrical system simply as a device which limits the flow of current as in the above explanation of transformer impedance is, in effect, assuming it to have an equivalent circuit as shown in *Figure 10.1(a)*. That is, it consists of an impedance having resistive and reactive components which has the effect of limiting the output voltage in proportion to the load. These components may be seen as being associated with the individual windings as indicated in the diagram. Indeed, the resistances of the individual windings R_1 and R_2 can be measured, but X_1 and X_2 cannot exist independently of each other and are thus not real values in the same way as the values

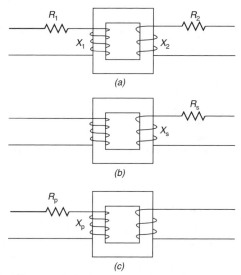

Figure 10.1 Different methods of representing the transformer circuits

of resistance are, although the individual windings will each have their own inductive reactances.

The equivalent circuit can be simplified further by lumping all the resistance and reactance together. The lumped values may be assumed to be all in the secondary (*Figure 10.1(b)*), or all in the primary (*Figure 10.1(c)*). It can be shown that the equivalent lumped values are:

$$R_P = R_1 + R_2 \left(\frac{N_1}{N_2}\right)^2$$

$$X_P = X_1 + X_2 \left(\frac{N_1}{N_2}\right)^2$$

when referred to the primary, and:

$$R_S = R_1 \left(\frac{N_2}{N_1}\right)^2 + R_2$$

$$X_S = X_1 \left(\frac{N_2}{N_1}\right)^2 + X_2$$

when referred to the secondary.

As indicated above, a transformer primary will carry some current simply by being connected to a supply even when its secondary is open-circuit, this is the magnetizing current. A more accurate equivalent circuit recognizes this by including the magnetizing branch as shown in *Figure 10.2*. Because

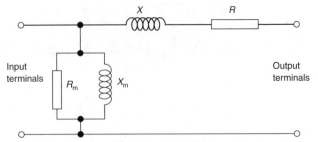

Figure 10.2 Transformer equivalent circuit including magnetizing branch

the magnetizing current is at low power factor, the magnetizing branch is largely reactive.

Transformer regulation. The transformer voltage ratio defined above applies at no load. It will be recognized from any of the equivalent circuits shown that when load current flows the output terminal voltage will fall. This is known as *regulation*. The approximate percentage regulation for a current loading of a times rated full-load current and an output power factor of $\cos\phi_2$ is given by the following expression:

$$\text{percentage regulation} = a(V_R\cos\phi_2 + V_X\sin\phi_2)$$
$$+ \frac{a^2}{200}(V_X\cos\phi_2 - V_R\sin\phi_2)^2$$

where V_R = percentage resistance voltage
V_X = percentage reactance voltage

This equation is sufficiently accurate for most practical transformers; however, for transformers having reactance values up to about 4% a further simplification may be made by omitting the a^2 term and for very high values of impedance, say 20% or over, a term in a^4 is sometimes necessary.

Transformer losses. Transformer losses are very small in relation to the power throughput, that is transformer efficiency, defined as output divided by input,

$$\frac{\text{input} - \text{losses}}{\text{input}}$$

or

$$\frac{\text{output}}{\text{output} + \text{losses}}$$

is very high.

Input and output are measured in VA, kVA or MVA and losses in W or kW.

For large transformers this ratio is over 99% and even for smaller transformers efficiencies of 98% or more are common.

There are basically two types of losses in transformers: *iron losses* and *copper losses*. Iron losses result from the magnetization of the core and themselves have two components; one results from taking the core steel through its successive cycles of magnetization, first to a maximum in one direction then back, through zero, and to a maximum in the opposite direction. The second component results from the flow of eddy currents in the core steel because the steel itself presents a closed current path linked by alternating flux. By building the core from thin laminations, see below, the cross-section of this current path is made as small as possible so as to increase its electrical resistance and thus reduce the eddy current loss component to a minimum. Iron losses are present whenever the transformer is energized, even if it is on open-circuit. They vary in accordance with the applied voltage but, since a transformer is designed to operate from a fixed voltage, within modest limits, normally $\pm 10\%$, iron losses are normally regarded as fixed whenever the transformer is energized.

Copper losses, sometimes referred to as *load losses*, arise because of the flow of load current through the windings. Load losses also have two components: one is a purely resistive component, the $I^2 R$ in the copper, the other is an eddy current component flowing within the copper windings at right angles to the main current flow. The resistance to the flow of current in this plane can be increased by subdividing the winding conductors into a number of strands or subconductors, and this has the effect of reducing winding eddy current losses.

Although small in relation to transformer throughput, losses create a considerable heating effect within the transformer core and windings and must therefore be removed by the provision of special cooling arrangements. Most power transformers, with the exception of some small distribution units, have their core and windings immersed in mineral oil. In addition to providing electrical insulation, this circulates to remove the heat. Circulation may be natural, produced by the thermal head resulting from the losses themselves, or it may be forced by means of a pump, or pumps. Many transformers have dual natural/forced oil circulation, the former being sufficient to cool up to half rated throughput with the latter providing cooling for up to full load. The oil is usually cooled by passing through heat exchangers, or radiators, for which the secondary coolant is the surrounding air. The air may circulate due to the natural thermal head or it may be blown by means of fans.

Transformer construction. A simple, single-phase, transformer can be represented diagrammatically as shown in *Figure 10.3* with its iron core linking primary and secondary windings. In practice most power transformers are three-phase and their construction is as shown in *Figure 10.4*. The core has three limbs, or legs, and each of these is wound with a primary and a secondary winding. The limbs are built from thin *laminations* of cold-rolled magnetic steel (Chapter 2) and these are varied in width, or *stepped* so that the core cross-section is approximately circular to enable the cylindrical windings to be fitted over the core with the minimum wasted space. Top and bottom *yokes* connect the three limbs together and complete the path for the magnetic flux. Because a system of three-phase fluxes produces no resultant at any instant in time, no return flux path is necessary. A three-phase transformer generally therefore only requires a three-limb core.

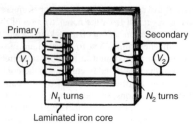

Figure 10.3 Single-phase transformer

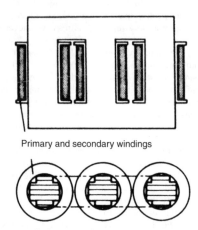

Figure 10.4 Core-type 3-phase transformer

Windings are normally made from copper. For smaller transformers this may be circular cross-section enamel insulated. For larger units the copper conductor is of rectangular cross-section. The copper conductor constituting a turn is subdivided into a number of separate strands, which may be insulated from each other either by enamel or by paper wrapping in order to reduce winding eddy current loss, as explained above. Where the bunch of strands making up a turn have only enamel insulation these are normally bound together by means of a paper outer wrapping which also provides the insulation to the adjacent turn. Windings are formed around paper or pressboard cylinders so that they may be assembled onto the core, or high voltage winding on to low voltage winding, after the winding operation.

For all but the smallest transformers which may be air cooled, cooling oil flows around the core and through the windings in ducts formed by means of pressboard strips. These run axially between the pressboard cylinder and winding turns, and also radially between individual turns or groups of turns. *Figure 10.5* shows a section through the core and windings of a typical medium sized transformer.

In the event of short-circuits on the external electrical systems, transformers are required to carry large short-circuit currents. These produce large

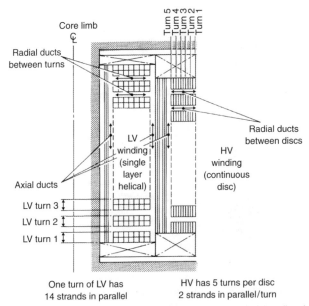

Figure 10.5 Section of LV and HV windings showing radial and axial cooling ducts

mechanical forces within the windings, both axially and radially. The radial forces are the easiest to withstand. An inner winding tends to be crushed further inwards by the winding outside, while an outer will tend to burst outwards. The inward crushing force is transmitted to the core and the outward bursting force is resisted by the tension in the copper conductors. Axial forces will tend to crush the winding turns together, but there are other axial components which, if the windings are not initially aligned, or *balanced*, will tend to increase the misalignment or unbalance even further.

It is necessary, therefore, to hold the windings very securely in position. This is done by means of core frames at the top and bottom of the limbs. These frames perform the combined functions of holding the core laminations together and clamping the windings to ensure that there can be no axial movement. This they achieve by transmitting the axial forces to the core limbs.

The core and windings of an oil-filled transformer are housed in a fabricated sheet-steel tank to contain the oil. Connections to the h.v. winding are frequently made via air/oil bushings mounted on the tank cover, although direct connection to gas insulated switchgear (GIS) is getting more common. L.v. winding connections are normally made via a cable box or boxes mounted on the side of the tank. The tank may have cooling radiators mounted on the sides not occupied by cable boxes or the radiators could be built into a separate free-standing cooler bank connected to the main tank by means of go and return pipework.

Figure 10.6 shows the complete core and windings of a fairly small oil-filled transformer and *Figure 10.7* shows a large three-phase transformer complete, having h.v. and l.v. bushings and separate cooler bank.

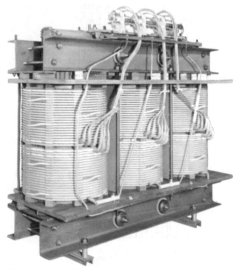

Figure 10.6 Three-phase 1500 kVA, 13.8/3.3 kV, 50 Hz core and windings. HV tappings at ±2.5% and ±5% taken from the HV disc type windings (Bonar Long Ltd)

Figure 10.7 Site installation of a 90 MVA, 132/33 kV, 50 Hz, three-phase transformer showing separate cooler bank (ABB Power T&D Ltd)

Transformer connections. In the above descriptions, mention has been made of the fact that most power transformers are three-phase. The three-phase windings can be connected in either *star* or *delta*. The arrangement selected depends on the following considerations.

Generally, in the UK, all electrical systems are required by the Electricity Supply Regulations to be connected to earth. In a three-phase system it is convenient to connect the system neutral to earth and the simplest way to do this is to provide a neutral on the incoming supply transformer. For this reason the transformer secondary windings are generally connected in star. (Occasionally, if the system is supplied directly from an alternating current generator, it will be the generator neutral which is earthed.)

It is very desirable in a three-phase system that the supply waveform should be as near as possible sinusoidal, that is it should have the minimum possible harmonic content since the presence of harmonics leads to increased stray losses and can also cause maloperation of some equipment. Harmonics in a supply arise from various causes but one of the most common is the non-linearity of the magnetization characteristics of electrical steels present, for example, in transformer cores. Third harmonic is one of the most prevalent of these, and the most effective way of limiting the third harmonic content in a transformer output current waveform is to connect one of the transformer windings in delta. The third harmonic voltages in a delta connected three-phase system are all in phase with each other so they are effectively acting in short-circuit around the delta. If the third harmonic voltage developed in the transformer is short-circuited in this way it will not appear in the output waveform and will not, therefore, give rise to any third harmonic output current – which was the object of the exercise.

A two-winding transformer can thus meet the requirements of providing an earth for the secondary system and elimination of third harmonic in the output if the secondary is connected in star with the primary in delta. The system supplying the primary of the transformer will generally be earthed at the neutral of its supply transformer.

Phase-shift in transformers. The connections of a delta/star transformer as described above are shown in *Figure 10.8(a)* and it will have a phasor diagram represented by *Figure 10.8(b)*. Considering the h.v. side of the transformer, the A-phase winding is represented by the phasor A_1A_2. The voltage induced in the l.v. winding must be in phase with and in the same sense as this since, as previously identified, a particular turn of a winding cannot behave any differently whether it is in the h.v. or l.v. winding. If the convention is followed that the higher numbered winding ends are connected to the line terminals, the h.v. winding connection A_2 to B_1 becomes terminal A, B_2 to C_1 terminal B, and C_2 to A_1 terminal C. L.v. line terminals are a_2, b_2, c_2 and a_1, b_1, c_1 the l.v. neutral. It can be seen that the transformer has produced a phase shift, or phase displacement, between h.v. and l.v. windings since, if the h.v. neutral is represented in the diagram by point N and the l.v. neutral by point n, whereas the h.v. A-phase phasor is at 12 o'clock with respect to the neutral, the l.v. a-phase has been moved forward to 1 o'clock. By simple geometry the angle of the phase shift can be deduced as $30°$ and the transformer is given the symbol Dyn1. The convention is that capital letters, D, Y, N, are used to indicate the h.v. winding and small letters, d, y, n, the l.v. The final digit indicates the phase shift as a number of multiples of $30°$.

190

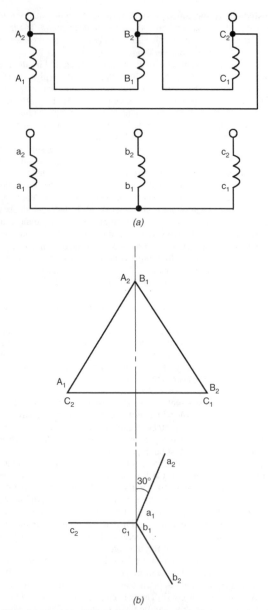

Figure 10.8 Winding connections, phasor and polarity diagram

Although it is preferable for all transformers to have one winding connected in delta, it is sometimes impracticable and both must be connected in star. One such occasion would be if it were important that there should be no phase shift through the transformer. It is possible to obtain no phase shift while still having one winding delta connected by using a delta/interstar connection as shown in *Figure 10.9*. The interstar arrangement (sometimes known as zigzag and abbreviated by the letter Z, or z) is achieved by internally subdividing one of the windings into two and connecting these two halves as shown in *Figure 10.9(b)*. Such a transformer effectively has a larger number of windings than a simple delta/star or star/delta connection and is therefore more costly to produce.

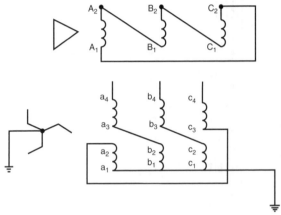

Figure 10.9 Transformer with delta secondary and interconnected-star earthing transformer with neutral connected to earth

Figure 10.10 is a table of the most common connection arrangements available for three-phase transformers and their associated phasor diagrams.

Parallel operation. Transformers may be paralleled in two ways. Two or more transformers may be arranged to supply a switchboard by totally different routes from a common source of supply. In this case the most important requirement is to ensure that the two supplies to be paralleled are in phase with each other. At the design stage this check may be carried by checking the theoretical phase displacements produced by each transformation from a common reference point; however, it is always important to prove any theoretical calculations by actually phasing out at the switchboard before closing the incoming circuit-breakers to operate in parallel for the first time.

If the transformers to be paralleled have their primaries connected to a common busbar it is desirable that their electrical characteristics match as closely as possible to ensure satisfactory load sharing. In this context 'matching' means:

- Having the same phase displacement.
- Having the same ratio on all tap positions and the same tapping range.
- Having the same impedance on all tap positions.

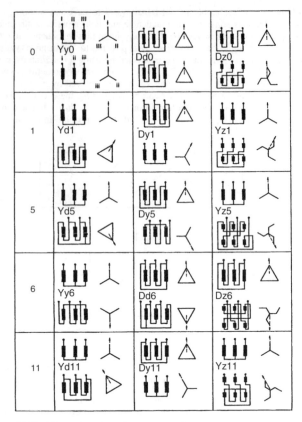

Figure 10.10 Three-phase transformer connections

It is, of course, still important to carry out phasing checks at the incoming circuit-breakers before connecting the transformers in parallel for the first time in order to confirm the correctness of all connecting cables.

Tapchanging in Transformers

Almost all transformers incorporate some means of adjusting their voltage ratio by means of the addition or removal of tapping turns. This adjustment may be made on-load, as is the case for many large transformers, by means of an off-circuit switch, or by the selection of bolted link positions which may be changed only with the transformer totally isolated. The degree of sophistication of the system of tap selection depends on the frequency with which it is required to change taps and the size and importance of the transformer.

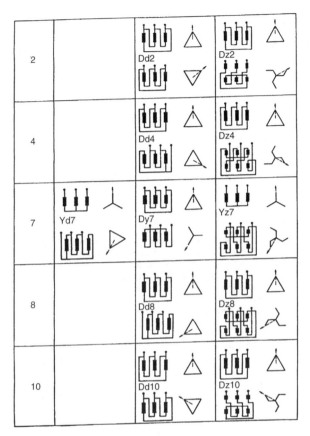

Figure 10.10 (continued)

Two definitions from the many which are set out in BS EN60076, Part 1, 'Power Transformers': *principal tapping* is the tapping to which the rated quantities are related and, in particular, the *rated voltage ratio*. This used to be known as normal tapping and the term is still occasionally used. It should be avoided since it can easily lead to confusion. It should also be noted that in most transformers tappings are *full-power tappings*, that is, the power capability of the tapping is equal to rated power so that on plus tappings the rated current for the tapped winding must be reduced and on minus tappings the rated current for the winding is increased. This usually means that at minus tappings, because losses are proportional to current squared, losses are increased, although this need not always be the case.

Uses of tapchangers. Before considering the operation and construction of tapchangers it is first necessary to examine the purpose and the way in which they are used. Transformer users require tappings for a number of reasons:

- To compensate for changes in the applied voltage on bulk supply and other system transformers.
- To compensate for regulation within the transformer and maintain the output voltage constant.
- On generator and interbus transformers to assist in the control of system VAr flows.
- To allow for compensation for factors not accurately known at the time of planning an electrical system.
- To allow for future changes in system conditions.

All the above represent sound reasons for the provision of tappings and, indeed, the use of tappings is so commonplace that most users are unlikely to consider whether or not they could dispense with tappings. However, transformers without taps are simpler, cheaper and more reliable. The presence of tappings increases the cost and complexity of the transformer and also reduces the reliability.

On-load tapchangers. One of the main requirements of any electrical system is that it should provide a voltage to the user which remains within closely defined limits regardless of the loading on the system. This is as important to domestic consumers as it is to industrial and commercial users. In many industrial systems, although the supply voltage must be high enough to ensure satisfactory starting of large motor drives, it must not be so high when the system is unloaded as to give rise to damaging overvoltages on, for example, sensitive electronic equipment. Some industrial processes will not operate correctly if the supply voltage is not high enough and some of these may even be protected by undervoltage relays which will shut down the process should the voltage become too low. Most domestic consumers require a supply voltage at all times of day and night which is high enough to ensure satisfactory operation of television sets, personal computers, washing machines and the like, but not so high as to shorten the life of filament lighting, which is often the first equipment to fail if the supply voltage is excessive.

This means that a method must be employed to obtain control of the voltage on transmission and distribution networks. Due to its comparatively low cost, reliability and ease of operation, on-load tapchanging on the transformer has become the accepted means of doing this.

By this means the turns ratio of the transformer windings is altered. Tappings are brought out from one of the windings and by appropriate connections the number of turns on that winding is altered. The tappings are nearly always on the high voltage winding to take advantage of the lower current conditions.

The principle of tapchanging on-load was developed in the late 1920s and requires a mechanism which will meet the following two conditions:

- The load current must not be interrupted during a tap change.
- No section of the transformer winding may be short-circuited during a tap change.

Early on-load tapchangers made use of reactors to achieve these ends but in modern on-load tapchangers these have been replaced by transition resistors which have many advantages. In fact, the first resistor-transition tapchanger made its appearance in 1929, but the system was not generally adopted in the UK until the 1950s. In the USA, the change to resistors only started to take place in the 1980s.

The tapchanger consists of two main components. The tap selector switch is the unit responsible for selecting the tap on the transformer windings, but does not make or break current. The diverter switch is where the actual switching of the load takes place.

The mechanical drive to these earlier tapchangers, both resistor or reactor types, was either direct drive or the stored energy type, the stored energy being contained in a flywheel or springs. But such drives were often associated with complicated gearing and shafting and the risk of failure had to be taken into account.

Most of these older designs have now been superseded by the introduction of the high-speed resistor type tapchanger. Reliability of operation has been greatly improved, largely by the practice of building the stored energy drive into close association with the actual switching mechanism thus eliminating many of the weaknesses of earlier designs. The introduction of copper tungsten alloy arcing tips has brought about a substantial improvement in contact life and a complete change in switching philosophy. It is recognized that long contact life is associated with short arcing time, and breaking at the first current zero is now the general rule.

A typical diagram showing the electrical switching sequence of a linear tapchanger is shown in *Figure 10.11*. For simplicity only a single phase is shown.

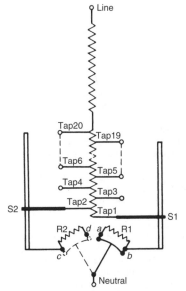

Figure 10.11 Electrical switching sequence resistor tap changing transformer used where load must not be interrupted

In the initial position selector switch S1 is on tap 1 and S2 on tap 2. The diverter switch connects tap 1 to the neutral point of the transformer winding. The sequence of operations in changing to tap 2 is as follows:

1. As the stored energy mechanism operates the moving contact starts to travel from one side of the diverter to the other; contact b is opened and the load current flows through resistor Rl to contact a.
2. The moving contact d then closes. Both resistors R1 and R2 are now in series across taps 1 and 2 and the load current flows through the mid-point of these resistors.
3. Further travel of the moving contact opens contact a and the load current then passes from tap 2 through resistor R2 and contact d.
4. Finally, when the moving contact reaches the other side of the diverter switch, contact c is closed and resistor R2 is shorted out. Load current from tap 2 now flows through contact c, the normal running position for tap 2.

The change from position 1 to 2 as described involves no movement of the selector switch. If any further change in the same direction is required, i.e. from 2 to 3, the selector switch S1 travels to tap 3 before the diverter switch moves and the diverter switch then repeats the above sequence but in the reverse order. If a change in the reverse direction is required, the selector switches remain stationary and the tap change is carried out by the movement of the diverter switch only.

Construction of on-load tapchangers. Although the speed of operation of diverter switches has been greatly increased in recent years, there is inevitably some arcing on contacts which are switching highly inductive circuits. Logically therefore these switches cannot be accommodated in the common oil with the transformer windings, since the contamination of the oil by arc products will rapidly lead to unacceptably poor dielectric properties. Selector switches, in theory, do not interrupt any current although in early on-load tapchangers it was considered desirable that their condition should be carefully monitored. In the UK, in order to allow this regular monitoring it was always practice to provide a further enclosure for the selector switches, separate from both main tank oil and the diverter switch oil.

As identified above, contact design and materials is now such that very long periods are possible without any inspection or maintenance of selector switches and for many years throughout mainland Europe it has been the practice to house the selector contacts in the main tank. This has the advantage that all tapping leads can be formed and connected to the appropriate selector switch contacts before the transformer is installed in the tank. With the separate compartment pattern, the usual practice is for selector switch contacts to be mounted on a baseboard of insulating material which is part of the main tank and forms the barrier between the oil in the main tank and that in the selector switch compartment. The tapping leads thus cannot be connected to the selector contacts until the core and windings have been installed in the tank. This is a difficult fitting task, requiring the tapping leads to be made up and run to a dummy selector switch base during erection of the transformer and then disconnected from this before tanking. Once the windings are within the tank, access for connection of the tapping leads is restricted and it is also difficult to ensure that the necessary electrical clearances between leads are maintained.

With in-tank tapchangers it is still necessary to keep the diverter switch oil separate from the main-tank oil. This is usually achieved by housing the diverter switches within a cylinder of glass-reinforced resin mounted above the selector switch assembly. When the transformer is installed within the tank, removal of the inspection cover which forms the top plate of this cylinder provides access to the diverter switches. These are usually removable via the top of the cylinder for maintenance and contact inspection. Such an arrangement is employed on the Reinhausen type M series which is a German design, also manufactured in France under licence by the Alstom group.

Single compartment tapchangers. Single compartment tapchangers have been developed to provide a more economic arrangement for medium sized, mainly 33/11 kV, transformers. This is achieved by combining the diverter switch and selector switch and mounting these in a separate compartment for bolting on to the side of the transformer tank. Tap selector contacts are brought from the tank interior through an insulating base and the contacts of each phase are arranged around the circumference of a circle. The transition resistors are mounted on the combined diverter/selector switch which rotates around a central shaft.

Figure 10.12 shows the switching sequence for a single compartment tapchanger which uses double resistor switching. Diagram 1 shows the condition with the transformer operating on tap position 1 with the load current carried by fixed and moving contacts. The first stage of the transition to tap position 2 is shown in diagram 2. Current has been transferred from the main contact to the left-hand transition resistor arcing contact and flows via resistor R_1. The next stage is shown in diagram 3 in which the right-hand transition contact has made contact with the tap 2 position. Load current is now shared between resistors R_1 and R_2 which also carry the tap circulating current. In diagram 4 the left-hand arcing contact has moved away from tap 1 interrupting the circulating current and all load current is now carried through the transition resistor R_2. The tap change is completed by the step shown in diagram 5 in which main and transition contacts are all fully made on tap 2. A single compartment tapchanger utilizing this arrangement is shown in *Figure 10.13*.

Moving coil voltage regulator. For many years the moving coil voltage regulator was used to provide a wide variation of output voltage for the operations such as battery charging or processes such as electroplating. Although now largely replaced by electronic equipment they were used in large numbers and may still be encountered. The basic winding arrangement of a typical moving coil voltage regulator consists of two fixed coils wound on the upper and lower halves of a magnetic core and connected in series opposition. A third coil of the same length is short-circuited upon itself and is free to move over the other two coils. The moving coil is entirely isolated electrically so that no flexible connections, slip-rings or sliding contacts are required.

The division of the voltage between the two fixed coils is determined by their relative impedances, and these are governed entirely by the position of the moving coil. With the moving coil in the position shown in *Figure 10.14* the impedance of coil *a* will be small and that of *b* large. If a voltage is then applied across the two coils connected in series the greater part of the voltage will appear across coil *b* and a small part across coil *a*.

When the moving coil is at the bottom of the leg as in *Figure 10.15* the relative impedances of coils *a* and *b* will be reversed, and the greater part of the voltage will now appear across coil *a*. With an arrangement similar to that

198

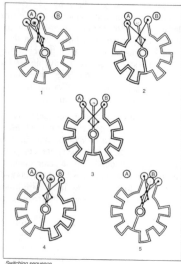

SWITCHING SEQUENCE

1. Shows in diagrammatic form the static condition of the tap-changer with the load current carried by copper fixed and moving contacts at tap A.

2. Current has been transferred from the main contact to the left-hand transition resistor arcing contact. The scissor action has increased the pitch between the two resistor moving contacts.

3. Right-hand transition resistor roller has made contact with tap B, establishing a circulating current between the two transition resistors.

4. Left-hand arcing contact has moved from tap A interrupting the circulating current and all power is carried through the right-hand transition resistor.

5. Tap-change is complete and power is carried through the main contact to tap B. The pitch of the transition resistor contacts is restored.

The same switching sequence applies with power flow in either direction.

Switching sequence.

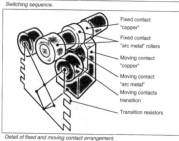

Fixed contact "copper"
Fixed contact "arc metal" rollers
Moving contact "copper"
Moving contact "arc metal"
Moving contacts transition
Transition resistors

Detail of fixed and moving contact arrangement.

Figure 10.12 Switching sequence for single compartment tapchanger (Associated Tapchangers)

shown in *Figures 10.14* and *10.15*, a range of voltage variation of practically 0–100% can be obtained with smooth infinitely variable control.

For applications on systems requiring voltage variations from 10–25% these can be obtained from a moving coil voltage regulator by the use of additional windings.

In these arrangements, the voltage variations obtained as shown in *Figures 10.14* and *10.15* are changed to the desired values by additional coils connected either to buck or boost a voltage in series with the line.

Figure 10.16 shows an arrangement incorporating two additional coils, r and l, which would be suitable for providing a constant output of 100% voltage while the input varied between 90% and 105% of the nominal value. Any desired value of buck or boost can be provided by choosing a suitable number of turns on the two coils. The position of the moving coil is altered by a small motor operated by a sensing device connected across the output.

Figure 10.13 A small single compartment tapchanger suitable for 300 A, 44 kV, 66 kV and 132 kV applications (Associated Tapchangers)

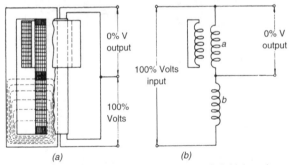

Figure 10.14 Typical moving-coil voltage regulator. Coil (a) impedance small; (b) impedance large

The Brentford linear regulating transformer. The Brentford voltage regulating transformer provides another alternative for the production of a stepless variable output voltage. This too has now ceased production but many are in service. The Brentford regulator is an autotransformer having a single layer coil on which carbon rollers make electrical contact with each successive turn of the winding. It can be designed for single- or three-phase operation and for either oil-immersed or dry-type construction. The winding is of the helical type which allows three-phase units to be built with a three-limb core as for a conventional transformer.

The helical winding permits a wide range of copper conductor sizes, winding diameter and length. The turns are insulated with glass tape and after

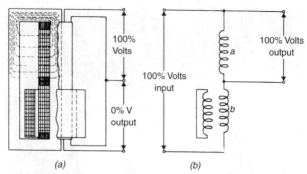

Figure 10.15 Typical moving coil voltage regulator. Coil (a) impedance large; coil (b) impedance small

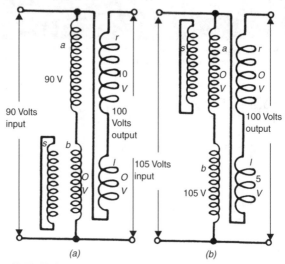

Figure 10.16 Typical moving coil voltage regulator with additional coils to buck or boost the line voltage

winding the coils are varnish impregnated and cured. A vertical track is then machined through the surface insulation to expose each turn of the winding. The chain driven carbon roller contacts supported on carriers operate over the full length of the winding to provide continuously variable tapping points for the output voltage.

As the contacts move they short-circuit a turn and a great deal of research was necessary to obtain the optimum current and heat transfer conditions at the coil surface. These conditions are related to the voltage between adjacent turns and the composition of the material of the carbon roller contacts.

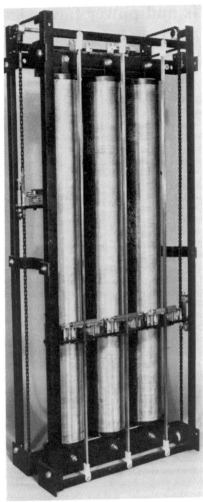

Figure 10.17 Three-phase, 100 A, 72 kVA, 415/0–415 V regulator (Allenwest Brentford Ltd)

The short-circuit current does not affect the life of the winding insulation or the winding conductor. The carbon rollers are carried in spring-loaded, self-aligning carriers and rotate as they travel along the coil face. Wear is minimal and the rolling action is superior to the sliding action of brush contacts. In normal use the contact life exceeds 100 km of travel with negligible wear on the winding surface. *Figure 10.17* shows a three-phase air insulated Brentford regulator.

11 Tariffs and power factor

Tariffs

Bulk electricity prices. With the privatization of electricity supply in 1989 came major changes in the tariff structure. This was, of course, to be expected since one of the main objectives of privatization was to introduce competition into the production and sale of electricity.

Since electricity is not a commodity which can be stored and used as required, the perceived difficulty in providing this competition was to do so while operating a system which enabled supply and demand to be balanced.

Before privatization it was the job of the CEGB, whose responsibility it was to provide the bulk supply of electricity, to maintain this balance. To do this they operated the generation plant on a 'merit order', in which the plant with the cheapest cost of production, normally based on thermal efficiency, was operated most, and the more expensive plant was operated in order of increasing cost only as the demand required. The majority of individual customers received their supplies from an Area Board who had a monopoly of supply over an area. The Area Boards received their supplies from the CEGB via a bulk supply tariff published annually by CEGB and calculated in accordance with a complex set of rules related to the cost of fuel and system operating costs.

On privatization the National Grid Company, NGC, was formed to own and operate the bulk transmission network and to manage a 'Pool' arrangement for electricity trading. Almost all electricity had to be sold into the Pool, and the distribution companies (termed 'suppliers' in this context) had to purchase from the Pool. The choice of plant to be operated effectively replicated the earlier, nationalized, system of central control, except that the costs of plant operation were declared a day ahead by the individual generators. There was little or no opportunity for suppliers to influence the prices by flexing their demand. The day was split into 48 half hour periods and Pool purchase and selling prices were set for each period. The purchase price was the price of the highest priced generating unit operating in the period and was paid to all generators for their generation in the period, regardless of their declared prices. The suppliers were all charged the Pool selling price, which was the Pool purchase price plus an amount to recover the additional costs of Pool operation.

After several years of operation, a number of problems were apparent with the Pool operation; the Pool prices had not shown any significant tendency to drop in line with the drops in the generators' costs, there was some evidence that generators were able to 'game play' in the Pool, with resulting sharp spikes in the Pool prices over some half hour periods and the suppliers were unable to influence the price of electricity by reducing demand at peak periods.

Following extensive consultation driven by the Government, revised New Electricity Trading Arrangements (NETA) were introduced in March 2000 to replace the Pool which ceased to operate on this date. Under NETA, generators and suppliers have to make bilateral contractual arrangements with each other for the supply of electricity, and the bulk of generation is traded through these bilateral contracts. Market participants are generally free to trade electricity with each other for forward periods. NGC continues to control the system, but its role is now one of ensuring that supply and demand are in balance second by second. The half hour periods are retained, and in each period and for some time previously (three and a half hours at the time of writing) only the System Operator can trade power, which he does to achieve balance. Generators and suppliers have to declare their contractual positions at the start of this period for each half hour, and commercial arrangements are in place to charge under- or oversuppliers or users for the costs of achieving an overall system balance. The charges are set at a fairly punitive rate to encourage participants to adhere to their contractual commitments.

The early indications are that NETA has been successful. It was brought into operation without any interruption of supply. After an initial settling down period the indications are that real downward pressure is being exerted on electricity prices which are now reflecting the production costs more than was the case under the Pool.

Electricity supplies to consumers. During the early phase of privatization only very large users of electricity were able to find their own suppliers and arrive at their own supply contracts. Since this initial phase, the size of consumer able to do this has been progressively reduced by government in accordance with a timetable published at the time of privatization, so that in 1999 all users were able to take their electricity from any supplier willing to provide them with a supply. Since the adoption of NETA, the suppliers may, but do not necessarily have to, purchase their electricity from generators. Whether they do so or not, in selling to individual consumers they must cover their own purchase costs and other operating cost, pay their profits, and pay 'use of system' charges. Use of system charges cover the cost of sending the electricity 'over the wires' connecting to the consumers' premises and, generally, include the costs associated with installing and maintaining those wires.

With such freedom available to suppliers and users, it is not surprising that many systems of tariffs are in operation. Most domestic users can choose from more than a dozen potential suppliers and it is not easy to decide which might be the most economic. For commercial and industrial users the situation can appear even more complex.

Despite this apparent complexity, however, there are factors which are common to most systems of tariffs and most of these factors are the same as those which applied before privatization. In providing a supply, there are two basic costs which must be met. One is the cost of the actual energy consumed, the other is the cost of providing the infrastructure – the wires – over which the supply is provided. Clearly, even if little or no electricity is used, the supplier would be faced with the cost of providing the connection. If large quantities of electricity are used then it will be the energy costs which will be predominant, but the supplier will still have the costs associated with providing the connection. This has led to the use by most suppliers of a *two-part tariff*.

The two-part tariff is based on these two costs of which the first part is covered by an annual amount and the second part by a charge per unit used. Block tariffs are related to maximum load, with different unit rates related to specified unit usage per quarter.

Industry generally operates on a maximum-demand system, while there are other forms of tariff for farms. Specialized tariffs include off-peak, Economy 7 and other time of day.

Load factor. This can be defined as the average load compared with the maximum load for any given period. It can be calculated as follows:

$$\frac{\text{Actual energy consumed}}{\text{Maximum demand} \times \text{Time in hours of period}}$$

The load factor of a consumer may vary from as low as 5% to as high as 80%, but usually it ranges from 10% (for lighting only) to 40% (for industrial or heating loads). Some industries are able to offer a 24-hour load and it is in these cases that very high load factor figures are obtained.

Owing to the two-part nature of the cost of supplying electrical energy, the actual load factor has a direct effect on the cost per unit since the fixed or standing charge to cover the first cost is divided into all the units used during that period. The more units used (and the higher the load factor), the less will be the fixed cost per unit. On this account it is the aim of every supply engineer to make his load factor as high as possible. As will be explained, special inducements are generally offered to consumers who will enable him to do this.

Diversity. The diversity of the supply load is given by the *diversity factor*, which is found from

$$\frac{\text{Sum of consumers' maximum demands}}{\text{Maximum demand on system}}$$

and it will be seen that

$$\frac{\text{System load factor}}{\text{Average consumer's load factor}} = \text{Diversity factor}$$

Note: The average consumer's load factor must be calculated with reference to actual consumption and not merely as a numerical average.

Tariffs. These are generally of three basic types: *industrial, commercial* and *domestic*. Despite privatization principles remain the same and the following examples are still generally valid. Most of the distribution companies have introduced two changes. They offer a seasonable time of day tariff which does not have a maximum demand component but has a higher unit cost in winter than in summer.

An industrial two-part tariff is almost invariably based on the maximum demand – either in kW or in kVA – and in many cases time of the year. A typical l.v. industrial tariff may therefore be: a fixed charge of £1.00 per kVA service capacity per month, £16.50 per month, £17.00 with night units, plus a maximum-demand charge per kW in each month as follows: April to October inclusive £0.11, November and February £4.50, December and January £7.70 and March £1.85. There are then unit charges which vary according to the time of day.

Most companies offer a wide range of options based on maximum demand and the supply can be taken at high voltage or low voltage. Typically an l.v. supply based on a monthly maximum demand if the load is over 10 kW might be as follows: fixed charge £12.80 per month, November–February, first 10 kW £9.50 per month. March–October, first 10 kW £0.35 per month with a day unit charge of 4.5p.

It is more usual today to base the fixed charge on a stated sum per month or per annum with a penalty charge for low power factor; it therefore pays the consumer to install power factor correction capacitors to lift a low power factor to a value in excess of 0.9.

The maximum demand figure is obtained by means of a maximum demand indicator which gives the highest load (either in kW or kVA according to the tariff) which occurs for a given period – such as 15 or 30 minutes. Special tariffs are in many cases offered to consumers with favourable loads.

Power Factor Correction

Many tariff charges encourage the user to maintain a high power factor (nearly unity) in his electrical network by penalizing a low power factor. The power factor can be improved by installing power factor correction equipment, the capital cost of which is often recovered in a few years by the savings made in reduced electricity bills.

Low power factors are caused mainly by induction motors and fluorescent lights and compensation may be applied to individual pieces of equipment, in stages by automatic switching, or in bulk at the supply intake position. Advice on the most economic system for a given installation is available from specialist firms.

There are a number of ways in which power factor correction can be provided and these are described below.

By capacitor. The kVA required for power factor correction may de determined graphically by the use of curves of the type shown in *Figure 11.1*. It is also possible to calculate the capacitance required as follows: referring to *Figure 11.2*, the load current is represented by OI_L lagging by angle ϕ_1 such that $\cos \phi_1$ is the power factor of the load.

Assuming that it is desired to improve the power factor to $\cos \phi_2$ by means of capacitors, the resultant current must be represented by OI_R in *Figure 11.2*. The method employed is the constant kW one.

To obtain this amount of correction the capacitor current of OI_C must equal $L_L - L_R$ and this value will be given by $OI_C = OI_L \sin \phi_1 - OI_R \sin \phi_2$.

The vector diagram is drawn for current, but is also applicable to kVA since the current is directly proportional to the kVA. Thus OI_L, OI_C and OI_R can be taken to represent the kVA of the load, the capacitor and the resultant kVA respectively.

In this case the initial conditions would be:

$$\cos \phi_1 = \frac{kW}{kVA_L}$$

$$\tan \phi_1 = \frac{kVAr_L}{kW}$$

The improved conditions would be:

$$\cos \phi_2 = \frac{kW}{kVAr_R}$$

$$\tan \phi_2 = \frac{kVAr_R}{kW}$$

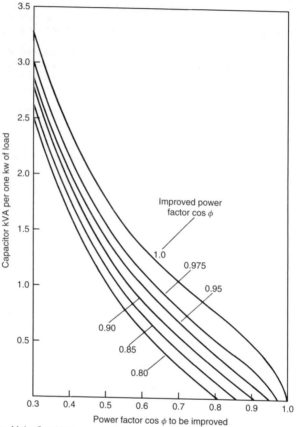

Figure 11.1 Graphical means of determining kVAr required when raising power factor from one value to a higher value

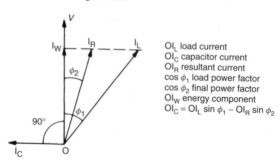

OI_L load current
OI_C capacitor current
OI_R resultant current
$\cos \phi_1$ load power factor
$\cos \phi_2$ final power factor
OI_W energy component
$OI_C = OI_L \sin \phi_1 - OI_R \sin \phi_2$

Figure 11.2 Diagram for capacitors

Capacitor kVAr required to improve factor from $\cos\phi_1$ to $\cos\phi_2$

$$= (\text{kVAr}_L - \text{kVAr}_R)$$
$$= \text{kW}(\tan\phi_1 - \tan\phi_2)$$

Actual capacitance required. It may be necessary to transform capacitor kVA to microfarad capacitance and the following relationship shows how this should be done.

Single-phase. Current in capacitor is given by

$$I_C = 2\pi f C V$$

where I_C = current in amperes

f = frequency

C = rating of capacitor in farads

V = voltage

(*Note*: 1 farad = $10^6\ \mu$F.)

Three-phase. The total line current taken by three capacitors in delta as shown in *Figure 11.3* is given by

Line current = $\sqrt{3}$ phase current in each capacitor

Total line current = $\sqrt{3}(2\pi f C V)$.

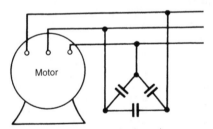

Figure 11.3 Capacitors connected in delta for three phase

The kVA is $\sqrt{3}VI \times 10^{-3}$ so that the kVA is given by

$$\text{kVA} = \frac{3(2\pi f C V^2)}{1000}$$

The C used in the above formula is the rating of one of the three capacitors forming the delta and so the total rating is $3C$. This gives us the formula:

Rating of each capacitor = $C = \dfrac{\text{kVA} \times 1000}{3(2\pi f V^2)}$

Total rating = $3C = \dfrac{\text{kVA} \times 1000}{2\pi f V^2}$ F

Synchronous motor correction. A synchronous motor can be made to take a leading current (a current at leading power factor) by overexciting it, and in so doing can be used to provide power factor correction.

Referring to *Figure 11.4* the current required for the synchronous motor cannot always be fixed by the desired amount of power factor correction, as in this case it is driving a load and the actual current will be fixed by the load on the synchronous motor and the power factor at which it is working.

It is impracticable to give formulae for working out these values as it is much better to start with the possible main load and the variable load which can be used for the synchronous motor.

Table 11.1 Wattless and power components for various power factors

Power factor $\cos \phi$	Angle (degrees)	Per kVA		Per kW	
		Power	Wattless	kVA	Wattless component
1.0	0	1.0	0	1.0	0
0.98	11.48	0.98	0.20	1.02	0.20
0.96	16.26	0.96	0.28	1.04	0.29
0.94	19.95	0.94	0.34	1.06	0.36
0.92	23.07	0.92	0.39	1.09	0.43
0.90	25.83	0.90	0.44	1.11	0.48
0.88	28.37	0.88	0.48	1.14	0.54
0.86	30.68	0.86	0.51	1.16	0.59
0.84	32.87	0.84	0.54	1.19	0.65
0.82	34.92	0.82	0.57	1.22	0.70
0.80	36.87	0.80	0.60	1.25	0.75
0.78	38.73	0.78	0.63	1.28	0.80
0.76	40.53	0.76	0.65	1.32	0.86
0.74	42.27	0.74	0.67	1.35	0.91
0.72	43.95	0.72	0.69	1.39	0.96
0.70	45.57	0.70	0.71	1.43	1.02
0.68	47.15	0.68	0.73	1.47	1.08
0.66	48.70	0.66	0.75	1.52	1.14
0.64	50.20	0.64	0.77	1.56	1.20
0.62	51.68	0.62	0.78	1.61	1.27
0.60	53.13	0.60	0.80	1.67	1.33
0.58	54.55	0.58	0.82	1.72	1.40
0.56	55.93	0.56	0.83	1.79	1.48
0.54	57.32	0.54	0.84	1.85	1.56
0.52	58.66	0.52	0.85	1.92	1.64
0.50	60	0.50	0.87	2.00	1.73

Referring to the phasor diagram, if values are taken either for currents as shown in the phasor diagram or their proportionate kVA, their resultant current or kVA can be obtained as follows:

$$OI_R = \sqrt{(OI_L \cos \phi_1 + OI_M \cos \phi_2)^2 + (OI_L \sin \phi_1 - OI_M \sin \phi_2)^2}$$

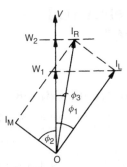

OI$_L$ main load current
OI$_M$ synch. motor current
OI$_R$ resultant current
cos ϕ_1 load power factor
cos ϕ_2 synch. motor power factor
cos ϕ_3 final power factor
OW$_1$ α original load, kW
OW α final load, kW

Figure 11.4 Diagram for synchronous motor

Resultant power factor can be obtained from

$$\tan \phi_2 = \frac{\text{OI}_L \sin \phi_1 - \text{OI}_M \sin \phi_2}{\text{OI}_L \cos \phi_1 + \text{OI}_M \cos \phi_2}$$

If in any given case there is a fixed main load at a stated power factor plus a given kW load for the synchronous motor it is advisable to calculate the resultant power factor by working this out for various leading power factors for the synchronous motor.

It should be borne in mind that synchronous and synchronous induction motors will not work satisfactorily at a very low power factor. Values between 0.6 and 0.9 leading are usually taken for satisfactory results.

12 Requirements for electrical installations (BS 7671)

IEE Wiring Regulations (Sixteenth Edition)

In 1992 the sixteenth edition of the IEE Wiring Regulations became a British Standard, BS 7671 : 1992. It retained the subtitle IEE Wiring Regulations Sixteenth Edition. Following a full review in accordance with BSI rules and procedures it has now become BS 7671 : 2001. This new edition, published in June 2001, came into effect on 1 January 2002, that is, installations designed after that date should comply with the new edition. Prior to that date designers have the choice of either edition.

This chapter was totally rewritten based on BS 7671 : 1992, and took into account all the changes up to and including Amendment No. 1, issued in December 1994. An outline of the changes introduced by the 2001 edition is given below and as far as possible the rest of the chapter has been updated to reflect these changes. However, this chapter is, like the rest of the book, intended as a handy reference source. It is not a guide to the Regulations and on all issues the Regulations themselves should be regarded as the definitive source of information relating to acceptable electrical installation design and practice.

Throughout this chapter references to the IEE Wiring Regulations are taken to mean BS 7671 : 2001. Information is also extracted from the Guidance Notes listed on page 260. Amendment No. 1 to BS 7671 : 2001, issued in February 2002, is also taken into account. Amendment No. 1 to the 1992 edition of BS 7671 coincided with the change in the nominal value of the supply voltages in the UK. By this the domestic single-phase supply was reduced from $240\,V \pm 6\%$ to $230\,V + 10\% - 6\%$; three-phase supplies were changed from $415\,V \pm 6\%$ to $400\,V + 10\% - 6\%$. In reality this has meant very little actual change because the new range encompasses the greater part of the previous one, but it now provides closer uniformity with Europe.

Notes on the layout of the 16th edition. In the numbering system used in the sixteenth edition (which is quite different to that of the previous editions), the first digit signifies a part, the second a chapter, the third a section, and the subsequent digits the regulation number. For example, the section number 413 is made up as follows:

Part 4 – Protection for safety
Chapter 41 (first chapter of Part 4) – Protection against electric shock
Section 413 (third section of Chapter 41) – Protection against indirect contact

The details of the six parts are as follows (a detailed list of contents is given later in this chapter):

Part No.	Subject
1	Fundamental principles for safety.
2	Definitions.

3   Assessment of general characteristics – identifies the characteristics of the installation that will need to be taken into account in choosing and applying the requirements of the subsequent Parts. These characteristics may vary from one part of an installation to another, and should be assessed for each location to be served by the installation.

4   Protection for safety – describes the basic measures that are available, for the protection of persons, property and livestock and against the hazards that may arise from the use of electricity.

           Chapters 41 to 46 each deal with a particular hazard. Chapter 47 deals in more detail with, and qualifies, the practical application of the basic protective measures, and is divided into sections whose numbering corresponds to the numbering of the preceding chapters, thus Section 471 needs to be read in conjunction with Chapter 41, Section 473 with Chapter 43, and Section 476 with Chapter 46.

5   Selection and erection of the equipment.

           Chapter 51 relates to equipment generally and Chapters 52 and 56 to particular types of equipment.

6   Special installations or locations – particular requirements.

7   Inspection and testing.

The sequence of the plan should be followed in considering the application of any particular requirement of the Regulations. The general index provides a ready reference to particular Regulations by subject, but in applying any one Regulation the requirements of related Regulations should be borne in mind. Cross-references are provided and the index is arranged to facilitate this. In many cases a group of associated Regulations is covered by a side heading which is identified by a two-part number, e.g. 547-03. Throughout the Regulations where reference is made to such a two-part number, that reference is to be taken to include all the individual Regulation numbers which are covered by that side heading and include that two-part number.

Changes Introduced by the 2001 Edition

Before looking at the Regulations in detail it is appropriate to examine the changes introduced by the 2001 edition. This still carries the subtitle 'IEE Wiring Regulations Sixteenth Edition' because the substance and philosophy remain as set out in the sixteenth edition, and it is only in certain specific areas of the detail that it has been changed. In some areas, particularly Part 1, the detail has been expanded. However, much of the revision is a tidying-up exercise.

First, the 2001 edition incorporates all the amendments to the previous edition including Amendment No. 3, dated April 2000, which has been further revised/corrected. The changes arise from three sources:

(i) Due to CENELEC harmonization (see below).
(ii) Due to changes in national requirements.
(iii) Due to new and obsolete product standards.

The resulting changes to the standard can be examined under six headings:

(i) Fundamental principles for safety.
(ii) Protection against overvoltage.
(iii) Precautions where risk of fire exists.
(iv) Special installations or locations.
(v) Inspection and testing.
(vi) Revised standards and new approaches.

Fundamental principles. The changes here have been implemented by a complete restructuring of Part 1. Historically the IEE Wiring Regulations were seen by many as applying only to domestic and similar applications.

Scope of Regulations. The new Chapter 11 – 'Scope, object and fundamental principles' – clearly sets out that the Regulations apply to all installations with the exception of specific applications where separate installation rules apply. Installations covered include those of residential, commercial, industrial, agricultural and horticultural, public premises, prefabricated buildings, caravans, caravan parks and similar sites, construction sites, exhibitions, fairs and other installations in temporary buildings, highway power supplies and street furniture, outdoor lighting.

Exclusions from scope. Regulation 110-02 excludes 'supplier's works', railway traction equipment, equipment of motor vehicles except caravans, ships, aircraft, mobile and fixed offshore, mines and quarries, radio interference suppression equipment except where it affects the safety of the installation, lightning protection of buildings covered by BS 6651 and those aspects of lift installations covered by BS 5655.

Relationship with statutory Regulations. It is also made clear, in Regulation 110-04, that the Regulations are not statutory Regulations. They may, however, be used in a court of law in evidence to claim compliance with statutory requirement. (However, see below those Regulations identified in BS 7671 which do have statutory force.)

The new Chapter 12 – 'Objects and Effects' – has been reduced in content. It now confines itself to setting out the objects of the Regulations. Some of its former contents, for example Regulations 110-04-01 and 110-05-01, have been moved to Chapter 11; some have been moved to Chapter 13. Chapter 12 now states that the Regulations contain the rules for design and erection of electrical installations so as to provide for safety and proper functioning for the intended use. It refers the reader to Chapter 13 for the fundamental principles.

The new Chapter 13 – 'Fundamental Principles' – details the principles underlying the subsequent parts dealing with safety, design, selection of equipment and inspection and testing. The requirements of this section are intended to provide for the safety of persons, livestock and property.

In electrical installations risk of injury may result from:

(i) shock currents
(ii) excessive temperatures
(iii) mechanical movement of electrically actuated equipment
(iv) explosion.

The electrical installation shall be designed for:

(i) the protection of persons, livestock and property
(ii) the proper functioning of the electrical installation.

Every item of equipment shall comply with the appropriate EN or HD or National Standard implementing the HD.

Erection includes good workmanship and proper materials, identification of conductors, satisfactory construction of joints and connections.

Verification of compliance includes inspection and testing on completion. The person carrying out the inspection and testing shall make a recommendation for subsequent periodic inspection and testing.

Protection against overvoltage. For the first time regulations concerning the use of surge protection are introduced. These are implemented by a new Chapter 44 – 'Protection Against Overvoltage' – the regulations do not apply where an installation is supplied by a low voltage network containing no overhead lines. Their application is also mitigated in locations which are subjected to a low level of thunderstorm days, or ceraunic level denoted by the symbol AQ, which is the criteria on which the use of surge protection is based. The level of surge protection required is nil for AQ1 ($\leq$25 thunderstorm days per year). As an alternative to the AQ criteria, the use of surge protection may be based on a risk assessment.

Precautions where risk of fire exist. These are covered in a new Chapter 48 although attention is drawn to the need to comply with the previously existing Chapter 42 and Section 527. This chapter applies to installations in locations where the risk of fire is due to the nature of the processed or stored materials, such as barns, paper mills, or textile manufacture. The chapter does not apply to locations with explosion risks which are separately covered by BS EN 50014 (see Chapter 23 of this book) or to installations in escape routes, for example those covered by BS 5266. Basically this chapter requires that electrical equipment for use in these locations shall be selected and erected such that its temperature in normal operation, and foreseeable temperature rise in the event of a fault, is unlikely to cause a fire.

In locations where there is a risk of fire due to the nature of the processed or stored material:

(i) electrical equipment shall have a degree of protection of at least IP5X
(ii) TN and TT systems (defined below), except MIC and Busbar Trunking systems, shall be protected by an RCD having $I_{\Delta n}$ not exceeding 300 mA
(iii) every circuit shall be capable of being isolated from live conductors by a linked circuit breaker or switch.

Regulations apply to a wide range of equipment including cables, busbar trunking, motors, luminaires, heating and ventilation, enclosures and distribution boards.

Special installations or locations. These changes affect Part 6. Section 601 incorporates Amendment No. 3 to BS 7671 : 1992 into the main body of the Standard (with corrections and additions). This covers locations containing a bath or shower. The new Section 604 covers construction site installations. New Regulations 604-01-02 and 604-01-03 identify where the requirements are applicable and not applicable. They apply to:

(i) the assembly comprising the main switchgear and the main protective devices
(ii) installations on the load side of the above, comprising mobile and transportable electrical equipment as part of the movable installation.

They do not apply to:

(i) the construction site offices
(ii) installations covered by BS 6907 (Equipment in opencast mines and quarries).

An important area of change is in Section 607, 'Earthing requirements for the installation of equipment having high protective conductor currents'. This has been brought about by recognition of the widespread use of information technology (IT) equipment. The first point to be noted is the use of the expression 'high protective conductor currents' rather than 'high earth leakage currents'. In the case of IT equipment the high protective conductor currents are not the result of earth leakage but are a design feature resulting from the need for suppression and the use of switch mode power supplies. The effect can, of course, be the same and the changes in Section 607 are of particular relevance to both appliance manufacturers and protective/distribution equipment manufacturers, where in many instances the advances in design and technology are diverging and leading to adverse reactions to unwanted tripping of protective devices and additional installation costs.

The amendment allows, perhaps even encourages, the use of twin socket outlets in ring circuits provided the socket outlets have two earth terminals and the two ends of the protective conductor at the fuseboard are secured in separate terminals of the earth bar.

In Section 611, 'Highway Power Supplies, Street Furniture and Street located Equipment', there are also amendments to Regulation 611-02-02 to include height limitations for access to electrical equipment. New Regulations 611-02-06 and 611-05-02 introduce requirements when using Class II equipment (defined below) and minimum degree of protection IP33, for electrical equipment, either by construction, or by method of installation.

Periodic inspection and testing. These revisions to Part 7, in particular Chapter 73, 'Periodic Inspection and Testing', place greater emphasis on periodic inspection and reporting. Particularly this will include the routine testing of RCDs, which is a requirement of the Electricity at Work Regulations and has been an initiative led by BEAMA. Regulation 731-01-03 now defines that the scope of the inspection and testing of an installation shall be decided by a competent person. Regulation 732-01-02 allows that periodic inspection may be replaced by a monitoring/maintenance programme with appropriate records.

Chapter 74, 'Certification and Reporting', includes a new Regulation 743-01-01 which requires that inspection schedules and reports shall be based on models given in Appendix 6.

Thermosetting and thermoplastic insulation. The use of an ever increasing number and variety of insulation materials has led to confusion as to their classification in what have hitherto been the two groups of materials, 'plastics' (PVC) and 'rubbers' and it has often been unclear as to their permitted temperature rating. This has been resolved by grouping the materials into 'thermoplastic' materials, which include PVC and 'thermosetting' which includes those hitherto regarded as rubber.

Thermoplastic materials are those which when heated become soft and can be moulded and reformed on any number of occasions, that is, they are 'plastic' in nature. Thermosetting materials are chemically cross-linked, that is, once the chemical bond has been formed, under the influence of heat, the material is set into its shape. Thermosetting materials have improved characteristics over thermoplastics such as resistance to deformation and higher operating temperatures.

The new method, to which rated temperature is added, gives a total and unambiguous method of classification, for example XLPE, 'cross-linked polyethylene', is neither a plastic nor a rubber, but is a thermosetting material rated at 90°C.

Initially, following the change, the old terms will be given in brackets following the new classification, e.g. 'thermoplastic' (PVC) and 'thermosetting' (rubber).

BS 7671 : 2001 Details of Regulations

In the new edition account has been taken of the technical substance of agreements reached in CENELEC. In addition it also takes into account the technical intent of the following CENELEC Harmonization Documents:

CENELEC Harmonization Document Reference		Part of the Regulation
HD 193	Voltage bands	Part 1 and Definitions
HD 384.1	Scope, object and fundamental principles	Part 1
HD 384.2	Definitions	Part 2
HD 384.3	Assessment of general characteristics	Part 3
HD 384.4.41	Protection against electric shock	Part 4, Chapter 41
HD 384.4.42	Protection against thermal effects	Part 4, Chapter 42
HD 384.4.43	Protection against overcurrent	Part 4, Chapter 43
HD 384.4.443	Protection against overvoltages of atmospheric origin or due to switching	Part 4, Section 443
HD 384.4.45	Protection against undervoltage	Part 4, Chapter 45
HD 384.4.46	Isolation and switching	Part 4, Chapter 46
HD 384.4.47	Application of measures for protection against electric shock	Part 4, Section 470

216

HD 384.4.473	Application of measures for protection against overcurrent	Part 4, Section 473
HD 384.4.482	Protection against fire where particular risks or danger exist	Part 4, Section 482
HD 384.5.51	Selection and erection of equipment, common rules	Part 5, Chapter 51
HD 384.5.52	Wiring systems	Part 5, Chapter 52 and Appendix 4
RD 384.5.523	Wiring systems, current-carrying capacities	Part 5, Section 523 and Appendix 4
HD 384.5.537	Switchgear and control gear, devices for isolation and switching	Part 5, Section 537
HD 384.5.54	Earthing arrangements and protective conductors	Part 5, Chapter 54
HD 384.5.551	Other equipment, low voltage generating sets	Part 5, Section 551
HD 384.5.56	Safety services	Part 5, Chapter 56
HD 384.6.61	Initial verification	Part 7, Chapter 71
HD 384.7.702	Special location – Swimming pools	Part 6, Section 602
HD 384.7.703	Special location – Locations containing a hot air sauna heater	Part 6, Section 603
HD 384.7.704	Construction and demolition site installations	Part 6, Section 604
HD 384.7.705	Special location – Agricultural and horticultural premises	Part 6, Section 605
HD 384.7.706	Special location – Restrictive conducting locations	Part 6, Section 606
HD 384.7.708	Special location – Caravan parks and caravans	Part 6, Section 608
HD 384.7.714	Outdoor lighting installations	Part 6, Section 611

Considerable reference is made throughout the Regulations to publications of the British Standards Institution, both specifications and codes of practice. Appendix 1 in the Regulations lists these publications and gives their full titles whereas throughout the Regulations they are referred to only by their numbers. Nearly 12 pages are included involving 110 different British Standards. Where they appear in the Regulations is also noted. Where a reference is made to a British Standard in the Regulations, and the British Standard concerned takes account of a CENELEC Harmonization Document, it is understood that the reference is to be read as relating also to any foreign standard similarly based on that Harmonization Document, provided it is verified that any differences between the two standards would not result in a lesser degree of safety than that achieved by compliance with the British Standard (see Section 511 of the Regulations).

A similar verification should be made in the case of a foreign standard based on an IEC standard but as national differences are not required to be listed in such standards, special care should be exercised.

In some cases the Regulations may need to be supplemented by requirements or recommendations of British Standards or of the person ordering the work. Installations falling into this category include emergency lighting to BS 5266, installations in explosive atmospheres to BS 5345 and fire detection and alarm systems in buildings to BS 5839. Other cases include installations subject to the Telecommunications Act 1984, BS 6701 Part I and electric surface heating systems to BS 6351. The Regulations do not apply to ten different types of installations and these are listed in BS 7671. They include railway traction equipment, installations on ships, and on mobile and fixed offshore installations.

Voltage ranges. Installations operating at the following levels are covered:

(i) Extra-low voltage – normally not exceeding 50 V a.c. or 120 V ripple free d.c. whether between conductors or to earth.
(ii) Low voltage – normally exceeding extra-low voltage but not exceeding 1000 V a.c. or 1500 V d.c. between conductors, or 600 V a.c. or 900 V d.c. between conductors and earth.

Equipment. The Regulations apply to items of electrical equipment only so far as selection and application of the equipment in the installation are concerned. They do not deal with requirements for the construction of prefabricated assemblies of electrical equipment, which are required to comply with appropriate specifications.

Contents of regulations

Part 1 – Scope, object and fundamental principles

Part 2 – Definitions

Part 3 – Assessment of general characteristics

Part 4 – Protection for safety

Part 5 – Selection and erection of equipment

Part 1. Scope, Object and Fundamental Principles

This has been discussed above under the heading 'Changes introduced by the 2001 edition'.

Part 2. Definitions

The Regulations include a number of definitions and some of these are included here. The well-known definitions, familiar to electrical contractors, are not reproduced. These definitions indicate the sense in which the terms defined are used in the Regulations. Some of the definitions are in line with those given in BS 4727 'Glossary of electrotechnical, power, telecommunications, electronics, lighting and colour terms'. Other terms that are not defined in the Regulations are used in the sense defined in that British Standard.

Arm's reach. Zone of accessibility to touch, extending from any point on a surface where persons usually stand or move about, to the limits a person can reach with a hand in any direction without assistance. Three diagrams in the Regulations illustrate the zone of accessibility.

Barrier. A part providing a defined degree of protection against contact with live parts, from any usual direction of access.

Basic insulation. Insulation applied to live parts to provide basic protection against electric shock and which does not necessarily include insulation used exclusively for functional purposes.

Bonding conductor. A protective conductor providing equipotential bonding.

Cable ducting. A manufactured enclosure of metal or insulating material other than conduit or cable trunking, intended for the protection of cables which are drawn in after erection of the ducting.

Cable trunking. A closed enclosure normally of rectangular cross-section, of which one side is removable or hinged, used for the protection of cables and for the accommodation of other electrical equipment.

Circuit. An assembly of electrical equipment supplied from the same origin and protected against overcurrent by the same protective device(s).

Circuit protective conductor (cpc). A protective conductor connecting exposed-conductive-parts of equipment to the main earthing terminal.

Class I equipment. Equipment in which protection against electric shock does not rely on basic insulation only, but which includes means for the connection of exposed-conductive-parts to a protective conductor in the fixed wiring of the insulation (see BS 2754).

Class II equipment. Equipment in which protection against electric shock does not rely on basic insulation only but in which additional safety precautions such as supplementary insulation are provided, there being no provision for the connection of exposed metalwork of the equipment to a protective conductor, and no reliance upon precautions to be taken in the fixed wiring of the installations (see BS 2754).

Class III equipment. Equipment in which protection against electric shock relies on supply at SELV and in which voltages higher than those of SELV are not generated (see BS 2754).

Double insulation. Insulation comprising both basic insulation and supplementary insulation.

Earthed concentric wiring. A wiring system in which one or more insulated conductors are completely surrounded throughout their length by a conductor, for example a metallic sheath, which acts as a PEN conductor.

Electrical installation (Abbreviated *Installation*). An assembly of associated electrical equipment supplied from a common origin to fulfil a specific purpose and having certain co-ordinated characteristics.

Extraneous-conductive-part. A conductive part liable to introduce a potential, generally earth potential, and not forming part of the electrical installation.

Final circuit. A circuit connected directly to current-using equipment, or to a socket-outlet or socket-outlets or other outlet points for the connection of such equipment.

Isolation. A function intended to cut off for reasons of safety the supply from all, or a discrete section, of the installation by separating the installation or section from every source of electrical energy.

Neutral conductor. A conductor connected to the neutral point of a system and contributing to the transmission of electrical energy. The term also means the equivalent conductor of an IT or d.c. system unless otherwise specified in the Regulations.

PEN conductor. A conductor combining the functions of both neutral conductor and protective conductor.

Protective conductor. A conductor used for some measures of protection against electric shock and intended for connecting together any of the following parts:

 (i) exposed-conductive parts
 (ii) extraneous-conductive parts
(iii) the main earthing terminal
(iv) earth electrode(s)
 (v) the earthed point of the source, or an artificial neutral.

A diagram in the Regulations (page 25) shows an example of earthing arrangements and protective conductors.

Reinforced insulation. Single insulation applied to live parts, which provides a degree of protection against electric shock equivalent to double insulation under the conditions specified in the relevant standard. The term 'single insulation' does not imply that the insulation must be one homogeneous piece. It may comprise several layers which cannot be tested singly as supplementary or basic insulation.

Residual current device. A mechanical switching device or association of devices intended to cause the opening of the contacts when the residual current attains a given value under specified conditions.

Residual operating current. Residual current which causes the residual current device to operate under specified conditions.

Ring final circuit. A final circuit arranged in the form of a ring and connected to a single point of supply.

Simultaneously accessible parts. Conductors or conductive parts that can be touched simultaneously by a person or, in locations specifically intended for them, by livestock.

Simultaneously accessible parts may be: live parts, exposed-conductive-parts, extraneous-conductive-parts, protective conductors or earth electrodes.

Skilled person. A person with technical knowledge or sufficient experience to enable him/her to avoid dangers which electricity may create.

Stationary equipment. Equipment which is either fixed, or equipment having a mass exceeding 18 kg and not provided with a carrying handle.

Supplementary insulation. Independent insulation applied in addition to basic insulation in order to provide protection against electric shock in the event of a failure of basic insulation.

Switch. A mechanical device capable of making, carrying and breaking current under normal circuit conditions, which may include specified operating overload conditions, and also of carrying for a specified time currents under specified abnormal circuit conditions such as those of short-circuit. It may be also capable of making, but not breaking, short-circuit currents.

Switchgear. An assembly of main and auxiliary switching apparatus for operation, regulation, protection or other control of an electrical installation.

System. An electrical system consisting of a single source of electrical energy and an installation. For certain purposes of the Regulations types of system are identified as follows, depending upon the relationship of the source, and of exposed-conductive-parts of the installation, to earth:

- **TN system**, a system having one or more points of the source of energy directly earthed, the exposed-conductive-parts of the installation being connected to that point by protective conductors.
- **TN-C system**, a system in which neutral and protective functions are combined in a single conductor throughout the system (*Figure 12.1*).

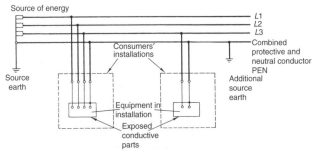

Figure 12.1 TN-C system. Neutral and protective functions combined in a single conductor throughout system. All exposed conductive parts of an installation are connected to the PEN conductor. An example of the TN-C arrangement is earthed concentric wiring but where it is intended to use this, special authorisation must be obtained from the appropriate authority. (Reproduced, by permission, from the 16th edition of the IEE Wiring Regulations)

- **TN-S system**, a system having separate neutral and protective conductors throughout the system (*Figure 12.2*).

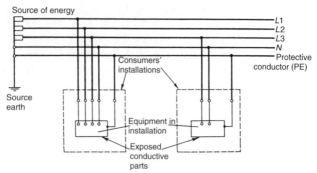

Figure 12.2 TN-S system. Separate neutral and protective conductors throughout the system. The protective conductor (PE) is the metallic covering of the cable supplying the installations or a separate conductor. All exposed conductive parts of an installation are connected to this protective conductor via the main earthing terminal of the installation. (Reproduced, by permission, from the 16th edition of the IEE Wiring Regulations)

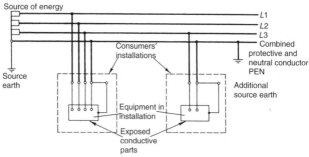

Figure 12.3 TN-C-S system. Neutral and protective functions combined in a single conductor in a part of the system. The usual form of a TN-C-S system is as shown, where the supply is TN-C and the arrangement in the installations is TN-S. This type of distribution is known also as Protective Multiple Earthing and the PEN conductor is referred to as the combined neutral and earth (CNE) conductor. The supply system PEN conductor is earthed at several points and an earth electrode may be necessary at or near a consumer's installation. All exposed conductive parts of an installation are connected to the PEN conductor via the main earthing terminal and the neutral terminal, these terminals being linked together. (Reproduced, by permission, from the 16th edition of the IEE Wiring Regulations)

- **TN-C-S system**, a system in which neutral and protective functions are combined in a single conductor in part of the system (*Figure 12.3*).
- **TT system**, a system having one point of the source of energy directly earthed, the exposed-conductive-parts of the installation being connected to earth electrodes electrically independent of the earth electrodes of the source (*Figure 12.4*).

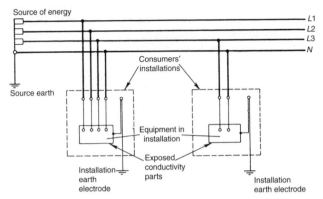

Figure 12.4 TT system. All exposed conductive parts of an installation are connected to an earth electrode which is electrically independent of the source earth. (Reproduced, by permission, from the 16th edition of the IEE Wiring Regulations)

- **IT system**, a system having no direct connection between live parts and earth, the exposed-conductive-parts of the electrical installation being earthed (*Figure 12.5*).

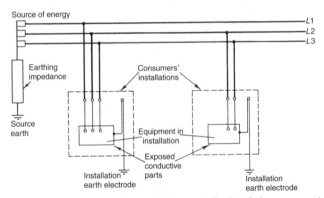

Figure 12.5 IT system. All exposed conductive parts of an installation are connected to an earth electrode. The source is either connected to earth through a deliberately introduced earthing impedance or is isolated from earth. (Reproduced, by permission, from the 16th edition of the IEE Wiring Regulations)

Part 3. Assessment of General Characteristics

Part 3 of the Regulations, 'Assessment of general characteristics', includes the purpose for which the installation is intended, supplies and structure

(Chapter 31); the external influences to which it is exposed (Chapter 32); compatibility of the equipment (Chapter 33); and its maintainability (Chapter 34).

Maximum demand. The maximum demand of an installation, expressed in amperes, has to be assessed. In determining this diversity may be taken into account (Reg. 311-01-01).

Live conductors and earthing. The number and type of live conductors and the method of earthing (depending on the type of system, i.e. TN-C, TN-S, TN-C-S, TT and IT) govern the method of protection for safety adopted in order to comply with Part 4 of the Regulations (Reg. 312-01-01).

Nature of supply. This relates to the nominal voltage; current; frequency; prospective short-circuit current at the origin of the installation; the earth fault loop impedance (Z_e) of that part of the system external to the installation; suitability of the installation to meet requirements, including the maximum demand; and the nature of the protective device acting at the origin of the installation (Reg. 313-01-01). Section 313-02 details requirements for safety services and standby purposes.

Circuit arrangement. Every installation must be divided into circuits as necessary to avoid danger and minimize inconvenience in the event of a fault and to facilitate safe operation, inspection, testing and maintenance (Reg. 314-01-01). See also other 314 Regulations.

Compatibility. The installer must check whether any equipment will have harmful effects upon other electrical equipment or other services. For an installation supplied from an external source of supply the supplier should be consulted about any equipment in the installation whose operation may have a significant influence on the supply.

Where an installation comprises more than one final circuit, each final circuit shall be connected to a separate way in a distribution board. The wiring of each final circuit must be electrically separate from that of every other final circuit to prevent indirect energization of a final circuit intended to be isolated (Reg. 314-01-04).

A requirement of the sixteenth edition is that of making an assessment of the frequency and quality of maintenance an installation will be expected to receive during its intended life. This implies that any periodic inspection, testing, maintenance and repairs likely to be necessary during the installation's intended life can be readily and safely carried out. Protective measures for safety must also be effective and the reliability of equipment for proper functioning must be appropriate (Reg. 341-01-01).

Part 4. Protection for Safety

Every installation, either as a whole or in its several parts, has to comply with the requirements of Part 4 by applying the protective measures outlined in Chapters 41 to 46 in the manner described in Chapter 47.

Protection against direct and indirect contact shall be provided by one of the following methods:

(i) Protection by safety extra-low voltage (Regs 411-02 and 471-02).
(ii) Protection by limitation of discharge of energy (Regs 411-04 and 471-03).

Functional extra-low voltage alone shall not be used as a protective measure (see Reg. 471-15).

The source to provide the safety extra-low voltage shall be one of the following: (a) an isolating transformer complying with BS 3535, in which there shall be no connection between the output winding and the body or the protective earthing circuit if any; or (b) a source of current such as a motor generator with windings providing electrical separation equivalent to that of the safety isolating transformer specified in (a) above or (c) an electrochemical source, i.e. a battery, or another source independent of a higher circuit voltage, i.e. an engine-driven generator; or (d) electronic devices complying with appropriate standards where measures have been taken so that even in the case of an internal fault the voltages at the outgoing terminals cannot exceed the value specified by Regulation 411-02-01. A higher voltage at the output terminals is permitted under specified circumstances.

Live parts of safety extra-low voltage equipment, other than cables, shall be electrically separate from those of higher voltage, neither should they be connected to earth or to a live part or a protective conductor forming part of another system (Reg. 411-02-05). Circuit conductors in these installations shall preferably be physically separated from those of any other circuit. There are provisions made where this is impracticable (Reg. 411-02-06). Plugs shall not be capable of being mated with socket-outlets of other voltage systems in use in the same premises; similarly socket-outlets must exclude plugs of other voltage systems. Socket-outlets shall not have a protective conductor contact (Reg. 411-02-10).

If the nominal voltage of an SELV system exceeds 25 V a.c. r.m.s. 50 Hz or 60 V ripple-free d.c. additional protection against direct contact is specified (Reg. 411-02-09). Where reduced or low body resistance is to be expected or where the risk of electric shock is increased by contact with earth potential (Regs 471-15, etc., as above) Regs 471-15 and 471-16 prescribe additional requirements. These conditions may be expected in situations where the hands and/or feet are likely to be wet or where the shock current path may not be through the extremities, or where a person is immersed in water or working in confined conductive locations.

Where extra-low voltage is used but not all the requirements regarding safety extra-low voltage can be met, appropriate measures are outlined (Reg. 471-14) to provide protection but the systems employing these measures are termed 'functional extra-low voltage systems' or FELV systems.

If the extra-low voltage system complies with the requirements of Reg. 411-02 for SELV except that the circuits are earthed at one point only, protection against direct contact shall be provided by either (i) barriers or enclosures affording a degree of protection of at least IP2X or IPXXB, or (ii) insulation capable of withstanding a test voltage of 500 V d.c. for 60 seconds (Reg. 471-14-02).

If the extra-low voltage system does not generally comply with the safety requirements, protection against direct contact shall be provided by barriers or enclosures or by insulation corresponding to the minimum test voltage for the primary circuit. Protection against indirect contact is also required (Reg. 471-14-03).

Circuits relying on protection against electric shock by limiting the discharge of energy, method (ii), shall be separated from other circuits in a manner similar to that specified in relation to safety extra-low voltage circuits (Regs 411-02-05 and 411-02-06).

Direct contact. One or more of the following basic protective measures for protection against direct contact shall be used:

(i) Protection by insulation of live parts (Regs 412-02 and 471-04).
(ii) Protection by a barrier or an enclosure (Regs 412-03 and 471-05).
(iii) Protection by obstacles (Regs 412-04 and 471-06).
(iv) Protection by placing out of reach (Regs 412-05 and 471-07).

When protecting live parts from direct contact by insulation it must be such that it can only be removed by destruction and it must be able to withstand the electrical, mechanical, thermal and chemical stresses to which it may be subjected in service (Reg. 412-02-01). If a barrier or enclosure is employed to prevent direct contact and the opening is larger than that allowed for by IP2X (i.e. for replacement of part or for functioning purposes) suitable precautions shall be taken to prevent persons or livestock from unintentionally touching live parts. As far as is practicable persons should be made aware that live parts can be touched through the opening and should not be touched (see Reg. 471-05-02).

Protection can also be provided by obstacles, but they shall be secured in such a way as to prevent unintentional removal but may be removable without using a key or tool.

Indirect contact. There are five basic protective measures against indirect contact specified in the Regulations and one or more should be adopted.

(i) Earthed equipotential bonding and automatic disconnection of supply (Regs 413-02 and 471-08).
(ii) Use of Class II equipment or equivalent insulation (Regs 413-03 and 471-09).
(iii) Non-conducting location (Regs 413-04 and 471-10).
(iv) Earth-free local equipotential bonding (Regs 413-05 and 471-11).
(v) Electrical separation (Regs 413-06 and 471-12).

When employing method (i) the bonding conductors shall connect to the main earthing terminal of the installation extraneous conductive parts including: main water and gas pipes, other service pipes and ducting, central heating and air conditioning systems, exposed metallic parts of the building structure and the lightning protective system. Bonding to the metalwork of other services may require the permission of the undertakings responsible. Bonding to any metallic sheath of a telecommunication cable is necessary; however, the consent of the owner shall be obtained. When an installation serves more than one building the requirements for earthed equipotential bonding apply to each building (Reg. 413-02-02).

The protective devices, earthing and relevant impedances of the circuits concerned shall be such that during an earth fault the voltages between simultaneously accessible exposed and extraneous conductive parts occurring anywhere in the installation shall be of such a magnitude and duration as not to cause danger (Reg. 413-02-04). Conventional means of compliance with this Regulation are given in Regulations 413-02-06 to 413-02-26 according to the type of system earthing, but other equally effective means shall not be excluded.

The Regulations provide comprehensive details of the methods of protection by earthed equipotential bonding and automatic disconnection of supply for the three types of earthing, i.e. TN, TT, and IT systems. In the UK the general practice is to employ the TN system of earthing in one of the three forms: (a) TN-C where the neutral and protective functions are combined in a single conductor (PEN) throughout the system; (b) TN-S where the neutral and protective conductors are separated throughout the system; and (c) TN-C-S where the two conductors' functions are combined in a single conductor in a part of the system. System (c) is very common as it allows single-phase loads to be supplied by live and neutral with a completely separate earth system connecting together all the exposed conductive parts before connecting them to the PEN conductor via a main earthing terminal which is also connected to the neutral terminal. In this chapter therefore we will confine our remarks to the TN system of earthing only.

Section 413 is the relevant part of the Regulations and it contains a number of tables related to protection to which the reader is referred. Reproduction of any of them is not permitted by the IEE.

Two types of protective device are allowed, overcurrent or residual current (r.c.d.) (Reg. 413-02-07). If an r.c.d. is employed in a TN-C-S system, a PEN conductor must not be used on the load side. Connection of the protective conductor to the PEN conductor must be made on the source side of the r.c.d.

Regulation 413-02-04 (see above) is considered to be satisfied if the characteristic of each protective device and the earth fault loop impedance of each circuit protected by it are such that automatic disconnection of the supply occurs within a specified time when a fault of negligible impedance occurs between phase conductor and a protective conductor or an exposed conductive part anywhere in the installation. The requirement is met where

$$Z_s \leq U_o / I_a$$

where Z_s = the earth fault loop impedance

I_a = the current causing the automatic operation of the disconnecting protective device within the time stated in Table 41A (of the Regulations) as a function of the nominal voltage U_o or, under the conditions stated in Regulations 413-02-12 and 413-02-13, within a time not exceeding 5 s

U_o = the nominal voltage to earth

The maximum disconnection times of Table 41A apply to a circuit supplying socket-outlets and to other final circuits which supply portable equipment intended for manual movement during use, or hand-held Class I equipment (Reg. 413-02-09). The requirement does not apply to a final circuit supplying an item of stationary equipment connected by means of a plug and socket-outlet where precautions are taken to prevent the use of the socket-outlet for supplying hand-held equipment, nor to the reduced low voltage circuits described in Regulation 471-15.

Where a fuse is used to meet Regulation 413-02-09, maximum values of earth fault loop impedance (Z_s) corresponding to a disconnection time of 0.4 s are stated in Table 41B1 (in the Regulations) for a nominal voltage to earth (U_o) of 230 V. For types and rated currents of general purpose (gG) fuses, other than those mentioned in Table 41Bl (of the Regulations), and for motor circuit fuses (gM) the reader is referred to the appropriate British Standard, to determine the value of I_a for compliance with Regulation 413-02-08.

Table 41B1 is in four parts and details the maximum earth fault loop impedance Z_s, for fuses, for 0.4 s disconnection time with U_o of 230 V: (a) for fuse ratings from 6 to 50 A (gG fuses to BS 88 Parts 2 and 6); (b) fuse ratings 5 to 45 A (fuses to BS 1361); (c) fuse ratings 5 to 45 A (fuses to BS 3036); and (d) 13 A fuses to BS 1362. Table 41B2 gives similar details covering miniature circuit-breakers.

Irrespective of the value of U_o, for a final circuit which supplies a socket-outlet or portable equipment intended for manual moving during use, or hand-held Class I equipment, the disconnection time can be increased to a value not exceeding 5 s for the types and ratings of the overcurrent protective devices and associated maximum impedance of the circuit protective conductors shown in Table 41C (of the Regulations). This table is in nine parts, the first four relating to the same types of fuses detailed in Table 41B1; the other five relate to miniature circuit-breakers to EN 60898, i.e. types 1, 2, B3 and D (Reg. 413-02-12).

For a distribution circuit the disconnection time must not exceed 5 s (Reg. 413-02-13). When a fuse is used to satisfy this Regulation, Table 41D (in the Regulations) specifies the maximum values of earth fault loop impedance (Z_s) corresponding to a disconnection time of 5 s for a nominal voltage to earth of 230 V. This table again has four parts similar to Table 41B1.

Regulation 413-02-16 covers the situation whereby an r.c.d. is used; the following condition has to be fulfilled:

$$Z_s I_{\Delta n} \leq 50 \, \text{V}$$

where Z_s is the earth fault loop impedance and $I_{\Delta n}$ is the residual operating current of the protective device in amperes. If the circuit protected by the r.c.d. extends beyond the earthed equipotential zone, exposed conductive parts need not be connected to the TN system protective conductors if they are connected to an earth electrode of the correct resistance value appropriate to the r.c.d.'s operating current (Reg. 413-02-17).

Class II or equivalent. Double or reinforced insulated equipment or low voltage switchgear assemblies having total insulation (see EN 60439) provide suitable protection (Reg. 413-03-01). Two other systems are mentioned relating to supplementary insulation applied to electrical equipment with only basic insulation and reinforced insulation applied to uninsulated live parts. When the electrical equipment is ready for operation, all conductive parts separated from live parts by basic insulation only shall be contained in an insulating enclosure affording at least the degree of protection IP2X or IPXXB (Reg. 413-03-04).

Where a circuit supplies items of Class II equipment a circuit protective conductor shall be run to and terminated at each point in the wiring and at each accessory except a lampholder having no exposed conductive parts and suspended from such a point (Reg. 471-09-02). This need not be observed where Regulation 471-09-03 applies. Exposed metalwork of Class II equipment should be mounted so that it is not in electrical contact with any part of the installation connected to a protective conductor. Such a contact might impair the Class II protection provided by the equipment specification.

Where a whole installation or circuit is intended to consist of Class II equipment (or equivalent) it must be verified that it is under effective supervision to ensure no change is made that would impair the effectiveness of the Class II (or equivalent) insulation. This measure cannot be applied to any circuits which include socket-outlets or where a user may change items of equipment without authorization (Reg. 471-09-03). Certain cables cannot be described as being of Class II construction. If these cables (they are specified) are installed in accordance with Chapter 52 they are considered to provide satisfactory protection (Reg. 471-09-04).

Protection against overcurrent. Live conductors shall be protected by one or more devices for automatic interruption of the supply in the event of an overload or short-circuit except where the source is incapable of supplying a current exceeding the current-carrying capacity of the conductors (Reg. 431-01-01).

A device protecting a circuit against overload has to satisfy a number of conditions:

(i) Its nominal current setting (I_n) shall not be less than the design current (I_b) of the circuit.
(ii) Its I_n shall not exceed the lowest of the current-carrying capacities (I_z) of any of the circuit conductors.
(iii) The current assuring effective operation of the protective device (I_2) shall not exceed 1.45 times the lowest of the current-carrying capacities (I_z) of any of the conductors of the circuit.

When the same protective device is used with conductors in parallel the value of I_z is the sum of the current-carrying capacities of those conductors. The conductors have to be of the same construction, cross-sectional area, length and disposition, having no branch circuits throughout their length and be arranged so as to carry substantially equal currents unless the suitability of the particular arrangement is verified (Reg. 473-01-06).

The omission of devices for protection against overload is permitted in circumstances where unexpected disconnection of the circuit could cause danger. Consideration should be given to the provision of an overload alarm in these circumstances.

Protection against fault current. Devices for this duty shall break any fault current flowing in the conductors of each circuit before this current causes danger due to thermal and mechanical effects produced in the conductors and connections (Reg. 434-01-01). Except under the conditions outlined in the next sentence, the breaking capacity rating of each device must not be less than the prospective short-circuit current or earth fault current at the point at which the device is installed. A lower breaking capacity than that at the point at which the device(s) is installed is permissible if it is backed up by another protective device(s), on the supply side, with the necessary breaking capacity. Co-ordination between the devices is important to ensure that no damage occurs to the protective device(s) on the load side of the conductors protected (Reg. 434-03-01).

Where a single device protects two or more conductors in parallel against fault current the operating characteristics of the device and the characteristics of the parallel conductors as installed shall be suitably co-ordinated. If the operation of a single protective device may not be effective, alternative measures identified in the Regulation may have to be taken (Reg. 473-02-05).

Protection according to the nature of circuits and distribution systems. A means of detection of overcurrent shall be provided and shall cause disconnection of the phase conductor. In a TT system, for a circuit supplied between phases and for which the neutral is not distributed, overcurrent protection need not be provided for one of the phase conductors provided certain conditions are met (Reg. 473-03-02). For an IT without a neutral conductor the overload protective device may be omitted in one of the phase conductors if an r.c.d. is installed in each circuit (Reg. 473-03-03).

Neutral conductor. In TN or TT systems where the neutral conductor has a cross-sectional area at least equal or equivalent to that of the phase conductor, it is not usually necessary to provide overcurrent detection or a disconnecting device for this conductor. Where the cross-sectional area is less than that of the phase conductors overcurrent detection for the neutral conductor shall be provided unless (i) it is protected against fault current by the protective device for the phase conductors of the circuit and (ii) the maximum current, including harmonics, likely to be carried by the neutral conductor in normal service, is significantly less than the value of the current-carrying capacity of that conductor.

Where either or both of the conditions are not met overcurrent detection shall be provided for the neutral conductor, appropriate to its cross-sectional area. Operation of the protective device causes disconnection of the phase conductors but not necessarily the neutral conductor (Regs 473-03-03 and 473-03-04).

The cross-sectional area of the neutral conductor may be smaller than the phase conductors depending on the degree of sustained imbalance except in circuits where the load is predominantly due to discharge lighting (Regs 524-02-01, 524-02-02 and 524-02-03).

Isolation and switching. Means shall be provided for non-automatic isolation and switching to prevent or remove hazards associated with the installation of electrically powered equipment and machines (Reg. 460-01-01). A main linked switch or linked circuit-breaker must be provided as near as practicable to the origin of every installation as a means of switching the supply on load and as a means of isolation. For d.c. systems all poles shall be provided with a means of isolation. There are further provisions if an installation is supplied from more than one source (Reg. 460-01-02).

Means of switching off for mechanical maintenance are necessary where such maintenance may involve a risk of physical injury (Reg. 462-01-01). Precautions must also be taken to prevent any equipment being unintentionally or inadvertently reactivated (Reg. 462-01-03).

Where it is necessary to disconnect rapidly an installation from the supply, emergency switching facilities shall be provided (Reg. 463-01-01). The means adopted shall be capable of cutting off the full-load current of the relevant part of the installation, due account being taken of stalled motor conditions where relevant (Reg. 537-04-01).

Where isolating devices for particular circuits are placed remotely from the equipment to be isolated, provision shall be made so that the means of isolation can be secured in the open position. Where this provision takes the form of a lock or removable handle, the key or handle shall be non-interchangeable with any others used for a similar purpose within the installation (Reg. 476-02-02).

Every motor circuit shall be provided with a disconnector which disconnects the motor and all equipment including any automatic circuit-breaker (Reg. 476-02-03). There are also stringent requirements relating to electric discharge lighting installations using an open-circuit voltage exceeding low voltage (Reg. 476-02-04).

Circuits outside an equipotential bonding zone. When a circuit in an equipotential bonding zone is specifically intended to supply fixed equipment outside the zone, and that equipment may be touched by a person in contact directly with the general mass of earth, the earth-fault loop impedance must be such that disconnection occurs within the time stated in Table 41A (of the Regulations) (Reg. 471-08-03).

Socket-outlet circuits which can be expected to supply portable equipment for use outdoors shall comply with Regulation 471-16-01. This regulation refers to socket-outlets rated at 32 A or less and lays down specific requirements with some exemptions.

Where automatic disconnection as protection against indirect contact is used in an installation forming part of a TT system, every socket-outlet circuit shall be protected by a residual current device (Reg. 471-08-06, see also Reg. 413-02-16).

Part 5. Selection and Erection of Equipment

Common rules. Chapter 51 covers the general rules applicable to selection of equipment. Every item of equipment shall comply with the requirements of the applicable British Standard, or Harmonized Standard. If equipment complying with a foreign standard is used it is the responsibility of the designer or person specifying the installation to verify there will be no lesser degree of safety than had the compliant British Standard equipment been used.

Equipment must be suitable for the voltage, current, frequency where this has an influence on the characteristics of the equipment, and power demanded by the duty.

Every item of equipment selected must be compatible with other equipment during normal service including switching.

Every item of equipment shall be of a design appropriate to the situation in which it is used. If it is not, by its construction, suitable for its operating environment, additional protection, which shall not adversely affect its operation, shall be provided.

Except for joints in cables, for which see Section 526, every item of equipment shall be accessible as necessary for operation, inspection, maintenance and connections.

Except where there is no possibility of confusion, every item of equipment shall be provided with suitable means of identification. Where the operation of switchgear or control gear cannot be observed by the operator and where this might lead to danger a suitable indicator, visible to the operator, shall be provided (Reg. 514-01-01).

Protective conductors. The colour combination of green and yellow is reserved exclusively for protective conductors and shall not be used for any other purpose. If a bare busbar is used as a protective conductor, it shall be identified throughout its length by an arrangement of green and yellow stripes (Reg. 514-03-01).

Diagrams. A legible diagram shall be provided indicating, in particular, type and composition of each circuit, the method used for compliance with

Regulation 413-01-01 (protection against indirect contact) and, where appropriate, the data required by Regulation 413-02-04 (characteristics of automatic protective devices), the information necessary for identification and location of protective devices.

Warning notices. Shall be provided for any enclosure within which a voltage exceeding 230 V exists (Reg. 514-10), and where there are live parts which are not capable of being isolated by a single device (Reg. 514-11). A permanent label with the words 'Safety Electrical Connection – Do Not Remove' shall be fixed in a visible position near to bonding conductors and earthing connections (Reg. 514-13). Notices shall be fixed near the origin of any installation on completion of inspection and testing (Reg. 514-12).

Electromagnetic compatibility. The requirements of BS EN 50082 and BS EN 50081 shall be taken into account when selecting equipment for an installation.

Wiring systems. Non-flexible or flexible cable or flexible cord on l.v. systems must comply with the appropriate British or Harmonized Standard. Non-flexible cable sheathed with lead, PVC or an elastomeric material for aerial use may incorporate a catenary wire or include hard-drawn copper conductors. But the Regulation does not apply to flexible cord forming part of a portable appliance or luminaire where these are the subject of and comply with a British Standard, or to special flexible cables/cords for combined power and telecommunications wiring (Reg. 521-01-01). The same Regulation permits flexible cable/cord to incorporate metallic armour, braid or screen. A busbar trunking system must comply with EN 60570 (Reg. 521-01-02).

Due to electromagnetic effects single-core cables armoured with steel wire or tape must not be used in a.c. circuits. Conductors of a.c. circuits installed in ferromagnetic enclosures have to be arranged so that all phase conductors and the neutral conductor (if any) together with the appropriate protective conductor of each circuit are in the same enclosure. Where such conductors enter a ferrous enclosure they have to be arranged so that the conductors are not individually surrounded by a ferrous material, or some other provision has to be made to prevent eddy currents (Reg. 521-02-01).

Enclosures. Conduits and conduit fittings, trunking, ducting and fittings must comply with their appropriate British Standards, which are specified in the Regulations. Where BS 4678 does not apply, non-metallic trunking, ducting and their fittings shall be of insulating material complying with the ignitability characteristic 'p' of BS 476 Part 5 (Regs 521-04-01 and 521-05-01).

Cable derating. Provision for increasing the section of a cable shall be made if a run is to be installed where thermal insulation may cover it. If installed in a thermally insulating wall or above a thermally insulated ceiling, with the cable being in contact on one side the current-carrying capacities are tabulated in Appendix 4 of the Regulations. If the cable is totally surrounded by thermally insulating material, derating factors have to be applied (Reg. 523-04-01).

Voltage drop. Under normal service conditions the voltage drop in a consumer's installation must be such that the voltage at the terminals of any fixed current-using equipment must be greater than the lower limit corresponding to the British Standard relevant to the equipment (Reg. 525-01-01). This Regulation is satisfied for a supply given in accordance with the Electricity Supply

Regulations 1988 (as amended) if the voltage drop between the supply terminals and the fixed current-using equipment does not exceed 4% of the nominal supply voltage.

Environmental conditions. The sixteenth edition contains a classification system relating to external influences developed for IEC Publication 364-3. The concise list of external influences occupies 12 pages and it introduces the characteristic letters AA, AD, etc. into the body of the text in Chapter 52. Appendix 5 of the Regulations explains the significance of these symbols.

Each condition of external influence is designated by a code comprising a group of two capital letters and a number, as follows.

The first letter relates to the general category of external influence:

> **A**   Environment
>
> **B**   Utilization
>
> **C**   Construction of buildings

The second letter relates to the nature of the external influence:

> ...   **A**
>
> ...   **B**
>
> ...   **C**

The number relates to the class within each external influence:

> **1**
>
> **2**
>
> **3**

For example, the code AA4 signifies:

> **A** = Environment
>
> **AA** = Environment − Ambient temperature
>
> **AA4** = Environment − Ambient temperature in the range of
>
> −5°C to +40°C

Note: The codification is not intended to be used for marking equipment.

Ambient temperature. A wiring system has to be suitable for the highest and lowest temperature likely to be encountered (Reg. 522-01-01). Similarly when handling or installing components of a wiring system they must be only at temperatures within the limits specified (Reg. 522-01-02). Wiring systems should be protected from the effects of heat from external sources, including solar gain by one of the following means, or an equally effective method; shielding, placing sufficiently far from the source of heat, selecting a system with due regard for the additional temperature rise which may occur, reductions in current-carrying capacity or local reinforcement or substitution of insulating material (Reg. 522-02-01).

Presence of water or high humidity. A wiring system has to be selected so that no damage can be caused by these two conditions. Where water may collect or condensation form in a wiring system, provision must be made for it to escape harmlessly. Where it may be subjected to waves, protection has to be provided (Section 522-03).

Corrosive/polluting substances. Where the presence of these is likely to affect any parts of a wiring system, these parts must be suitably protected or made from materials resistant to such substances (Reg. 522-05-01).

Identification. The colour combination green and yellow is reserved exclusively for the identification of protective conductors. Where electrical conduits are required to be distinguished from pipelines of other services, orange (to BS 1710) shall be used (Regs 514-02-01 and 514-03-01). Every single-core non-flexible cable and every core of non-flexible cable for use as fixed wiring shall be identifiable at its terminations and preferably throughout its length, by the appropriate method described in items (i) to (v) below:

(i) for thermosetting (rubber) and thermoplastic (PVC) insulated cables, the use of core colours is specified in *Table 12.1*, or the application at terminations of tapes, sleeves or discs of the appropriate colours in the table

Table 12.1 Colour identification of cores of non-flexible cables and bare conductors for fixed wiring

Function	Colour identification
Protective (including earthing) conductor	green-and-yellow
Phase of a.c. single-phase circuit	red[†]
Neutral of a.c. single- or three-phase circuit	black
Phase R of 3-phase a.c. circuit	red
Phase Y of 3-phase a.c. circuit	yellow
Phase B of 3-phase a.c. circuit	blue
Positive of d.c. 2-wire circuit	red
Negative of d.c. 2-wire circuit	black
Outer (positive or negative) of d.c. 2-wire circuit derived from 3-wire system	red
Positive of 3-wire d.c. circuit	red
Middle wire of 3-wire d.c. circuit	black[*]
Negative of 3-wire d.c. circuit	blue
Functional earth	cream

For armoured PVC-insulated cables and paper-insulated cables, see Regulation 514-06-01 (ii) and (iii)

[*]Only the middle wire of three-wire circuits may be earthed

[†]As alternatives to the use of red, if desired, in large installations, on the supply side of the final distribution board, yellow and blue may also be used.

(ii) for armoured auxiliary thermoplastic (PVC) cables, either (i) above or the use of sequentially numbered cores commencing with number 1

(iii) for paper-insulated cables, the use of numbered cores provided that the numbers 1, 2 and 3 signify phase conductors, number 0 the neutral conductor and number 4 for the fifth ('special purpose') core if any

(iv) for cables with thermosetting insulation, the use of core colours as in *Table 12.1* or alternatively sequentially numbered cores commencing with the number 1

(v) for mineral insulated cables, the application at terminations of tapes, sleeves or discs of the appropriate colours prescribed in *Table 12.1*.

Binding and sleeving for cables should comply with BS 3858 where appropriate (Reg. 514-06-01).

Bare conductors should be identified in a similar manner by tapes, sleeves and discs or by painting with the appropriate colours specified in *Table 12.1* (Reg. 514-06-03).

Flexible cable and cord shall be identifiable throughout its length as appropriate to its functions as indicated in *Table 12.2*. Such cables shall not use the

Table 12.2 Colour identification of cores of flexible cables and flexible cords

Number of cores	Function of core	Colour(s) of core
1	Phase	Brown (Note 1)
	Neutral	Blue
	Protective	Green-and-yellow
2	Phase	Brown
	Neutral	Blue (Note 2)
3	Phase	Brown (Note 3)
	Neutral	Blue (Note 2)
	Protective	Green-and-yellow
4 or 5	Phase	Brown or black (Note 4)
	Neutral	Blue (Note 2)
	Protective	Green-and-yellow

Notes to Table 12.2.

(1) Or any other colour not prohibited by Regulations 514-03-01 and 514-07-02, except blue.

(2) The blue core can be used for functions other than the neutral in circuits which do not incorporate a neutral conductor, in which case its function must be appropriately identified during installation; provided that the blue core is not in any event used as a protective conductor. If the blue core is used for other functions, the coding L1, L2, L3, or other coding where appropriate should be used.

(3) In three-core flexible cables or cords not incorporating a green and yellow core, a brown core and a black core may be used as phase conductors.

(4) Where an indication of phase rotation is desired, or it is desired to distinguish the functions or more than one phase core of the same colour, this must be by use of numbered or lettered (not coloured) sleeves to the core, preferably using the coding L1, L2, L3 or other coding where appropriate. (Reg. 514-07-01).

following core colours: green alone; yellow alone; or any bi-colour other than the colour combination green and yellow (Regs 514-07-01 and 514-07-02).

Fire alarm and emergency lighting circuits (Category 3 circuits) must be segregated from all other cables and from each other in accordance with BS 5266 and BS 5839. Telecommunication circuits have to be segregated in accordance with BS 6701 as appropriate (Reg. 528-01-04).

Bends. The internal radius of every bend in a non-flexible cable must be such as not to cause damage to the conductors and cable (Reg. 522-08-03).

Switchgear. No fuse or, excepting where linked, switch or circuit-breaker shall be connected in the neutral conductor of TN or TT systems (Reg. 530-01-02).

When using a residual current device its operating current shall comply with Section 413 as appropriate to the type of system earthing.

In a TN system where for certain equipment in a certain part of the installation, one or more of the conditions in Regulation 413-02-08 cannot be satisfied, an r.c.d. can be used as a protective device. There are certain other conditions which have to be fulfilled (Reg. 531-03-01). Where a single r.c.d. is used to protect part of a TT installation, it must be placed at the origin of the installation, unless the part between the origin and the device complies with the requirements for protection by the use of Class II equipment or equivalent insulation. The Regulation applies to each origin if there is more than one (Reg. 531-04-01).

For an IT system where an r.c.d. is used for protection and disconnection following a first fault is not envisaged, the non-operating residual current of the device has to be at least equal to the current which circulates on the first fault to earth of negligible impedance affecting a phase conductor (Reg. 531-05-01).

Earthing arrangements. The types of earth electrode recognized by the Regulations are: earth rods or pipes; earth tapes or wires; earth plates; underground structural metalwork embedded in foundations; welded metal reinforcement of concrete (except pre-stressed concrete) embedded in the earth; lead sheaths and other metallic coverings of cables not precluded by the Regulations; other suitable underground metalwork (Reg. 542-02-01). Certain precautions are specified when using some of these systems (Regs 542-02-02 to 02-05).

Cross-sectional areas of earthing conductors where buried in the ground shall not be less than the figures in Table 54A of the Regulations (Reg. 542-03-01). Every earthing conductor also has to comply with Section 543.

Regulations 542-03-02 and 547-01-01 banning the use of aluminium and copperclad aluminium for final connections to earth electrodes or for bonding connections have been deleted by Amendment No. 1.

Protective conductors. The cross-sectional area of every protective conductor, other than an equipotential bonding conductor, shall be calculated in accordance with Regulation 543-01-03 or selected in accordance with Regulation 543-01-04. If this conductor is not an integral part of a cable, and is not formed by conduit, ducting or trunking, and is not contained in an enclosure formed by a wiring system, the cross-sectional area shall not be less than 2.5 mm^2 copper equivalent if protection against mechanical damage is provided and 4 mm^2 copper equivalent if mechanical protection is not provided (Reg. 543-01-01). The reader is also referred to Regulation 543-03-01. For an earthing conductor buried in the ground Regulation 542-03-01 applies.

The formula for calculating the cross-sectional area is $S = \sqrt{(I^2 t)}/k$ mm^2, where S is cross-sectional area in mm^2, I is the value (a.c., r.m.s.) of the fault current for a fault of negligible impedance that can flow through the protective device in amperes (with certain provisions), t is the operating time of the disconnecting device in seconds corresponding to the fault current I amperes, and k is a factor taking account of the resistivity, temperature coefficient and heat capacity of the conductor material, and the appropriate initial and final temperatures. The values of k are tabulated in the Regulations in Tables 54B–F (Reg. 543-01-03).

When it is desired not to calculate the minimum cross-sectional area of a protective conductor the value may be selected in accordance with *Table 12.3*. Where the application of *Table 12.3* produces a non-standard size, a conductor having the nearest larger standard cross-sectional area shall be used (Reg. 543-01-04).

Table 12.3 Minimum cross-sectional area of protective conductor in relation to the cross-sectional area of associated phase conductor

Cross-sectional area of phase conductor (S)	Minimum cross-sectional area of the corresponding protective conductor	
	If the protective conductor is of the same material as the phase conductor	If the protective conductor is not the same material as the phase conductor
mm^2	mm^2	mm^2
$S \leq 16$	S	$\dfrac{k_1 S}{k_2}$
$16 < S \leq 35$	16	$\dfrac{k_1 16}{k_2}$
$S > 35$	$\dfrac{S}{2}$	$\dfrac{k_1 S}{k_2 2}$

Types of protective conductors. Flexible or pliable conduit cannot be used as a protective conductor; neither can a gas or oil pipe (Reg. 543-02-01). It may consist of one or more of the following: a single-core cable; a conductor in a cable; an insulated or bare conductor in a common enclosure with insulated live conductors; a fixed bare or insulated conductor; a metal covering, for example, the sheath, screen or armouring of a cable; a metal conduit or other enclosure or electrically continuous support system for conductors; and an extraneous conductive part complying with Regulation 543-02-06. (Reg. 543-02-02).

A protective conductor of the types described above and of cross-sectional area 10 mm^2 or less, shall be of copper (Reg. 543-02-03). Where a metal enclosure or frame of an l.v. switchgear or control gear assembly or busbar trunking system is employed as a protective conductor it has to satisfy three requirements which are laid down (Reg. 543-02-04).

Where the protective conductor is formed by conduit, trunking, ducting, or the metal sheath and/or armour of cables, the earthing terminal of each accessory shall be connected by a separate protective conductor to an earthing terminal incorporated in the associated box or other enclosure (Reg. 543-02-07).

PEN conductors. The provisions on combined protective and neutral conductors forming the section of Regulation 546 are applicable only under certain conditions including that of privately owned generating plant, or where the installation is supplied by a privately owned transformer or converter in such a way that there is no metallic connection with the general public supply, or where special authorization has been granted (Reg. 546-02-01).

Conductors of the following types may serve as PEN conductors provided that the part of the installation concerned is not supplied through a residual current device:

(i) for fixed installations, conductors of cables not subject to flexing and having a cross-sectional area not less than $10 \, mm^2$ for copper, or $16 \, mm^2$ for aluminium

(ii) the outer conductor of concentric cables where that conductor has a cross-sectional area not less than $4 \, mm^2$ in a cable complying with an appropriate British Standard and selected and erected in accordance with Regulations 546-02-03 to 02-08 (Reg. 546-02-02).

The conductance of the outer conductor of a concentric cable is specified in Regulation 546-02-04.

Dimensions of equipotential bonding conductors. Except where PME conditions apply the cross-sectional area of main equipotential bonding conductors shall not be less than half that of the earthing conductor of the installation, subject to a minimum of $6 \, mm^2$. The cross-sectional area need not exceed $25 \, mm^2$ if the bonding conductor is of copper or a cross-sectional area affording equivalent conductance in other materials. Where PME conditions apply the main equipotential bonding conductor must be selected in accordance with the neutral conductor of the supply and Table 54H of the Regulations (Reg. 547-02-01). The position of the main bonding connections to water or gas services is specified in Regulation 547-02-02.

Plugs and socket-outlets. In low voltage circuits these must conform to British Standards listed in *Table 12.4* (Regs 553-01-02 to 553-02-02).

Table 12.4 Plugs and socket-outlets for low voltage circuits

Type of plug and socket-outlet	Rating, amperes	British Standard
Fused plugs and shuttered socket-outlets, 2-pole and earth for a.c.	13	BS 1363 (fuses to BS 1362)
Plugs, fused or non-fused and socket-outlets, 2-pole and earth	2, 5, 15, 30	BS 546 (fuses, if any to BS 646)
Plugs, fused or non-fused, and socket-outlets, protected type, 2-pole with earthing contact	5, 15, 30	BS 196
Plugs and socket-outlets (theatre type)	15	BS 5550 subsection 7.3.1
Plugs and socket-outlets (industrial type)	16, 32, 63, 125	EN 60309-2

Part 6. Special Installations or Locations

A completely new section was introduced into the sixteenth edition (BS 7671 : 1992) as Part 6 with the title as above. It covers what are termed 'Special installations or locations'. Protective measures appropriate to these special cases are described in the section dealing with Part 6, below. These are locations containing a bath tub or shower basin; swimming pools; locations containing a hot air sauna heater; construction site installations; electrical installations of agricultural and horticultural premises; restrictive conductive locations; equipment having high protective conductor currents; electrical installations in caravans and motor caravans (Division one) and in caravan parks (Division two); and highway power supplies and street furniture. This has been amended and added to in the 2001 edition of BS 7671. A section is reserved for marinas and there is a further unnamed section reserved for future use.

The particular requirements for these special installations or locations supplement or modify the general requirements contained in other parts of the Regulations (Reg. 600-01). The absence of references to the exclusion of a chapter, section or clause means that the corresponding general Regulations are applicable.

Locations containing a bath or shower. This section highlights some of the Regulations relating to bath tubs, shower basins and their surroundings where the risk of electric shock is increased by a reduction in body resistance and contact of the body with earth potential. Such conditions apply to most domestic dwellings. The Regulations do not apply to emergency facilities in industrial areas and laboratories. Special requirements may be necessary for a location containing a bath for medical treatment.

In April 2000, Amendment No. 3 to BS 7671 : 1992 effected a complete revision of the existing Section 601, 'Locations containing a bath tub or shower basin'. BS 7671 : 2001 incorporated this amendment with further minor revisions and editorial corrections.

The major change brought about by the new (in 2000) Section 601 was the introduction of the concept of zones. Four zones: zone 0, zone 1, zone 2 and zone 3 are defined in Regulation 601-02-01. These take into account walls, doors, fixed partitions and floors, where these effectively limit the extent of a zone. Diagrams, 601A and 601B (in the Regulations), are given as examples to assist interpretation of the definitions. These are reproduced here as *Figures 12.6* and *12.7*.

It is worthwhile, and avoids the risk of any ambiguity, if these definitions are examined in full.

Zone 0 is the interior of the bath tub or shower basin.

In a location containing a shower without a basin, zone 0 is limited by the floor and by the plane 0.05 m above the floor. In this case:

(i) where the shower head is demountable and able to be moved around in use, zone 0 is limited by the vertical plane(s) at a radius of 1.2 m horizontally from the water outlet at the wall, or
(ii) where the shower head is not demountable, zone 0 is limited by the vertical plane(s) at a radius of 0.60 m from the shower head.

Zone 1 is limited by:

(i) the upper plane of zone 0 and the horizontal plane 2.25 m above the floor, and

240

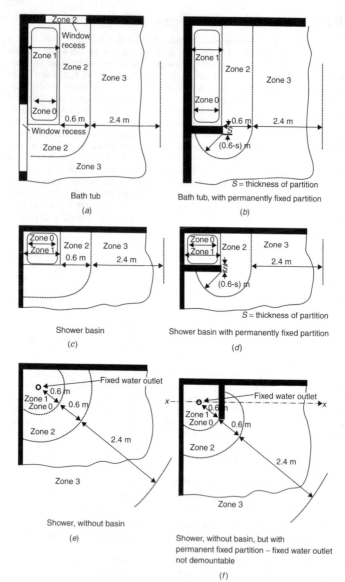

Figure 12.6 Examples of zone dimensions (plan) NOT TO SCALE (See Regulation 601-02-01 for definitions of zones)

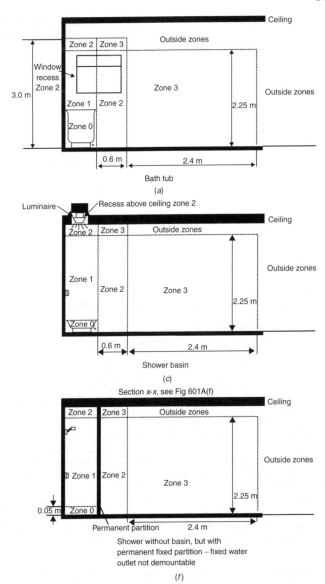

(a) Bath tub

(c) Shower basin

Section *x-x*, see Fig 601A(f)

(f) Shower without basin, but with permanent fixed partition – fixed water outlet not demountable

* Zone 1 if the space is accessible without the use of a tool.

Spaces under the bath, accessible only with the use of a tool, are outside the zones.

Figure 12.7 Examples of zone dimensions (elevation) NOT TO SCALE (See Regulation 601-02-01 for definitions of zones)

(ii) (a) by the vertical plane(s) circumscribing the bath tub or shower basin and includes the space below the bath tub or shower basin where that space is accessible without the use of a tool, or

    (b) for a shower without a basin and with a demountable shower head able to be moved around in use, the vertical plane(s) at a radius of 1.2 m from the water outlet at the wall, or

    (c) for a shower without a basin and with a shower head which is not demountable, the vertical plane(s) at a radius 0.60 m from the shower head.

Zone 2 is limited by:

(i) the vertical plane(s) external to zone 1, and parallel vertical plane(s) 0.60 m external to zone 1, and
(ii) the floor and the horizontal plane 2.25 m above the floor.

In addition, where the ceiling height exceeds 2.25 m above the floor, the space above zone 1 up to the ceiling or a height of 3.0 m above the floor, whichever is lower, is zone 2.

Zone 3 is limited by:

(i) the vertical plane(s) external to zone 2 and the parallel vertical plane(s) 2.40 m external to zone 2, and
(ii) the floor and the horizontal plane 2.25 m above the floor.

In addition, where the ceiling height exceeds 2.25 m above the floor, the space above zone 2 up to the ceiling or a height of 3.0 m above the floor, whichever is lower, is zone 3.

The protective measures to be applied are then determined according to the zone. In zone 0, only protection by SELV at a nominal voltage not exceeding 12 V a.c. r.m.s. or 30 V ripple-free d.c. is permitted (Reg. 601-05-01). The measures of protection by obstacles, placing out of reach, by non-conducting location and by earth-free local equipotential bonding are not permitted (Regs 601-05-02 and 601-05-03).

No switchgear or accessories may be installed in zone 0.

Where SELV or PELV is used, whatever the nominal voltage, protection against direct contact is required either by barriers or enclosures affording at least IP2X or IPXXB degree of protection or insulation capable of withstanding a type-test voltage of 500 V r.m.s. a.c. for 1 minute.

Swimming pools. This covers swimming and paddling pools and their surrounding zones; special requirements may be necessary for swimming pools for medical use. Three zones are identified in Regulation 602-02. Zone A is the interior of the basin, chute or flume and includes the portions of essential apertures in its walls and floor which are accessible to persons in the basin.

Zone B is limited (a) by the vertical plane 2 m from the rim of the basin, and (b) by the floor or surface expected to be accessible to persons, and (c) by the horizontal plane 2.5 m above that floor or surface, except where the basin is above ground, when it shall be 2.5 m above the level of the rim of the basin. This Zone also includes other areas where applicable like diving and spring boards, starting blocks or a chute and the immediate vicinity around such areas.

Zone C is limited by (a) the vertical plane circumscribing Zone B and the parallel vertical plane 1.5 m external to Zone B, and (b) by the floor or surface expected to be occupied by persons and the horizontal plane 2.5 m above that floor or surface.

Regulations cover protection for safety and selection and erection of equipment. For example, where SELV is used as protection against electric shock, irrespective of the nominal voltage, protection against direct contact has to be provided either by barriers and enclosures affording at least IP2X or IPXXB degree of protection, or by insulation able to stand a test voltage of 500 V d.c. for 1 minute (Reg. 602-03-01). In Zones A and B only the protective measure against electric shock by SELV (Reg. 411-02) at a nominal voltage not exceeding 12 V a.c. r.m.s. or 30 V d.c. can be used, the safety source being installed outside the Zones A, B and C with certain specified exceptions (Reg. 602-04-01).

Some protective measures cannot be used in any zone: they include protection by means of obstacles (Reg. 412-05); by means of placing out of reach (Reg. 412-05); by means of a non-conducting location (Reg. 413-04); and by means of earth-free local equipotential bonding (Reg. 413-05). These prohibitions form Regulation 602-04-02.

In Zones A and B surface wiring systems must not employ metallic conduit or metal trunking or an exposed metallic cable sheath or an exposed earthing or bonding conductor (Reg. 602-06-01). Zones A and B shall contain only wiring necessary to supply equipment located in those zones (Reg. 602-06-02). Accessible metal junction boxes must not be installed in Zones A and B (Reg. 602-06-03).

A socket-outlet, switch or accessory is permitted in Zone C only under certain conditions which are specified in Regulation 602-07-02. Socket-outlets must comply with EN 60309-2.

Hot air saunas. The particular requirements of this section apply to locations in which hot air sauna heating equipment to EN 60335-2-53 is installed (Reg. 603-01-01). Four temperature classification zones are illustrated in Figure 603A (in the Regulations).

Where SELV is used to protect against electric shock, irrespective of the nominal voltage, one has to provide protection against electric shock by insulation able to withstand a test voltage of 500 V d.c. for 1 minute or barriers or enclosures affording at least an IP24 or IPX4B degree of protection (Reg. 603-03-01). Two methods of protection against direct contact are prohibited, by using obstacles or placing out of reach (Reg. 603-04-01). Two methods of protection against indirect contact are also prohibited, non-conducting location and earth-free local equipotential bonding (Reg. 603-05-01). All equipment used must have at least an IP24 degree of protection (Reg. 603-06-01). Equipment characteristics for the four temperature zones A, B, C and D are indicated (Reg. 603-06-02). As regards wiring systems only flexible cords complying with BS 6141 having 150°C rubber insulation can be employed and it must be mechanically protected with material which complies with Regulation 413-03-01 (Reg. 603-07-01).

Construction site. This section applies to installations provided for the purpose of electricity supply for the following works: new building construction, repair, alterations, extensions or demolition of existing buildings; engineering construction; earthworks; and similar works (Reg. 604-01-01). These requirements do not apply to installations in construction site offices, cloakrooms, meeting rooms, canteens, restaurants, dormitories, toilets, etc. where the general requirements of the Regulations apply. Construction site fixed installations are limited to the assembly of the main switchgear and principal protective devices. Installations on the load side of such assemblies and transportable electrical equipment are covered by the requirements of this section (Reg. 604-01-02).

There is a limit to the voltages one can use which are as follows: portable hand lamps in confined or damp locations – 25 V single-phase SELV or 50 V single-phase SELV; portable hand lamps for general use and portable hand-held tools and local lighting up to 2 kW – 110 V single-phase centre point earthed; small mobile plant up to 3.75 kW – 110 V three-phase star point earthed; fixed floodlighting – 230 V single-phase; and fixed and movable equipment above 3.75 kW – 400 V three-phase. High voltage supplies can be used for large equipment where necessary (Reg. 604-02-02).

Considerable space is devoted to the requirements for protection for safety against indirect contact for TN, TF and IT systems. An important Regulation 604-03-01 states that where an alternative system is available an IT system shall not be used. If it is, then permanent earth monitoring must be provided. Where the protection against indirect contact is by earthed equipotential bonding and automatic disconnection of supply (Regs 413-02 and 471-08 as appropriate to the type of earthing system) then Regulations 604-06 to 604-08 apply.

For a TN system many of the Regulations in Section 413 of the fifteenth edition are replaced by new Regulations contained in Section 604. For example, Regulation 413-02-08 Table 41A is replaced by a new table, Table 604A, where the maximum disconnection times are appreciably reduced. For example, for U_o 120 V the time drops from 0.8 to 0.35 s (Reg. 604-04-01). For a TT system the formula in Regulation 413-02-20 is replaced by $R_a I_a \leq 25$ V (Reg. 604-05-01). This is the only change for such a system. For an IT system (see the preceding paragraph), the formula in Regulation 413-02-03 is replaced by $R_b I_a \leq 25$ V (Reg. 604-06-01). Also Table 41E in Regulation 413-02-26 is replaced by Table 604E and the definition of I_a is altered to relate to the new table (Reg. 604-06-02).

As regards wiring systems there is a requirement to protect cables that run across site roads or walkways (Reg. 604-10-02). All assemblies used for site distribution networks must comply with BS 4363 and EN 60439-4 (Reg. 604-09-01). Except for assemblies covered by Regulation 604-09-01 equipment must have a degree of protection appropriate to the external influence; every socket-outlet must also form part of an assembly complying with this Regulation (Regs 604-09-02 and 604-12-01). All plugs and socket-outlets and cable couplers must comply with EN 60309-2 (Regs 604-12-02 and 604-13-01).

Agricultural and horticultural premises. The requirements laid down in Section 605 apply to all parts of fixed installations of these types of premises both outdoors and indoors, and to where livestock is kept. Where the premises include dwellings solely for human habitation they are excluded from the scope of this section.

Where SELV is used, irrespective of the nominal voltage, protection against direct contact must be provided either by barriers or enclosures to at least IP2X or IPXXB insulation capable of withstanding a test voltage of 500 V d.c. for 1 minute (Reg. 605-02-02). All socket-outlet circuits, except those supplied from an SELV supply, must be protected by an r.c.d. complying with the appropriate British Standard and having the characteristics specified in Regulation 412-06-02(ii) (Reg. 605-03-01).

Where protection against indirect contact for livestock is provided by earthed equipotential bonding and automatic disconnection Regulations 605-05 to 605-09 apply and these cover TN systems (Reg. 605-04-01). In these Regulations many of the requirements of the Regulations in Section 413 are replaced including provision of new tables. For TT and IT systems the same conditions apply about the formulas $R_a I_a$ and $R_b I_d$ as for construction sites (Regs 605-06-01 and 605-07-01). For an IT system Table 41E is replaced by Table 605E and the definition of I_a is altered similar to that for construction sites (Reg. 605-07-02).

Restrictive conductive locations. Section 606 deals only with installations of this nature and not to locations in which freedom of movement is not physically constrained. Regulation 606-02-01 specifies that where protection by the use of SELV is used (Reg. 411-02) regardless of the voltage, protection against direct contact must be provided by a barrier or enclosure affording at least IP2X or IPXXB degree of protection or insulation able to withstand a test voltage of 500 V d.c. for 1 minute. One cannot use obstacles or placing out of reach as a means of protection against direct contact (Reg. 606-03-01). Protection against indirect contact can be provided by (i) SELV (Reg. 606-2); (ii) electrical separation; (iii) automatic disconnection in which case a supplementary equipotential bonding conductor shall be provided; (iv) use of Class II equipment. Certain other requirements are specified for some of these methods (Reg. 606-04-01).

Equipment having high protective conductor currents. The requirements of Section 607 apply to installations where a piece of equipment has protective conductor current exceeding 3.5 mA. It also applies to circuits for which the total protective conductor current exceeds 10 mA. It includes information technology (IT) equipment, industrial telecommunications equipment with r.f. interference suppression filtering and heating elements (Reg. 607-01-01). Generally no special measures are necessary where the protective conductor current is less than 3.5 mA.

The strategy in providing protection for safety in the situations where high protective conductor currents exist is wherever possible to ensure that the equipment is supplied via a permanent connection rather than plugs and sockets (Reg. 607-02-03). Flexible cables are permitted, but where these are used the plug and socket must comply with BS EN 60309-2. The integrity of the protective conductor must be ensured by using $2.5 \, mm^2$ cross-section minimum for plugs rated 16 A and not less than $4 \, mm^2$ for plugs rated above 16 A or the protective conductor must be not less than half the cross-section of the phase conductor. Alternatively the equipment may be connected to the supply using a protective conductor complying with Section 543 having an earth monitoring system to BS 4444 which will disconnect the equipment in the event of a continuity fault occurring in the earth conductor (Reg. 607-02-03).

For every final circuit and distribution circuit intended to supply equipment where the protective conductor current is likely to exceed 10 mA the integrity of the protective conductor shall be ensured by using either a single conductor of not less than $10\,mm^2$ or not less than $4\,mm^2$ and enclosed to provide additional protection against damage. In both cases the protective conductor must comply with 543-02 and 543-03. Two protective conductors of different types may be used, say, a metallic conduit plus an additional conductor. If two conductors of a multicore cable are used, their total cross-section must be not less than $10\,mm^2$, alternatively metallic sheath, armour or wire braid complying with 543-02-05 may be used. Other options include use of an earth monitoring system complying with BS 4444, as above, or connection of the equipment by means of a double wound transformer with the protective conductor of the incoming supply being connected to the exposed conductive parts of the equipment and to a point of the secondary winding of the transformer.

Socket-outlet final circuits. For a final circuit with a number of socket outlets connected in a ring where it is known or reasonably expected that the total protective conductor current will exceed 10 mA the circuit shall be provided with a high integrity protective conductor as described previously.

In all instances where two protective conductors are used in order to comply with the above requirements, the ends of the protective conductors must be terminated independently of each other at all connection points throughout the circuit. This requires all accessories to be provided with two separate earth terminals.

Caravans, motor caravans and caravan parks. Section 608 dealing with caravans and their parks is divided into two divisions, the first dealing with caravans themselves and the second with parks. Division one restricts the Regulations to caravans and motor caravans where the supply does not exceed 250/440 V. Electrical circuits and equipment covered by Road Vehicles Lighting Regulations 1989 and installations covered by BS EN 1648, Parts 1 and 2, are not covered (Reg. 608-01-01). Certain requirements of Section 601 (bath tubs and shower basins) also apply to caravans and motor caravans. They do not apply to certain installations which are identified in BS 7671.

Protection against direct contact cannot be provided by obstacles or by placing out of reach (Reg. 608-02-01); and similarly for protection against indirect contact, non-conducting locations, earth-free equipotential bonding or electrical separation are prohibited (Reg. 608-03-01). If automatic disconnection of supply is employed a double pole r.c.d. must be provided and the wiring system must include a circuit protective conductor which must be connected to (i) the protective contact of the inlet; and (ii) the exposed conductive parts of the electrical equipment; and (iii) the protective contacts of the socket-outlets (Reg. 608-03-02). A final circuit must be provided with overcurrent protection capable of disconnecting all live conductors of that circuit (Reg. 608-04-01).

Wiring systems can use (i) flexible single-core insulated conductors in non-metallic conduits; (ii) stranded insulated conductors with a minimum of seven strands in non-metallic conduits; or (iii) sheathed flexible cables (Reg. 608-06-01). Pliable polyethylene conduits shall not be used. Minimum conductor size is $1.5\,mm^2$. The limit of $6\,mm^2$ in Regulation 543-03-02 does not apply and all protective conductors, irrespective of cross-sectional area, have to be insulated (Reg. 608-06-03). No electrical equipment must be installed in a compartment intended for fuel storage. A notice of durable material has to be fixed near the main isolating switch inside the caravan or motor caravan and carry specified text headed 'Instructions for electricity supply' (Reg. 608-07-05). There are a number of Regulations dealing with socket-outlets and luminaires.

The means of connection to the caravan or motor caravan switch socket-outlet are specified (Reg. 608-08-08). This Regulation also includes Table 608A giving details of the cross-sectional areas of flexible cords and cables used for caravan connectors from 16 A up to 100 A.

Division two of Section 608 dealing with caravan parks also prohibits the use of protection by obstacles, placing out of reach, non-conducting locations, earth-free equipotential bonding and electrical separation (Regs 608-10-01 and 608-11-01). Underground cables should, as far as possible, be used to connect caravan pitch supply equipment. Overhead conductors can be employed subject to certain conditions (Reg. 608-12-03).

Installation of highway power supplies, street furniture and street located equipment. These Regulations do not apply to supplier's works as defined by the Electricity Supply Regulations 1988 (as amended) in accordance with Regulation 110-02. This comparatively short Section 611 deals with protection against electric shock; devices for isolation and switching; identification of cables; external influences; and temporary supplies. There are a number of protective measures which cannot be used with highway power supplies and street furniture.

Part 7. Inspection and Testing

Every installation must be inspected and tested to verify that the Regulations have been met. This should be an ongoing activity during erection and on completion. Precautions must be taken to avoid danger to persons and damage to property and to installed equipment during inspection and testing.

Regulation 712-01-03 provides a checklist for this inspection.

Regulations 713-02 to 713-13 list the tests to be carried out and the order in which these should be performed. These include: continuity of protective conductors and equipotential bonding, continuity of final ring circuits, insulation resistance, site applied insulation, separation of circuits, integrity of barriers, insulation of non-conducting floors and walls, correct polarity of fuses, switches and lampholders, earth electrode resistance and earth loop impedance. Prospective fault current should be measured, calculated or determined by another method. Finally equipment must be subjected to the appropriate functional testing.

The above requirements also apply to alterations and additions to an existing installation.

Periodic inspection and testing. Periodic inspection and testing should be carried out so far as is reasonably practicable to determine whether an installation is in a satisfactory condition for continued service. This should be without dismantling or with partial dismantling, as required. The frequency of inspection and testing should be determined having regard to the type of installation, its use and operation, quality of maintenance and external influences.

Continuous monitoring by suitably skilled persons with the keeping of appropriate records could serve in lieu of periodic inspection and testing.

Certification and reporting. On completion of the verification of a new installation or changes to an existing installation an Electrical Certificate should be produced. Appendix 6 of the Regulations provides a model certificate. The certificate should adequately identify the circuits covered and the tests carried out.

An inspection report should be raised upon completion of periodic inspection and testing of an existing installation.

All such documentation should be signed and authenticated by a competent person or persons.

Conventional Circuit Arrangements

This section gives details of conventional circuit arrangements which satisfy the requirements of Chapter 46 for isolation and switching, together with the requirements as regards current-carrying capacities of conductors prescribed in Chapter 52 and Appendix 4. It is the responsibility of the designer and installer when adopting them to take the appropriate measures to comply with the other chapters or sections of the Regulations that are appropriate. The following information has been extracted from Appendix E of Guidance Note No. 1.

Circuit arrangements other than those detailed below are not precluded where they are specified by a suitably qualified electrical engineer, in accordance with the general requirements of the Regulation 314-01-03. The conventional circuit arrangements are:

A. Final circuits using socket-outlets complying with BS 1363.
B. Final circuits using socket-outlets complying with BS 196.
C. Final radial circuits using socket-outlets complying with BS 4343.
D. Cooker final circuits in household premises.

Circuits under A. A ring or radial circuit, with spurs if any, feeds permanently connected equipment and an unlimited number of socket-outlets. The floor area served by the circuit is determined by the known or estimated load but does not exceed the value given in *Table 12.5*.

For household installations a single 30 A ring circuit may serve a floor area of up to 100 m² but consideration should be given to the loading in kitchens which may require a separate circuit. For other types of premises, final circuits complying with *Table 12.5* may be used subject to conditions laid down in the Appendix.

The number of socket-outlets shall be such as to ensure compliance with the Regulation which states that every portable appliance must be able to be fed from an adjacent and conveniently accessible socket-outlet. Account should be taken of the length of flexible cord normally fitted to the majority of appliances and luminaires. Each socket-outlet of a twin or multiple socket-outlet unit is regarded as one socket-outlet.

Conductor size (*circuits under A*). The minimum size of conductor in the circuit and in non-fused spurs is given in *Table 12.5*. If cables of more than two circuits are bunched together or the ambient temperature exceeds

Table 12.5 Final circuits using BS 1363 socket-outlets

Type of circuit		Overcurrent protective device		Minimum conductor size*, mm²		Maximum floor area served, m²
		Rating, A	Type	Copper conductor rubber- or PVC-insulated cables mm²	Copper conductor mineral-insulated cables mm²	
A1	Ring	30 or 32	Any	2.5	1.5	100
A2	Radial	30 or 32	Cartridge fuse or circuit-breaker	4	2.5	50
A3	Radial	20	Any	2.5	1.5	20

*The tabulated values of conductor size may be reduced for fused spurs.

$30°C$, the size of the conductor shall be increased by applying the appropriate correction factors from Appendix 4 of the Regulations such that the size then corresponds to a current-carrying capacity not less than: 20 A for circuit A1; 30 A or 32 A for circuit A2; 20 A for circuit A3.

The conductor size for a fused spur is determined from the total current demand served by that spur, which is limited to a maximum of 13 A. The minimum conductor size for a spur serving socket-outlets is: $1.5 \, mm^2$ for rubber- or PVC-insulated cables with copper conductors; $2.5 \, mm^2$ for rubber- or PVC-insulated cables with copperclad aluminium conductors; $1 \, mm^2$ for mineral-insulated cables with copper conductors.

Spurs (*circuits under A*). The total number of fused spurs is unlimited but the number of non-fused spurs shall not exceed the total number of socket-outlets and stationary equipment connected directly in the circuit.

A non-fused spur feeds only one single or one twin socket-outlet or one permanently connected item of equipment. Such a spur is connected to a ring circuit at the terminals of socket-outlets, or at joint boxes, or at the origin of the circuit in the distribution board.

A fused spur is connected to a ring circuit through a fused connection unit; the rating of the fuse in the unit not exceeding that of the cable forming the spur, and in any event, not exceeding 13 A.

Circuits under B. A ring or radial circuit, with fused spurs if any, feeds equipment the maximum demand of which, allowing for diversity, is known or estimated not to exceed the rating of the overcurrent protective device and in any event does not exceed 32 A. No diversity is allowed for permanently connected equipment. The number of socket-outlets is unlimited and the total current demands of points served by a fused spur shall not exceed 16 A. The overcurrent protective device rating shall not exceed 32 A.

The size of conductor is obtained by applying the appropriate correction factors from Appendix 4 of the Regulations and is such that it then corresponds to a current-carrying capacity of:

(i) For ring circuits – not less than 0.67 times the rating of the overcurrent device.
(ii) For radial circuits – not less than the rating of the overcurrent protective device.

The conductor size for a fused spur is determined from the total demand served by that spur which is limited to a maximum of 16 A.

Spurs (*circuits under B*). A fused spur is connected to a ring circuit through a fuse connection unit, the rating of the fuse in the unit not exceeding that of the cable forming the spur and in any event not exceeding 16 A. Non-fused spurs are not used.

Circuits under C. A radial circuit feeds equipment the maximum demand of which, having allowed for diversity, is known or estimated not to exceed the rating of the overcurrent protective device, and in any event does not exceed 20 A. The number of socket-outlets is unlimited. The overcurrent protective device has a rating not exceeding 20 A.

Conductor size is determined by applying appropriate correction factors from Appendix 4 of the Regulations and is such that it then corresponds to a current-carrying capacity not less than the rating of the overcurrent protective device. Socket-outlets have a rated current of 16 A.

Circuits under D. The circuit supplies a control switch or a cooker control unit which may incorporate a socket-outlet. The rating of the circuit is determined by the assessment of the current demand of the cooking appliance(s) and control unit socket-outlet, if any, in accordance with Table J1 of Appendix J.

A circuit of rating exceeding 15 A but not exceeding 50 A may supply two or more cooking appliances where these are installed in one room, subject to certain conditions.

Limitation of Earth Fault Loop Impedance

Guidance Note No. 1 details how to calculate the earth fault loop impedance Z_s in order to comply with the Regulation dealing with a.c. circuits. For cables having conductors of cross-sectional area not exceeding 35 mm^2 the inductance can be ignored so that Z_s is given by:

(i) For radial circuits

$$Z_s = Z_E + R_1 + R_2 \text{ ohms}$$

Where Z_E is that part of the earth fault loop impedance external to the circuit concerned; R_1 is the resistance of the phase conductor from the origin of the circuit to the most distant socket-outlet or other point of utilization; R_2 is the resistance of the protective conductor from the origin of the circuit to the most distant socket-outlet or other point of utilization.

(ii) For ring circuits without spurs

$$Z_s = Z_E + 0.25R_1 + 0.25R_2 \text{ ohms}$$

where Z_E is as described under (i) above; R_1 is now the total resistance of the phase conductor between its ends prior to them being connected together to complete the ring; R_2 is similarly the total resistance of the protective conductor.

Having determined Z_s the earth fault current I_F is given by

$$I_F = \frac{U_o}{Z_s} \text{ amperes}$$

where U_o is the nominal voltage to earth (phase to neutral voltage). From the relevant time/current characteristics the time for disconnection (t) corresponding to this fault current is obtained. Substituting for I_F, t and the appropriate k value in the equation given in Regulation 543-01-03 gives the minimum cross-sectional area of the protective conductor and this must be equal to or less than the size chosen.

When the cables are to Table 5 of BS 6004 or are other PVC-insulated cables to that standard, the Guidance Note gives in tabular form the maximum earth loop impedance for circuits having protective conductors of copper ranging from 1 to 16 mm^2 cross-section area and the overcurrent protective device is a fuse to BS 88 Part 2, BS 1361 or BS 3036. These tables also apply if the protective conductor is bare copper and in contact with cable insulated with PVC. For each type of fuse, two tables are given:

(i) Where the circuit concerned feeds socket-outlets and the disconnection time for compliance is 0.4 seconds, and

(ii) Where the circuit concerned feeds fixed equipment and the disconnection time for compliance is 5 seconds.

The graphs and tables are not reproduced here, being too numerous (totalling six); the reader is referred to the Guidance Note for details.

Cable Current-Carrying Capacities

Current-carrying capacities and voltage drops for cables and flexible cords are contained in Appendix 4. Typical methods of installation of cables and conductors are contained in a table ranging from single and multicore sheathed cables clipped direct to or lying on a non-metallic surface (No. 1 method) to single and multicore cables in an enclosed trench 600 mm wide by 760 mm deep (minimum dimensions) including 100 mm cover (No. 20 method). Correction factors for groups of more than one circuit of single-core cables or more than one multicore cable, and for cables in trenches and for mineral insulated cables installed on perforated trays are specified in other tables. These correction factors are applied as necessary to the current-carrying capacities and associated voltage drops of the types of cables covered.

There are many tables for copper conductor cables and somewhat fewer for cables with aluminium conductors. Cables with copper conductors include PVC-insulated, thermosetting, 85°C rubber-insulated, flexible cables and cords and mineral-insulated cables. Cables with aluminium conductors include PVC-insulated and thermosetting cables.

The current-carrying capacities take account of IEC Publication 364.5.523 (1983) so far as the latter is applicable. For types of cable not treated in the IEC Publication (e.g. armoured cables) the current-carrying capacities of the Appendix are based on data provided by ERA Technology and the British Cable Makers Confederation. Readers are referred to the Regulations for fuller details.

To comply with the requirements of Chapter 52 of the Regulations for the selection and erection of wiring systems in relation to risks of mechanical damage and corrosion the IEE On-site Guide to BS 7671 tabulates types of cable and flexible cord for particular uses and external influences. The tables are not intended to be exhaustive and other limitations may be imposed by the relevant regulations, in particular those concerning maximum permissible operating temperature.

An example of the type of information contained in these tables is as follows: thermoplastic (PVC)- or thermosetting (rubber)-insulated non-sheathed cables for fixed wiring can be used in conduits, cable ducting or trunking, but not when the enclosures are buried underground. Mineral-insulated cables are suitable for general use but additional precautions in the form of a PVC sheath are necessary when such cables are exposed to the weather or risk of corrosion, or where installed underground, or in concrete ducts.

Methods of Cable Support

Examples of methods of support for cables, conductors and wiring systems are described in Guidance Note No. 1 but other methods are not precluded when specified by a suitably qualified electrical engineer.

Cables. For cables supported on structures which are only subject to vibration of low severity and a low risk of mechanical impact the following conditions should be observed:

(i) For cables of any construction, installation in conduit, trunking or ducting without further fixing of the cables, precautions being taken against undue compression or other mechanical stressing of the insulation at the top of any vertical runs exceeding 5 m in length.

(ii) For sheathed and/or armoured cables installed in accessible positions, support by clips at spacings not exceeding the appropriate value stated in *Table 12.6*.

(iii) For sheathed and/or armoured cables in vertical runs which are inaccessible and unlikely to be disturbed (e.g. installations in an inaccessible cavity), supported at the top of a run by a clip and a rounded support of a radius not less than the appropriate value stated in Appendix I, Table I1 (of Guidance Note No. 1). For unarmoured cables, the length of run without intermediate support should not exceed 2 m for a lead-sheathed cable or 5 m for a rubber- or plastics-sheathed cable. Where these figures are intended to be exceeded the advice of the manufacturer of the cable should be obtained. Care should be taken that the cable does not bridge a cavity between external and internal faces in a manner which would transmit moisture.

In this Note there are recommendations for the support of cables in caravans, for flexible cords used as pendants and overhead wiring. For overhead wiring cables sheathed with rubber or PVC, support by a catenary wire is required either continuously bound up with the cable or attached thereto at intervals, these intervals not to exceed those given in Appendix I, Table 13 (of Guidance Note No. 1). Three other overhead wiring systems are also contained in the same Table.

Conduit and cable trunking. *Table 12.7* gives supporting spacings for rigid conduit and cable trunking. Conduit, embedded in the material of the building and pliable conduit embedded in the material of the building or in the ground need no further support. Otherwise the support of pliable conduit must be in accordance with *Table 12.7*. Cable trays should be supported in accordance with the installation designer's requirements.

Cables installed in conduit and trunking. This subject is dealt with in Appendix A of Guidance Note No. 1.

Methods of Testing

The standard methods of testing described in a Guidance Note are suitable for the corresponding testing prescribed in Part 7 of the Regulations. They are given as examples, and the use of other methods giving no less effective results is not precluded.

254

Table 12.6 Spacing of supports for cable in accessible positions the entire support derived from the clips

Overall diameter of cable*, mm	Maximum spacing of clips, mm							
	Non-armoured rubber-, plastics- or lead-sheathed cables				Armoured cables		Mineral-insulated copper-sheathed or aluminium-sheathed cables	
	Generally		In caravans					
	Horizontal†	Vertical†	Horizontal†	Vertical†	Horizontal†	Vertical†	Horizontal†	Vertical†
≤ 9	250	400			–	–	600	800
> 9 ≤ 15	300	400	150 (for all sizes)	250 (for all sizes)	350	450	900	1200
> 15 ≤ 20	350	450			400	550	1500	2000
> 20 ≤ 40	400	550			450	600	–	–

Note: For the spacing of supports for cables of overall diameter exceeding 40 mm, and for single-core cables having conductors of cross-sectional area 300 mm² and larger, the manufacturer's recommendations should be observed.

*For flat cables taken as the measurement of the major axis.

†The spacings stated for horizontal runs may be applied also to runs at an angle of more than 30° from the vertical. For runs at an angle of 30° or less from the vertical, the vertical spacings are applicable.

Table 12.7 Spacings of supports for conduits and cable trunking

(a) Conduits

Nominal size of conduit, mm	Maximum distance between supports, m					
	Rigid metal		Rigid insulation		Pliable	
	Horizontal	Vertical	Horizontal	Vertical	Horizontal	Vertical
Not exceeding 16	0.75	1.0	0.75	1.0	0.3	0.5
Exceeding 16 and not exceeding 25	1.75	2.0	1.5	1.75	0.4	0.6
Exceeding 25 and not exceeding 40	2.0	2.25	1.75	2.0	0.6	0.8
Exceeding 40	2.25	2.5	2.0	2.0	0.8	1.0

(b) Cable trunking

Cross-sectional area, mm^2	Maximum distance between supports, m			
	Metal		Insulating	
	Horizontal	Vertical	Horizontal	Vertical
Up to 700	0.75	1.0	0.5	0.5
Exceeding 700 and not exceeding 1500	1.25	1.5	0.5	0.5
Exceeding 1500 and not exceeding 2500	1.75	2.0	1.25	1.25
Exceeding 2500 and not exceeding 5000	3.0	3.0	1.5	2.0
Exceeding 5000	3.0	3.0	1.75	2.0

Note 1. The spacings tabulated allow for maximum fill of cables to the Wiring Regulations and the thermal limits specified in the relevant British Standards. They assume that the conduit or trunking is not exposed to other mechanical stress.

2. The above figures do not apply to lighting suspension trunking, or where special strengthening couplers are used. A flexible conduit is not normally required to be supported in its run. Supports should be positioned within 300 mm of bends or fittings.

Each leg of the ring circuit is identified. The phase conductor of one leg and the neutral conductor of the other leg are temporarily bridged. The resistance is measured between the remaining phase and neutral conductors, a finite reading confirms that there is no open-circuit on the ring conductors under test. These remaining conductors are then temporarily bridged together. The connections and instrument are indicated in the Guidance Note. The resistance between phase and neutral contacts at each socket-outlet around the ring is measured and noted. The readings obtained should be substantially the same, provided that no multiple loops exist over the length of the ring. Where the protective conductor is in the form of a ring, the test is repeated, transposing the circuit protective conductor with the phase conductors. The phase conductor from one leg of the ring is temporarily bridged with the circuit protective conductor of the other leg of the ring. The resistance is measured between the remaining phase conductor and the remaining unconnected circuit protective conductor at the origin of the circuit. A finite reading confirms that there is no open-circuit on the ring conductors under test. The remaining circuit protective conductor and phase conductor are then temporarily bridged together. The resistance is measured between the circuit protective conductor and phase conductor contacts at each socket-outlet around the ring. The readings obtained should be substantially the same, provided that no multiple loops exist, and the readings at the centre point of the ring are approximately equal to $(R_1 + R_2)$ for the circuit. This value should be recorded and, when corrected for temperature, may be used to calculate the earth fault loop impedance of the circuit Z_s to verify compliance with the requirements of Table 41B1 and Table 41B2 of the Wiring Regulations.

Where single-core cables are used, special precautions should be taken and these are outlined in the Guidance Note.

Continuity of protective conductors' main and supplementary bonding. Every protective conductor including any bonding conductor must be tested to ensure it is electrically sound and correctly connected.

For cables having conductors of cross-sectional area not exceeding 35 mm^2 their inductance can be ignored. Above that figure the inductance becomes significant and an appropriate a.c. instrument should be used for the measurement. The test methods detailed below, as well as checking the continuity of the protective conductor, also measure $(R_1 + R_2)$ which, when corrected for temperature, enables the designer to verify the calculated earth fault loop impedance Z_s. Use a low resistance ohmmeter for these tests.

Test method 1. Strap the phase conductor to the protective conductor at the distribution switchboard to include all the circuit. Then test between phase and earth terminals at each outlet in the circuit. The measurement $(R_1 + R_2)$ at the circuit's extremity should be recorded to verify compliance with the Wiring Regulations.

When the testing of ring circuit continuity including protective conductors is required, the test should be made prior to connecting supplementary bonds to the protective conductors.

Test method 2. Connect one terminal of the continuity tester to one test lead and connect this to the consumer's earth terminal. Then connect the other terminal of the continuity tester to another test lead and use this to make contact with the protective conductors at various points on the circuit, such as light fittings, switches, spur outlets, etc.

The resistance reading obtained by the above method includes the resistance of the test leads. The resistance of the test leads should be measured and deducted from any resistance reading obtained using this method. To test the bonding conductors' continuity use Test method 2.

Earth electrode resistance. After an earth electrode has been installed it is necessary to verify that the resistance meets the conditions of the Wiring Regulations for TT and IT installations. Use an earth electrode resistance tester for this test. Refer to Section 16 of Guidance Note No. 1.

Two test methods are described but we only cover one in this book. The test requires the use of two test spikes (electrodes), and is carried out in the following manner.

Connection to the earth electrode is made using terminals C1 and P1 of a four-terminal earth tester. Details are given to exclude test lead resistance. Connection to the temporary spikes is made as shown in *Figure 12.8*. The distance between the test spikes is important. In general, reliable results may be expected if the distance between the electrode under test and the current spike is at least ten times the maximum dimension of the electrode system, e.g. 30 m for a 3 m long electrode.

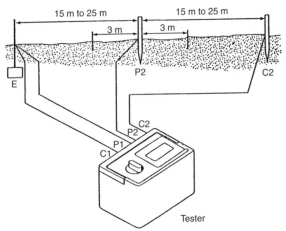

Figure 12.8 Measurement of earth electrode resistance

Three readings are taken: with the potential spike initially midway between the electrode and current spike, second, at a position 10% of the electrode-to-current spike distance back towards the electrode, and finally at a position 10% of the distance towards the current spike. By comparing the three readings, a percentage deviation can be determined. This is calculated by taking the average of the three readings, finding the maximum deviation of the readings from this average in ohms, and expressing this as a percentage of the average. The accuracy of the measurement using this technique is typically 1.2 times the percentage deviation of the readings. It is difficult to aim for a measurement accuracy better than 2%, and inadvisable to accept readings that differ by more than 5%. To improve the accuracy of the measurement to acceptable levels, the test must be repeated with a larger separation between the electrode and the current spike.

The instrument output current may be a.c. or reversed d.c. to overcome electrolytic effects. Because these testers employ phase-sensitive detectors the errors associated with stray currents are eliminated.

If the temporary spike resistances are too high, measures to reduce these will be necessary, such as driving the spikes deeper into the ground or watering with brine to improve contact resistance.

Earth fault loop impedance. The earth fault current loop (phase to earth loop) comprises the following parts, starting at the point of fault:

- the circuit protective conductor
- the main earthing terminal and earthing conductor
- for TN systems the metallic return path, or for TT and IT systems the earth return path
- the path through the earth and neutral point of the transformer
- the transformer winding and the phase conductor from the transformer to the point of fault.

The earth fault impedance Z_s should be determined at the farthest point of each circuit including socket-outlets, lighting points, submain cables and any other fixed equipment. The value obtained, after adjustment to take into account the effects of fault current, should not exceed that detailed in Table 41B or Table 41D, or should not exceed a value which might prevent conformity with the Regulations for r.c.d. protected circuits.

Z_e is measured using a phase earth loop impedance tester at the source of the installation supply.

The impedance measurement is made between the main phase supply and the main means of earthing with the main switch open or with all the circuits isolated. The means of earthing will be isolated from the installation earthed equipotential bonding for the duration of the test. Care should be taken to avoid any shock hazard to the testing personnel and other persons on the site while both establishing contact and performing the test. Use a loop impedance tester for this test. Refer to Section 16 of this Guidance Note.

While testing the continuity of protective conductors of radial circuits, or while testing the continuity of ring final circuits, a value of $(R_1 + R_2)$ will have been measured at ambient temperature. The measured values of $(R_1 + R_2)$ for the final circuit should be added to the appropriate values of $(R_1 + R_2)$ for any distribution circuits.

An alternative means to determine $(R_1 + R_2)$ is to measure the loop impedance value at the extremity of the final circuit, taking care to use the correct phase supply and protective conductor return for the circuit. This loop impedance less the Z_e value measured earlier can be taken as the value of $(R_1 + R_2)$. These $(R_1 + R_2)$ values will be used to determine the value of Z_s of the circuit under fault conditions, but must be first corrected for conductor temperature. This requires the use of correction factors dependent on the type of cable used in the circuit and ambient temperature at the time of continuity testing. Details of how to obtain them are given.

Residual current operated general purpose (non-delayed) devices to BS 4293. Prior to these r.c.d. tests it is essential for safety reasons that the earth loop impedance is tested to check the requirements have been met. Use an r.c.d. tester for these tests.

Tests are made on the load side of the r.c.d. between the phase conductor of the circuit protected and the associated c.p.c. The load supplied should be disconnected during the test. Since r.c.d. testers require a few milliamperes to operate the instrument, these are normally obtained from the phase and neutral of the circuit under test. When testing a three-phase r.c.d. protecting a 3-wire circuit, its neutral is required to be connected to earth. This means that the test current will be increased by the instrument supply current and will cause some devices to operate during the 50% test at a time when they should not operate. It is necessary to check the operating parameters of the r.c.d. with the manufacturers before failing the device.

Under certain conditions (which are given) these tests can result in a potentially dangerous voltage on exposed and extraneous-conductive-parts. Precautions must therefore be taken to prevent contact of persons or livestock with such parts.

Tests are as follows. (i) With a fault current flowing equivalent to 50% of the rated tripping current of the r.c.d. for a period of 2 s, the device should not open. (ii) With a fault current flowing equivalent to 100% of the rated tripping current of the r.c.d., the device will open in less than 200 ms. Where the r.c.d. incorporates an intentional time delay it should trip within the time range of 50% of the rated time delay plus 200 ms and 100% of the rated time range plus 200 ms. Because of the variability of the time delay it is not possible to specify a maximum test time. It is therefore imperative that the circuit protective conductor does not rise more than 50 V above earth potential. It is suggested that in practice a 2 s maximum test time is sufficient. (iii) Where the r.c.d. is used to provide supplementary protection against direct contact, with a test current of 150 mA the device should open in less than 40 ms. The maximum test time must not be longer than 50 ms. An integral test device is incorporated in each r.c.d. This device enables the mechanical parts of the r.c.d. to be checked. Tripping the r.c.d. by means of the above electrical tests and the integral test device establishes the following. (i) That the r.c.d is operating with the correct sensitivity. (ii) The integrity of the electrical and mechanical elements of the tripping device. Operation of the integral test device does not provide a means of checking: (i) the continuity of the earthing conductor or the associated circuit protective conductors, or (ii) any earth electrode or other means of earthing, or (iii) any other part of the associated installation earthing.

Battery insulation resistance testers. In the insulation resistance test covered by a Guidance Note the testing voltage may be derived from a hand-driven generator. There are now a number of battery-powered insulation testers available in which the high voltage is derived from special circuitry inside the unit.

The BM101, *Figure 12.9*, from Avo Megger Instruments Ltd has four scales, one reading insulation resistance values in megohms from 0 to 200 and infinity; the other three reading resistance from 0 to 1 MΩ, 0 to 200 Ω and 0 to 2 Ω respectively. The latter scale is suitable for continuity testing. A taut-band suspension indicator is used which is inherently robust so that the instrument is ideal for field use. A battery condition indicator is incorporated and a pushbutton is used to energize the circuit. A single 9 V dry battery provides the power source and the insulation testing voltage is 500 V d.c.

Figure 12.9 Type BM101 insulation and continuity tester (Avo Megger Instruments Ltd)

Intended for use where high values of insulation resistance may be met, such as h.v. cable, or where tests may be of long duration, is the BM8/2 battery-operated multi-voltage tester. It has four testing voltages selected by a rotary switch: 100 V, 250 V, 500 V and 1000 V. At 1000 V the range is 0–20 000 megohms and infinity. At 100 V the range is 0–2000 megohms and infinity. A black and white band on the scale indicates the battery condition. Like the BM101 it is pushbutton operated and has a power source provided by six 1.5 V cells.

Guidance Notes

1. Selection and erection
2. Isolation and switching
3. Inspection and testing
4. Protection against fire
5. Protection against electric shock
6. Protection against overcurrent
7. Special locations (not yet published)

13 Lighting

Virtually all buildings have electric lighting which serves two purposes. It helps us to recognize objects quickly and in sufficient detail to learn all we need to know about them, and it contributes to making buildings safe and pleasant places in which to work or take part in other activities. There must always be enough light to make objects visible but other factors are no less important. The directions from which light comes, the brightness and colour contrasts created between details of interest and their background, the presence or absence of bright reflections in the part of the object being looked at, and changes in colour resulting from the type of lamp used can all affect ease of recognition.

CIBSE. The Chartered Institute of Building Services Engineers, Lighting Division (formerly the Illuminating Engineering Society), has produced a code for interior lighting and some of the information contained in this chapter is taken from this code which was published in January 1994. Some of the more common terms used in lighting design and the associated units are given below.

Luminous flux. The light emitted by a source, or received by a surface. It is expressed in lumens. Symbol: ϕ.

Lumen. This is the SI unit of luminous flux. An ordinary 100 W lamp, for example, emits about 1200 lumens. One lumen is the luminous flux emitted within unit solid angle (one steradian) by a point source having a uniform luminous intensity of one candela. Symbol: lm.

Luminous intensity. The quantity which describes the power of a source or illuminated surface to emit light in a given direction. It is the luminous flux emitted in a very narrow cone containing the given direction divided by the solid angle of the cone. The result is expressed in candelas. Symbol: I.

Candela. The SI unit of intensity. It is 1 lumen per steradian. Symbol: cd.

Illuminance. The luminous flux density at a surface, i.e. the luminous flux incident per unit area. The quantity was formerly known as the illumination value or illumination level. It is expressed in lux (lumens/m^2 or lm/m^2). Symbol: E.

Standard service illuminance. The service illuminance throughout the life of an installation and averaged over the relevant area. This area may be the whole area of the working plane in an interior, or the area of the visual task and its immediate surround. It is expressed in lux. The tables that are contained in this chapter specifying the illuminance for specific areas are based on the standard service illuminance.

Lux. SI unit of illuminance. It is equal to one lumen per square metre.

Foot-candle. An obsolete unit of illuminance. It has the same value as the lumen per square foot.

Mean spherical illuminance: scalar illuminance. The average illuminance over the whole surface of a very small sphere located at a given point. It is expressed in lux. Symbol: E_s.

Mean cylindrical illuminance. The average illuminance over the curved surface of a very small cylinder located at a given point. Unless otherwise stated, the axis of the cylinder is taken to be vertical. It is expressed in lux. Symbol: E_c.

Luminance. A term which expresses the intensity of the light emitted in a given direction by unit area of a luminous or reflecting surface. It is the luminous flux emitted in the given direction from a surface element, divided by the product of the projected area of that element perpendicular to the prescribed direction and the solid angle containing the direction. It is expressed in lumens per square metre per steradian which is equivalent to candelas per square metre. In interior lighting design it is the product of the illuminance and the luminance factor (q.v.) for the particular conditions of illumination and viewing. If the surface can be assumed without too much error to be perfectly matt, its luminance in any direction is the product of the surface illuminance and its reflectance, and can be expressed in lm/m^2 (apostilbs). Symbol: L.

Candela per square metre. SI unit of luminance in lumens per square metre per steradian. Unit luminance in this system is that of a uniform plane diffuser emitting π lumens per square metre. Symbol: cd/m^2. To convert the luminance from apostilbs to the SI unit divide apostilbs by π.

Apostilb. A metric unit of luminance. Unit luminance is expressed in this system as that of a uniform diffuser emitting $1 \, lm/m^2$. Symbol: asb. $1 asb = 1/\pi \, cd/m^2$.

Luminosity. Visual sensation associated with the amount of light emitted from a given area. The term brightness is used colloquially.

Glare. The discomfort or impairment of vision experienced when parts of the visual field are excessively bright in relation to the general surroundings. There are a number of terms to express the amount of glare. For example, disability glare which prevents seeing detail; discomfort glare which causes visual discomfort but might not impair ability to see detail; direct glare caused when excessively bright parts of the visual field are seen directly, i.e. unshielded light sources.

I.E.S. glare index. A numerical index calculated according to the method described in I.E.S. Technical Report No. 10. It enables the discomfort glare to be ranked in order of severity and the permissible limit of this from an installation to be prescribed quantitatively. This Technical Report has been updated and is now available as Technical Memorandum No. 10 of the CIBSE.

Luminous efficacy. The ratio of the luminous flux emitted by a lamp to the power consumed by it. Expressed in lm/W.

Reflectance. The ratio of the flux reflected from a surface to the flux incident upon it. The value is always less than unity. Symbol: ρ.

British zonal system (BZ). A system for classifying luminaires according to their downward light distribution. The BZ class number denotes the classification of a luminaire in terms of the flux from a conventional installation directly incident on the working plane, relative to the total flux emitted below the horizontal. Although sometimes still quoted it has now been replaced, for the purpose of flux calculations by the methods outlined in CIBSE Technical Memorandum No. 5.

Room index. An index related to the dimensions of a room, and used when calculating the utilization factor and the characteristics of the lighting installation. It is given by:

$$\frac{lw}{h_{\mathrm{m}}(l + w)}$$

where l is the length and w the width of the room and h_{m} the height of the luminaires above the working plane.

Utilization factor. The total flux reaching the working plane divided by the total lamp flux.

PSALI. An abbreviation of 'permanent supplementary artificial lighting of interiors'.

PAL. An abbreviation of 'permanent artificial lighting'.

Reflection. Whenever light falls on to a surface, some of it is absorbed and the remainder is either reflected or transmitted. If the surface is opaque and smoothly polished, the *specularly* reflected light leaves the surface at the same angle as it arrived (as a billiard ball striking a cushion), and by suitably shaping the surface it is possible to redirect the light in any desired direction (e.g. a motor-car headlight, with lamp placed at the focal point of a polished parabolic mirror directing most of the light forward).

Diffuse reflection occurs from matt surfaces. The light is reflected most strongly at right angles to the surface (whatever the direction from which the light arrives) and progressively more weakly at other angles. Matt surfaces show no highlights. Most painted and many other surfaces are partly specular and partly diffuse reflectors of light and are classified according to which type of reflection predominates.

Diffusion. Light passes straight through a transparent material, but is scattered or diffused to a greater or lesser extent in a translucent material. Flashed opal glass or its plastics equivalent scatters it completely so that it emerges in all directions, and complete concealment of lamps behind a panel of this material is easily achieved. Frosted glass diffuses the light less perfectly, so that it emerges mainly in the same general direction as when it entered the glass; in effect, it is usually possible to see vaguely the positions of lighted lamps behind frosted panels. Hammered and rolled glasses and clear plastics with a similar finish generally have less diffusing and concealing power than frosted glass but have a sparkle that may be preferred in many cases.

Refraction. If light passes through a transparent material which does not have parallel sides, it will be bent away from its original direction by a process known as refraction. Ribbed glass or plastic fittings in which each rib is a carefully designed prism can therefore be made to control light very accurately in a required direction, and this principle is very widely used in electric street lighting fittings.

Shadows. When an obstruction completely masks the only source of light from a point on, say, the floor, the shadow at that point will be complete, but where the source is only partially masked there will be only a partial shadow. It follows, then, that a concentrated light source tends to promote deep shadows with hard edges, while physically large sources promote soft and faint shadows; also that the greater the number of sources in a room, and the more light that is reflected from the ceilings and walls, the softer and fainter shadows become.

Fluorescent lighting is often called shadowless. This does not apply to single lamps, for though the length of the lamp tends to reduce shadows of linear objects at right angles to its main axis, its small width tends to promote those of objects parallel to it. Thus the shadows cast depend partly on the shape of the obstruction and partly on the orientation of the fluorescent lamp with regard to it. When reduction of shadow from a particular fitting is desired, the aim should be to ensure that every important point on the working area can see at least part of the fitting; thus, for lighting a kitchen sink, a fluorescent lamp should be placed above and parallel to the front edge of the sink, for in this position a person's head and shoulders obstruct less light than if the lamp were placed at right angles to this position.

Glare. Glare has been described as 'light out of place'. Properly applied light makes it easy and comfortable to see, glare has the reverse effect and means that money is being wasted reducing visual effectiveness. Broadly, glare is caused either by too much light entering the eye from the wrong directions, or by some things being too bright in relation to other surfaces in the normal field of view, and may be prevented or at least minimized by applying the following rules:

1. Use lighting fittings which put the downward light mainly where it is wanted – on the work – with relatively little escaping in the direction of the worker's eye. It is just as necessary to screen fluorescent lamps from normal angle of view as any other.
2. Make the actual detail being looked at a little brighter than anything else seen at the same time (e.g. white paper against a light-coloured desktop). If dark cloth were viewed against a light-coloured surface, a local light would be required to make the cloth look bright.
3. Use light-coloured decorations and ensure that sufficient light from the fittings goes upwards and sideways to illuminate the ceiling and walls to make them fairly bright, and thus reduce the contrast of brightness between the fittings and their background.
4. Avoid if possible the use of glossy working surfaces (e.g. polished wood or glass tabletops) which mirror the lighting fittings.
5. The more extensive the installation, and the higher the illumination, the more carefully should these rules be followed.

The CIBSE Code for interior lighting published in 1994 giving recommendations for good interior lighting makes reference to glare index. This factor should be applied in all illumination calculations. Due to the complicated nature of the subject the reader should refer to the Code for a full explanation of its use.

Electric Lamps

Filament lamps. Almost all filament lamps for general lighting service are made to last an average of at least 1000 hours. This does not imply that every individual lamp will do so, but that the short-life ones will be balanced by the long-life ones; with British lamps the precision and uniformity of manufacture now ensures that the spread of life is small, most individual lamps in service lasting more or just less than 1000 hours when used as they are intended to be used.

In general, vacuum lamps, which are mainly of the tubular and fancy shapes, can be used in any position without affecting their performance. The ordinary pear-shaped gas-filled lamps are designed to be used in the cap-up position in which little or no blackening of the bulb becomes apparent in late life. The smaller sizes, up to 150 W, may be mounted horizontally or upside-down, but as the lamp ages in these positions the bulb becomes blackened immediately above the filament and absorbs some of the light. Also vibration may have a more serious effect on lamp life in these positions. Over the 150 W size, burning in the wrong position leads to serious shortening of life.

Coiled-coil lamps. By double coiling of the filament in a lamp of given wattage a longer and thicker filament can be employed, and additional light output is obtained from the greater surface area of the coil, which is maintained at the same temperature thus avoiding sacrificing life. The extra light obtained varies from 20% in the 40 W size to 10% in the 100 W size.

Effect of voltage variation. Filament lamps are very sensitive to voltage variation. A 5% overvoltage halves lamp life due to overrunning of the filament. A 5% undervoltage prolongs lamp life but leads to the lamp giving much less than its proper light output while still consuming nearly its rated wattage. The rated lamp voltage should correspond with the supply voltage. Complaints of short lamp life very often arise directly from the fact that mains voltage is on the high side of the declared value, possibly because the complainant happens to live near a substation.

Reflector lamps. For display purposes reflector lamps are available in sizes of 25 W to 150 W. They have an internally mirrored bulb of parabolic section with the filament at its focus, and a lightly or strongly diffusing front glass, so that the beam of light emitted is either wide or fairly narrow according to type. The pressed-glass (PAR) type of reflector lamp gives a good light output with longer life than a blown glass lamp. Since it is made of borosilicate glass, it can be used out of doors without protection. The very low efficacy of incandescent lamps is often the factor that precludes their selection for commercial lighting projects.

Tungsten halogen lamps. The life of an incandescent lamp depends on the rate of evaporation of the filament, which is partly a function of its temperature and partly of the pressure exerted on it by the gas filling. Increasing the pressure slows the rate of evaporation and allows the filament to be run at a higher temperature thus producing more light for the same life.

If a smaller bulb is used, the gas pressure can be increased, but blackening of the bulb by tungsten atoms carried from the filament to it by the gas rapidly reduces light output. The addition of a very small quantity of a halide, iodine or bromine to the gas filling overcomes this difficulty, as near the bulb wall at a temperature of about $300°C$ this combines with the free tungsten atoms to form a gas. The tungsten and the halide separate again when the gas is carried back to the filament by convection currents, so that the halide is freed to repeat the cycle.

Tungsten halogen lamps have a longer life, give more light and are much smaller than their conventional equivalents, and, since there is no bulb blackening, maintain their colour throughout their lives. Mains-voltage lamps of the tubular type should be operated within $5°$ of the horizontal. A 1000 W tungsten halogen lamp gives 21 000 lumens and has a life of 2000 hours. They are used extensively in the automotive industry. They are also making inroads into shop display and similar areas in the form of l.v. (12 V) single-ended dichroic lamps.

Discharge lamps. In discharge lamps the light output is obtained, not from incandescence, but by 'exciting' the gas or vapour content of the discharge tube. The excited gas emits energy of a characteristic wavelength and this may appear as several disconnected spectral lines. Light emitted in such a discontinuous manner may result in serious distortion of the colour appearance of objects seen in it. The main advantage of discharge lamps is their high luminous efficacy with very long life.

In any discharge lamp, once started, the current tends to increase instantaneously to a destructive value so to protect the lamp and wiring a device such as a choke has to be incorporated in the circuit to limit the current to a designed safe value. On d.c. supplies a resistance may be used which, in most cases, consumes about as much power as the lamp itself and thus lowers the luminous efficacy compared with operation on a.c.

The choke is connected in series with the lamp, and a power-factor capacitor is placed across the mains on the mains side of the choke. The lamp takes a few minutes, according to type, to reach full brightness. If switched off when hot, it will not restart until it has cooled down, but since, in general, it will withstand a sudden voltage drop of about 30 V, ordinary fluctuations of mains voltage do not affect it seriously.

Fluorescent tubes. A tube is a mercury discharge lamp in which the inside of the discharge tube is coated with fluorescent phosphors. Because the vapour pressure is low a higher proportion of u.v. radiation than of visible light is emitted and this radiation excites the fluorescent powders which then give off visible light. By choosing the phosphors correctly any colour can be produced. Because the light comes from the phosphors, the proportion of visible light to radiant heat is much greater in fluorescent tubes than in incandescent lamps. Consequently the lamps have a high efficacy and also 'run cool' so that they can be used in situations where filaments lamps would generate too much heat.

It used to be that the efficacies of the 'white' lamps in the range were closely related to their colour rendering properties. This is because the eye is less sensitive to the red and blue light at the ends of the spectrum than to the green and yellow in the middle and so more electrical energy must be used to produce the sensation of red light than is needed for the same apparent value of green and yellow. Consequently lamps with a high efficacy (white or warm white) have poor colour rendering properties while lamps which show objects in their true colours, i.e. Natural, Kolor-rite or Northlight, have a lower efficacy. In about 1980 narrow band or so-called 'rare earth' phosphors were introduced into fluorescent tubes. These enable good colour rendering lamps to be produced with high efficacy by having three or more narrow bands of energy in the red, green and blue wavelengths, which the eye interprets as 'white light'. These phosphors can operate at higher temperatures and have permitted the development of a new generation of compact lamps.

A range of higher efficacy narrower diameter (26 mm instead of 38 mm) lamps has been introduced. A change of gas fill has enabled the tubes to have similar electrical characteristics to conventional fluorescent tubes but with a 5% reduction in power consumption. Combined with the rare earth phosphors, lamps of higher efficacy and better colour rendering are now available. These tubes can be operated on the existing circuits with a reduction in power.

Recently a new range of lamps (T5s – in ratings from 14 W to 35 W) has been developed with lengths that will fit modern metric luminaires. These operate only from special electronic control gear and are even slimmer than standard tubes (16 mm instead of 26 mm). Lamp efficiency is better than existing lamps. Phosphors are pre-coated triphosphors so the lamps have excellent lumen maintenance.

Numerous compact fluorescent lamps have been introduced recently, aimed at eventually replacing the g.l.s. design and, in many instances, linear fluorescents. The colour rendering of these is nearly as good as the g.l.s. lamps and typically they are five times more efficient. The GE 2D is an example and its characteristics have been included in *Table 13.5*.

Nearly all present-day fluorescent tubes are of the 'hot cathode' type. This means that the electrodes at the ends of the lamp must be heated by passing an electric current through them before an arc will strike between them. These cathodes consist of tungsten wire or braid covered with or encasing a slug of electron emitting material. When the cathode is heated, a cloud of electrons forms around the cathodes at either end of the tube, ionizing the argon within it. A choke is normally placed in series with the tube to control the current in the arc and the heating current passes through it and both the cathodes.

The various circuits used to start fluorescent tubes are described below.

Glow starter circuit. As soon as the main circuit switch is closed, the full mains voltage is applied across the electrodes of the glow starting switch (*Figure 13.1*). The voltage is sufficient to cause a glow discharge in the starting switch bulb. This has the effect of warming up the bi-metallic strips on which the switch contacts are mounted. The heating of these bi-metallic strips causes them to bend towards each other until the contacts touch. The glow discharge in the starter switch then disappears. The heater elements which form the electrodes in the fluorescent tube are heated by the current which now passes through them. In the meantime the bi-metallic strips not being heated by the glow discharge cool down and spring away from each other.

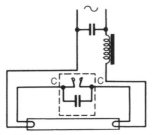

Figure 13.1 Glow starter circuit

This sudden interruption of the circuit, which contains a choke, causes a voltage surge across the fluorescent lamp electrodes which starts the discharge in the fluorescent lamp.

Electronic ignitors. A number of electronic ignitors for fluorescent tubes are now on the market. These may take the form of a simple replacement for a conventional starter switch or be incorporated in a complete ballast circuit. Most of them are solid-state devices that provide a cathode-heating current for a controlled period, followed by a starting pulse voltage which may utilize the choke winding as does the conventional glow starter. They operate at temperatures down to −5°C, require only a simple choke ballast, and have the same low watts loss as a glow switch. They are a 'first time' device and eliminate the blinking which sometimes occurs with conventional starters and shortens lamp life. The 'stuck starter' condition cannot occur and they are permanent devices and do not need to be replaced. They are available for all sizes of fluorescent tubes.

The series circuit. Fluorescent lamps more than 600 mm long require one set of control gear each; lamps of 600 mm length and less can either be run singly on 100–130 V or on 200–260 V (except 40 W 600 mm) or, as is generally the case, two in series with a single choke on 200–260 V. The circuit is shown in *Figure 13.2* and it will be seen that the lamps will start one after the other. Both must be of the same wattage but not necessarily of the same colour. Series running is economical since only one choke is necessary per pair of lamps, and choke losses are small, but if one lamp or starter misbehaves it may also affect the other lamp; therefore trouble in this circuit should be investigated as soon as possible. If series-operated lamps are to be run on a Quickstart circuit a double-wound transformer is needed as shown in *Figure 13.3* and both lamps must have the earthed metallic strip.

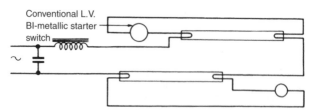

Figure 13.2 Series operation of short fluorescent lamps with switched starting

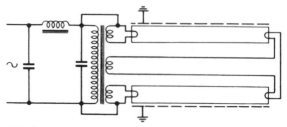

Figure 13.3 Series operation of short fluorescent lamps with Quickstart circuits

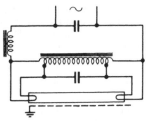

Figure 13.4 Quickstart circuit for single-lamp operation

Quickstart circuit. A fluorescent tube circuit designed to give a rapid start without flickering is shown in *Figure 13.4*. The unit consists of an auto-transformer, the primary winding of which is connected across the fluorescent tube, with the secondary winding in two separate sections, one across each cathode. This method is not normally used today but is included because it may be found in old installations.

Semi-resonant start circuit. In this circuit the place of the choke is taken by a specially wound transformer. Current flows through the primary coil to one cathode of the lamp, and thence through the secondary coil which is wound in opposition to the primary. The other end of the secondary is connected to a fairly large capacitor, and thence through the second cathode of the lamp to neutral (*Figure 13.5*). This method is not normally used today but is included because it may be found in old installations.

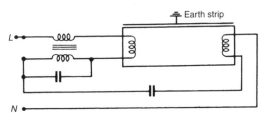

Figure 13.5 Semi-resonant start circuit

Electronic ballast circuit. The recently introduced high frequency (h.f.) electronic ballast provides silent, instant flicker-free starting for single and twin standard fluorescent lamps up to 1800 mm in length, both 26 mm and 38 mm in diameter. The new ballasts eliminate the need for a separate starter, choke and capacitor. The basic construction of an h.f. ballast involves a harmonic filter, rectifier and inverter, similar to those used in emergency lighting. The principle of operation is to convert 50 or 60 Hz mains input into a d.c. voltage and then convert this to a high frequency, around 32 kHz, for operating fluorescent lamps. The h.f. ballast takes advantage of a characteristic of fluorescent lamps whereby greater efficiency is obtained at high frequency. The overall lighting system efficacy can be increased by 20 to 30% due to:

1. Improved lamp efficacy at high frequency operation.
2. Reduced circuit power losses.
3. Lamp operates closer to optimum performance in most enclosed luminaires.

For instance, an 1800 mm lamp normally rated at 70 W with standard control gear can now be run at 62 W for the same light output. Furthermore, ballast losses are reduced – in the case of a twin 1800 mm circuit from 26 W (starter switch circuit) to only 8 W with an HF ballast. The overall achievement then for a twin 70 W 1800 mm circuit is some 20% being 132 W (that is 2 × 62 W plus 8 W) compared with 166 W (2 × 70 W plus 26 W). Together these energy saving features enable lighting levels to be maintained with a dramatic cut in electricity costs.

D.C. working. If desired, fluorescent lamps may be used on d.c. supplies by omitting the normal power-factor capacitor and inserting a suitable resistor in series with the choke. This resistor consumes a wattage generally roughly equal to that of the lamp, so that luminous efficacy of the circuit is about half of that on a.c. Suitable resistance values are given in *Table 13.1*.

Table 13.1 Resistors for D.C. operation

Mains voltage	Resistor values (ohms)					
	1–80 W	1–40 W 1200 mm	1–30 W	2–40 W 600 mm	2–20 W	2–15 W
200	103	208	264	116	182	235
210	116	235	293	128	208	264
220	128	264	330	147	235	293
230	147	293	380	147	264	330
240	166	330	420	166	293	380
250	166	330	420	166	330	380

After a period of working the positive end of the tube may darken owing to migration of the mercury to the negative end, and to counteract this tendency it is usual to fit a polarity-reversing switch to the subcircuit, preferably of the unidirectional rotary type so that polarity is changed at every switching. A thermal starter is normally used.

Colour appearance and colour rendering. The light output of a fluorescent tube is not necessarily the only consideration. Good colour rendering can be most important and there are many situations where the use of a de-luxe colour such as Polylux can produce a stimulating atmosphere which far outweighs the small additional cost. In some cases where critical colour matching is important the old Kolor-rite or Natural tubes may be appropriate.

Operating characteristics. Fluorescent lamps emit about one third as much total heat as filament lamps giving the same amount of light, but only about one fifth as much radiated heat. Their light output varies by about 1% for each 1% change in mains voltage (compared with 4% variation in light of filament lamps). In general, light output after 5000 hours' burning is about 85% of that at 100 hours.

Effect of operating temperature. Fluorescent lamps are designed to operate at ambients of about 25°C. If they are used at elevated temperatures, a drop in light output occurs due to reduction in u.v. emission and enhancement of the visible mercury lines caused by an increase in vapour pressure. One way to overcome this is to use 'Amalgam' lamps in which a ring of indium

Table 13.2 Tubular fluorescent lamps

Lamp rating, W	Nominal lamp length, mm	Actual length, mm	Diameter, mm	Nominal tube voltage	Nominal lamp current, A
125	2400	2389.1	38	152	0.94
100	2400	2389.1	38	128	0.96
85	2400	2389.1	38	185	0.54
70	1800	1778.0	26	128	0.70
75/85	1800	1778.0	38	123	0.77
80/65	1500	1514.3	38	100/110	0.87/0.67
58	1500	1514.3	26	113	0.63
40	1200	1213.6	38	102	0.44
36	1200	1213.6	26	104	0.42
30	900	908.8	26	101	0.36
40	600	604.0	38	104	0.42
20	600	604.0	38	58	0.38
18	600	604.0	26	58	0.38
15	450	451.6	26	57	0.34
13	525	531.0	16	92	0.17
8	300	302.4	16	55	0.17
6	225	226.2	16	43	0.16
4	150	150.0	16	30	0.15

2D lamps	Lamp width and length, mm	Depth, mm			
38	205	35	110	0.49	
28	205	35	107	0.32	
16	140	27	97	0.20	

near the end of the lamps absorbs mercury and thus reduces pressure. The other is to use ventilated or ventilating fittings.

Low ambient temperatures also cause a fall-off in light output due to the condensation of mercury on the tube walls. Enclosing lamps in a clear acrylic sheath usually maintains them at the correct working temperature in low temperature surroundings.

Table 13.3 T5 lamp information

Lamp rating, W	Lamp lumens (lm)	Luminous efficacy (lm/W)	Colour temp. (K)	Colour rendering group	Average life (hours)	Length (mm)
14	1350	96	3000/4000	1B	16 000	549
21	2100	100	3000/4000	1B	16 000	849
28	2900	104	3000/4000	1B	16 000	1149
35	3650	104	3000/4000	1B	16 000	1449

Fault finding. A fault in some part of the circuit may often be traced by observing the following symptoms:

Lamp glows continuously at both ends, or one end, but makes no effort to start. In the first case, faulty starter; in the second, an earth in some part of the circuit and possibly a faulty starter also. With the instant-start circuit, lamp has no earthing strip, or earth is poor, or low mains voltage; if glowing at one end only faulty transformer or leads to one end of the lamp short- or open-circuited.

Lamp flashes repeatedly but cannot start. Faulty starter giving insufficient pre-heating time; or an old lamp. In the latter case one or both ends of the lamp may have become blackened and the lamp may light normally for a few moments, then die away with a shimmering effect.

Lamp lights normally but extinguishes after a few seconds, then repeats. Probably abnormally low mains voltage. May also be due to faulty starter.

Swirling effect of light in lamp. Probably disappears after a few switching operations on a new lamp. If it persists, change starter. If it still persists, renew lamp.

Cold cathode fluorescent tubes. Another type of fluorescent tube – the cold cathode type – is usually made of 20 mm tubing coated with fluorescent powder and with either a mercury and argon or neon filling. It is of the high voltage type, operated from a step-up transformer and there is no delay period in switching on. A life of 10 000 hours or more may be expected and is unaffected by the frequency of switching. Mercury-filled tubes show the usual drop in efficiency throughout life, but that of neon-filled lamps remains constant. Both types remain alight through severe reductions in mains voltage, and can be dimmed by suitable apparatus.

Cold cathode lamps are manufactured in a wide range of standard colours including daylight, warm white, blue, green, gold and red. The colours can be used separately or mixed to give any desired result, the gold and red lamps being particularly useful for providing a warm-toned light. Dimming of one of the mixed colours will provide a colour change. As the tubes can be manufactured in a variety of shapes and curves, they are suitable for decorative illumination in restaurants, etc. in situations where the higher voltage required for operating them will not be objectionable.

The main problems are their relative inefficiency as compared to hot cathode lamps and the high voltages needed for starting and operation.

Mercury and metal halide lamps. The mercury spectrum has four well-defined lines in the visible area and two in the invisible ultraviolet region. This radiation is used to excite fluorescence in certain phosphors, by which means some of the missing colours can be restored to the spectrum. The proportion of visible light to u.v. increases as the vapour pressure in the discharge tube, so that colour correction is less effective in a high pressure mercury lamp than in a low pressure (fluorescent) tube.

High pressure mercury lamps are designated MBF and the outer bulb is coated with a fluorescent powder. MBF lamps are now commonly used in offices, shops and indoor situations where previously they were considered unsuitable. Better colour rendering lamps have recently been introduced with a slight increase in efficacy. They are designated MBF de-luxe or MBF-DL lamps and are at present slightly more expensive than ordinary MBF lamps.

Table 13.4 Light output of standard fluorescent tubes (GE lamp data)

Lamp		T5 16 mm Ø				T8 26 mm							
	watts length mm	**4** 150	**6** 225	**8** 300	**13** 525	**15** 450	**18** 600	**30** 900	**36** 970	**36** 1200	**38** 1050	**58** 1500	**70** 1800
Triphosphor	**827 Polylux XL**					1050	1450	2500		3450			
Polylux XL	**830 Polylux XL**					1050	1450	2500		3450			
	835 Polylux XL						1450			3450			
	840 Polylux XL					1050	1450	2500		3450			
	860 Polylux XL						1450			3250			
Triphosphor	**827 Polylux**												
Polylux	**830 Polylux**			460					3100		3300		
	835 Polylux												
	840 Polylux			460					3100		3300		
5 Band	**930 Polylux Deluxe**											3750	
Phosphor	**940 Polylux Deluxe**											3750	
Standard	Warm white	150	300	400	850	950	1225	2300		3000	3050	4800	5800
Halophosphate	White	150	300	400	850	950	1225	2300		3000	3050	4800	5800
	Natural/universal white											4100	
	Cool White	150	290	380	800	900	1200	2250	2600	3000		4700	5700
	Daylight	150	290	380	800	730	950	1700	1900	2400		3580	

Table 13.4 (continued)

	Lamp	T12 38 mm Ø							Mod-U-Line®		Circline®			
	watts / length mm	**20** / 600	**40** / 1200	**65** / 1500	**75** / 1800	**85** / 2400	**100** / 2400	**125** / 2400	**20** / 265	**40** / 525	**22** / 209.5	**32** / 311.5	**40** / 412.6	**60** / 412.6
Triphosphor Polylux XL	**827 Polylux XL**													
	830 Polylux XL													
	835 Polylux XL													
	840 Polylux XL													
	860 Polylux XL													
Triphosphor Polylux	**827 Polylux**	1450												
	830 Polylux		3350	5300	6700	8450	9400	10550						
	835 Polylux													
	840 Polylux	1450	3350	5300	6700	8450	9400			3250				
5 Band Phosphor	**930 Polylux Deluxe**													
	940 Polylux Deluxe													
Standard Halophosphate	Warm white	1225	3050	5000	5850					2875	1050			3700
	White	1225	3050	5000	5850	7350	8600	9500		2875		1875	2800	
	Natural/universal white	1000	2375	3775	4650									
	Cool White	1200	3000	4850	5700		8450	9300		2875	1000	1825	2700	
	Daylight	950	2400	3700							875	1550	2500	

Table 13.5 Colour appearance and colour rendering properties of fluorescent lamps

Tube colour	Colour rendering quality	Colour appearance	Applications
Polylux XL 827	Very good	Warm	Tubes of various efficacies for
Warm white	Fair	Warm	use in social residential and domestic situations
Polylux XL 835	Very good	Intermediate	Tubes of various efficacies for
White	Fair	Intermediate	general illumination of work
Plus white	Good	Intermediate	areas – shops, factories, warehouses, etc.
Polylux XL 840	Very good	Cool	Tubes of various efficacies for
Cool white	Fair	Cool	work areas requiring illumination
Natural	Very good	Cool	to blend with natural daylight – offices, shops, etc.
Deluxe natural	Good	Intermediate	Butchers, fishmongers, supermarkets. Enhances appearance of red objects
Kolor-rite	Excellent	Cool	Complies with DHSS requirements for hospital lighting
Northlight/Colour matching	Very good	Cool	Areas for matching materials, etc. Any application where a wintery effect or an impression of coolness is required
Artificial daylight	Excellent	Cool	Areas for exact colour matching. Best colour rendering with cool appearance. Meets BS 950 Part 1
Colours	Poor	–	Saturated for display, floodlighting, stage lighting

A more fundamental solution to the problem of colour rendering is to add the halides of various metals to mercury in the discharge tube. In metal halide lamps (designated MBI) the number of spectral lines is so much increased that a virtually continuous spectrum of light is achieved, and colour rendering is thus much improved. The addition of fluorescent powders to the outer jacket (MBIF) still further improves the colour rendering properties of the lamp, which is similar to that of a de-luxe natural fluorescent tube.

Metal halide lamps are also made in a compact linear form for floodlighting (MBIL) in which case the enclosed floodlighting projector takes the place of the outer jacket and in a very compact form (CSI) with a short arc length which is used for projectors and encapsulated in a pressed glass reflector, for long range floodlighting of sports arenas, etc. In addition, single-ended low wattage (typically 35 W – 150 W) metal halide lamps (MBI-T) have been developed offering excellent colour rendering for display lighting, floodlighting and uplighting of commercial interiors.

Table 13.6 Electric discharge lamps

Type	Watts	Bulb shape	Length, mm	Dia., mm	Design lm	Cap
Mercury MBF	50	Oval	129	56	1 900	E27 (ES)
	80	Oval	154	71	3 650	E27 (ES)
	125	Oval	175	76	5 800	E27 (ES)
	250	Oval	227	91	12 500	E40 (GES)
	400	Oval	286	122	21 500	E40 (GES)
	700	Oval	328	143	38 000	E40 (GES)
	1000	Oval	410	167	58 000	E40 (GES)
MBF-DL	80	Oval	154	71	3 650	E27 (ES)
	125	Oval	175	76	6 200	E27 (ES)
	250	Oval	227	91	13 300	E40 (GES)
	400	Oval	286	122	22 800	E40 (GES)
Mercury- reflector MBFR	250	Parabolic	260	166	10 500	E40 (GES)
	400	Parabolic	300	181	18 000	E40 (GES)
	700	Parabolic	328	202	32 500	E40 (GES)
	1000	Parabolic	380	221	48 000	E40 (GES)
Mercury- tungsten MBFT	160	Oval	175	76	2 560	B22 (BC)
	250	Oval	227	91	4 840	E40 (GES)
	500	Oval	286	122	11 500	E40 (GES)
Metal- halide MBI-T CSI	750	Linear	254	21	58 500	R × 75
	1000	Linear	254	21	110 000	R × 75
	150	Linear (SE)	80	22	12 000	G12
	1000	Parabolic	175	205	67 000	G38
MBIF	250	Oval	227	91	16 000	E40 (GES)
	400	Oval	286	122	24 000	E40 (GES)
	1000	Oval	410	167	85 000	E40 (GES)
High-pressure sodium SON	50	Oval	154	71	3 100	E27 (ES)
	70	Oval	154	71	5 300	E27 (ES)
	150	Oval	227	91	15 000	E40 (GES)
	250	Oval	227	91	25 500	E40 (GES)
	400	Oval	286	122	45 000	E40 (GES)
	1000	Oval	410	167	110 000	E40 (GES)
SONDL	150	Oval	227	91	11 000	E40 (GES)
	250	Oval	227	91	19 000	E40 (GES)
	400	Oval	286	122	33 000	E40 (GES)
SON-T	50	Cylindrical	154	39	3 100	E27 (ES)
	70	Cylindrical	154	39	5 500	E27 (ES)
	150	Cylindrical	210	47	15 500	E40 (GES)
	250	Cylindrical	257	47	27 000	E40 (GES)
	400	Cylindrical	285	47	47 000	E40 (GES)
	1000	Cylindrical	380	67	120 000	E40 (GES)
SONDL-T	150	Cylindrical	210	47	11 500	E40 (GES)
	250	Cylindrical	257	47	20 500	E40 (GES)
	400	Cylindrical	285	47	34 000	E40 (GES)
SON.TD	250	Linear	189	24	25 000	2R × 7s
	400	Linear	254	24	46 000	2R × 7s

Table 13.6 (continued)

Type	Watts	Bulb shape	Length, mm	Dia., mm	Design lm	Cap
Low-pressure	18	Linear (SE)	210	53	1 750	B22 (BC)
sodium SOX	35	Linear (SE)	311	53	4 500	B22 (BC)
	55	Linear (SE)	425	53	7 500	B22 (BC)
	90	Linear (SE)	528	67	12 500	B22 (BC)
	135	Linear (SE)	775	67	21 500	B22 (BC)

No attempt should ever be made to keep an MB or MBF lamp in operation if the outer bulb becomes accidentally broken, for in these types the inner discharge tube of quartz does not absorb potentially dangerous radiations which are normally blocked by the outer glass bulb.

Sodium lamps. Low pressure sodium lamps give light which is virtually monochromatic; that is, they emit yellow light at one wavelength only, all other colours of light being absent. Thus white and yellow objects look yellow, and other colours appear in varying shades of grey and black.

However, they have a very high efficacy and are widely used for streets where the primary aim is to provide light for visibility at minimum cost; also for floodlighting where a yellow light is acceptable or preferred.

The discharge U-tube is contained within a vacuum glass jacket which conserves the heat and enables the metallic sodium in the tube to become sufficiently vaporized. The arc is initially struck in neon, giving a characteristic red glow; the sodium then becomes vaporized and takes over the discharge.

Sometimes leakage transformers are used to provide the relatively high voltage required for starting, and the lower voltage required as the lamp runs up to full brightness – a process taking up to about 15 minutes. Modern practice is to use electronic ignitors to start the lamp which then continues to operate on a conventional choke ballast. A power-factor correction capacitor should be used on the mains side of the transformer primary.

A linear sodium lamp (SLI/H) with an efficacy of 150 lm/W is available and in the past was used for motorway lighting. The outer tube is similar to that of a fluorescent lamp and has an internal coating of indium to conserve heat in the arc. Mainly because of its size the SLI/H lamp has been replaced with the bigger versions of SOX lamps as described above.

Metallic sodium may burn if brought into contact with moisture, therefore care is necessary when disposing of discarded sodium lamps; a sound plan is to break the lamps in a bucket in the open and pour water on them, then after a short while the residue can be disposed of in the ordinary way. The normal life of all sodium lamps has recently been increased to 4000 hours with an objective average of 6000 hours.

SON high-pressure sodium lamps. In this type of lamp, the vapour pressure in the discharge tube is raised resulting in a widening of the spectral distribution of the light, with consequent improvement in its colour-rendering qualities. Although still biased towards the yellow, the light is quite acceptable for most general lighting purposes and allows colours to be readily distinguished. The luminous efficacy of these lamps is high, in the region of 100 lm per watt, and they consequently find a considerable application in industrial situations, for street lighting in city centres and for floodlighting. Three types of lamp are available: an elliptical type (SON) in which the outer bulb is coated with a fine diffusing powder, intended for general lighting; a single-ended cylindrical type with a clear glass outer bulb, used for floodlighting (SON.T); and a double-ended tubular lamp (SON.TD) also designed for floodlighting and dimensioned so that it can be used in linear parabolic reflectors designed for tungsten halogen lamps. This type must always be used in an enclosed fitting.

The critical feature of the SON lamp is the discharge tube. This is made of sintered aluminium oxide to withstand the chemical action of hot ionized sodium vapour, a material that is very difficult to work. Recent research in the UK has resulted in improved methods of sealing the electrodes into the tubes, leading to the production of lower lamp ratings, down to 50 W, much extending the usefulness of the lamps.

Most types of lamps require some form of starting device which can take the form of an external electric pulse ignitor or an internal starter. At least one manufacturer offers a range of EPS lamps with internal starters and another range that can be used as direct replacements for MBF lamps of similar rating. They may require small changes in respect of ballast tapping, values of p.f. correction capacitor and upgrading of the wiring insulation to withstand the starting pulse voltage. Lamps with internal starters may take up to 20 minutes to restart where lamps with electronic ignition allow hot restart in about 1 minute.

Considerable research is being made into the efficacy and colour rendering properties of these lamps and improvements continue to be introduced.

Recent developments have led to the introduction of SON de-luxe lamps. At the expense of some efficacy and a small reduction in life far better colour rendering has been obtained. They are increasingly being used in offices and shops as well as for industrial applications.

Induction lamps. An induction lamp is like a fluorescent tubular lamp. It contains a low pressure of mercury which when excited radiates u.v. which in turn is absorbed by a phosphor coating and re-radiated as visible light. However, unlike a fluorescent, the induction lamp has no electrodes and the discharge is created by a magnetic field generated externally to the bulb. The bulb has a hollow central stem and the antenna, like a small radio transmitting aerial, is in the centre of the lamp. Because there are no electrodes the lamp life is not dependent upon the initial ion emission from the cathode coating. They operate at high frequency so there is little flicker. The principle finds its application in the Philips QL lamp which operates from separate dedicated electronic control gear. The phosphor coating is sealed into the bulb surface so does not degrade due to mercury contamination. Average lamp life is 60 000 hours.

'Life' and light output of lamps. In an incandescent lamp failure occurs long before its light output has fallen a substantial amount, indeed in a tungsten halogen lamp there is virtually no difference between the light output of a new lamp and that of one at the point of failure. This is by no means the case with discharge lamps, including fluorescent tubes, which will often 'run' for many hours longer than their rated 'life' and may finally prove quite uneconomical in use, since, although they still consume about the same amount of power, their light output may be reduced to a fraction of its original value.

In consequence, all British and most European manufacturers no longer publish 'life' figures for these types of lamp, but issue lumens maintenance

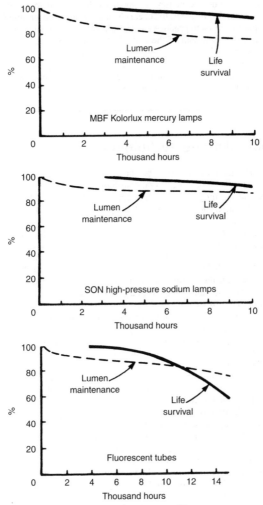

Figure 13.6 Typical performance in laboratory conditions

and lamp mortality curves to show the rate at which light output may be expected to diminish during use and the percentage of lamp failures which may occur within a representative batch. These factors taken in conjunction with the likely deterioration of reflecting surfaces of ceilings, walls and floors and the accumulation of dirt on the reflectors and diffusers of luminaires, make it possible to determine the point at which lamps should be replaced. Typical curves for various types of discharge lamps are shown in *Figure 13.6*.

Failure of a fluorescent tube is not normally related to fracture of the filament or cathode. It generally occurs when the electron emitting material on the cathodes ceases to produce electrons in sufficient quantity to permit the lamp to strike. A symptom of the end of life is a strong yellow glow at one end and severe blackening of both ends of the tube.

Before this occurs, however, the reduction of light output of the phosphors will have brought the total light output down to an uneconomic figure. For this reason tubes are normally replaced before actual failure.

Interior Lighting Techniques

Types of fittings. Fittings should be chosen according to their qualities in the following respects: appearance; light distribution; brightness; ease of erection, relamping and cleaning; cost; and luminous efficacy. The order of importance changes with different applications, but light distribution should always be considered first.

Fittings classed as 'mainly direct' give most of their light in one direction. With any given type of lamp they will give the highest illuminance per watt of electrical input, with some likelihood of glare either direct from the lamp or by reflection from polished working surfaces, with a tendency to hard and deep shadows, but with the possibility of ensuring that at any particular working position most of the light comes from a desired direction – as may be required in, say, a drawing office.

On the other hand, 'mainly indirect' fittings, which give all or nearly all their light upwards, have the opposite characteristics. Indirect lighting, also known as uplighting, can be an efficient system measured in illuminance per watt, and there is no likelihood of glare except from the ceiling at high illuminance; shadows are very soft and weak, and the light at every working point is received more or less equally from all directions. Midway between these extremes is the 'general diffusing' class of fitting which, of course, has intermediate characteristics.

'Direct' lighting is almost always used for industry and for display purposes in shops; it gives a brisk, lively effect which emphasizes light and shade and reveals shape well. Indirect lighting is usually considered more restful and is used mainly for restaurants, hotels and other 'social' interiors, and in combination with direct lighting for many classes of shops although it is becoming increasingly popular for office lighting using so-called 'uplights' fitted with SON-DL or MBIF lamps. 'General diffusing', or 'direct', lighting is generally used in offices and schools.

Spacing of fittings. In most interiors which contain several lighting fittings the aim is usually to provide 'general lighting', i.e. a comparatively even illumination at working level all over the room. This allows furniture and plant to be moved or added to without altering the lighting installation. Too wide a spacing of fittings would give the effect of comparative darkness between pools of light, whereas an unnecessarily close spacing would be uneconomic. As a general rule, applying both to fluorescent lamp fittings and other types normally used for general lighting, a satisfactorily even illumination results when the fittings are spaced no more than the maximum allowed spacing to mounting ratio times the distance above the working plane in both transverse and axial directions.

Interior lighting design. It is not the purpose of this section to provide enough information to design a complete lighting installation but to give general guidance on the matter. One design method is based on calculating the number of lamp lumens required and from this, based on the light source, the number of fittings and the mounting conditions, the layout of the installation.

Multiplying the area to be lit by the illuminance required gives the total lumens needed at the working place, taken at a point 0.85 m above floor level. For example, if a room 10 m by 25 m is to have an illuminance of 1000 lux, the total lumens required at the working plane is $10 \times 25 \times 1000 = 250\,000$ lumens. Because of the absorption of light by the walls, ceiling and floor and by the lighting fittings themselves the number of lumens emitted by the lamps must be in excess of 250 000. The ratio of the actual number of lumens reaching the working plane in a particular area to the number of lumens emitted by the lamp is called the utilance or coefficient of utilization. In addition allowance in the calculation must be made for the effect of dirt or dust on the fittings themselves, the maintenance factor. The importance of maintenance is stressed on page 282.

The formula for calculating the total lamp lumens is:

$$F = \frac{E_{av} \times A}{UF \times M \times Abs}$$

where F is the total lamp lumens; E_{av} the average illuminance (in lux); A the area to be lighted in square metres; UF the utilance factor; M the maintenance factor; Abs the absorption factor, which can be ignored except in very dusty atmospheres.

The next step is to calculate the number of lamps and fittings required. There is a close relationship between the spacing and mounting height of the lighting fittings. Too wide a spacing results in a fall-off in illuminance between the fittings. The actual maximum spacing to mounting height ratio that a fitting can be used at is normally incorporated with the UF tables for the fitting. This value should never be exceeded in normal circumstances. The example below is based on Thorn Diffusalux fittings with a maximum spacing to height ratio of 1.67.

The majority of lighting engineers arrive at the minimum number of fittings required by scaling them off on the plan. Alternatively it can be derived from the formula:

$$\frac{L}{MS} \times \frac{W}{MS}$$

where L and W are the length and width of the room (m) and MS is the maximum space between fittings (m). Each part of the equation is worked out separately to the nearest whole number above the actual answer.

The lamp lumens per fitting having been calculated, the number and rating of lamps or tubes are determined from the lamp characteristic tables (see pages 271, 273, 274, 276).

A practical example illustrates the use of the information above. It is proposed to light an office 15 m long by 10 m wide to an illuminance of 400 lux using fluorescent tubes. The ceiling height is 3 m and its reflectance 80%. Desk height is 1 m. Wall reflectance is 30% but owing to windows running along both long sides the average reflectance is dropped to 10% with floor reflectance of 20%. The ceiling is unobstructed by beams. Maximum spacing allowed is 3.34 m (SHR max . = 1.67 times height above working plane, which is 2 m: $1.67 \times 2 = 3.34$). This gives a requirement of 12/9 fittings from the formula but the best arrangement that fits into a room of that shape is the three rows of fluorescent fittings, making a total of nine. The room index is 3 if the height of the working plane above floor level is taken as 1 m. (*Note*: Room index is calculated from the formula given above.)

If Thorn 'Diffusalux' fittings with prismatic diffusers are employed and mounted on the ceiling a utilization factor of 0.72 obtains for a single lamp and 0.62 for a twin lamp fitting.

From the formula $F = E_{av} \times A/UF \times M \times Abs$, we see that the total lumens required is 60 000 (floor area 150 m^2 × 400 lux) divided by the utilance (0.62 for a twin-lamp fitting) and a maintenance factor of 0.8. (We can ignore *Abs* as this is a clean atmosphere.) This gives us a figure of 121 000 lumens approximately which, divided by the 12 fittings, shows 10 083 lumens per fitting or 5042 lumens per lamp. A 1500 mm 58 W Polylux white tube gives 5400 lumens, so we can use 12 fittings, each carrying two lamps of this rating. Note that occasionally the number of fittings installed may have to be higher than the calculated minimum. This can happen where fittings of the required lumen output are not available or where a certain rating of lamp or tube is specified to allow standardization of replacement lamps throughout an installation.

Maintenance factor. Lighting levels within a building decrease progressively due to the accumulation of dirt on windows, fittings and room surfaces, and to the fall in lumen output of the lamps themselves. Illuminance values, therefore, vary continuously, decreasing because of depreciation but being restored by cleaning, redecorating and relamping.

When natural and artificial lighting installations are designed, allowance is made for depreciation by incorporating appropriate factors in the design formulae as indicated earlier. The tables in this chapter giving recommended illuminances are for average service conditions and so the factors adopted to take account of the lamp, fitting and room surface depreciation relate to these average conditions. The maintenance factor M is defined as the ratio of the illuminance provided by an installation in the average condition of dirtiness expected in service, to the illuminance from the same installation when clean. Selection of maintenance factor for an installation is rather complicated and the reader is referred to IES Technical Report No. 9, 1967. A modified and shortened version of this is reprinted in Table 4.8 of the 1994 CIBSE Code.

Cleaning of fittings. The fall in light output due to dirt collecting on the light controlling surfaces of a fitting can usually be almost completely recovered by cleaning. The optimum economic cleaning interval (T) is that at which the cost of the light loss by dirt on the fitting equals the cost of cleaning the fitting. It is given by:

$$T = \frac{-C_c}{C_a} + \sqrt{\frac{2C_c}{C_a \Delta}}$$

where C_c is the cost of cleaning the fittings once; C_a is the annual cost of operating the fittings without cleaning; Δ is the notional average rate of decrease of luminous flux caused by dirt deposition and diminution of light from the tube – Δ can be calculated from the formula:

$$\Delta = (E_0 - E_1)/(E_0 \times T)$$

where E_0 is the initial illumination; E_1 is the minimum illumination after time T in years.

Maintenance techniques. Techniques for lighting maintenance fall into three categories: relamping, cleaning of fittings and cleaning and redecoration of room surfaces. Spot replacement of filament lamps calls for no comment but discharge lamps should be changed as quickly as possible after failure to prevent damage to associated control gear. If bulk replacement of lamps is adopted then this should take place when failures have reached about 20% of the total. Alternatively bulk replacement can take place when the illuminance falls below an agreed level which should not be less than the minimum value as defined in IES Technical Report No. 9.

All inside walls, partitions and ceilings should be washed at least once every 14 months and either whitewashed every 14 months or repainted every seven years. Surfaces decorated with a washable water paint should be repainted every three years.

Floodlighting Techniques

This form of outdoor illumination has three main applications: (a) for industrial purposes, i.e. the lighting of railway sidings and goods yards and for outdoor constructional work, (b) for decorative purposes, i.e. the illumination of buildings, monuments and gardens on special occasions, and (c) outdoor sports.

Industrial floodlighting. All the high pressure discharge lamps described in the preceding pages are suitable for this type of application. The SON lamp, for example, is commonly used for area floodlighting and SON.TD or MBIL/H lamps used in accurately designed parabolic reflectors are also used for long distance industrial floodlighting.

Docks loading wharves. For illumination of both docks and wharves floodlight projectors have many advantages over the old-fashioned 'cluster' fittings. Their flexibility and much greater efficiency ensures that adequate illumination is readily available over a working area. As a general rule, the projectors should be mounted high up on the crane platform or runway, and facilities provided for the spreading and training of the beams to meet the exigencies of the work.

Railway shunting yards, sidings, etc. The introduction of floodlighting has enabled a much more even distribution of light to be achieved than was possible with fittings mounted on short posts and has to a large extent eliminated the accident hazard. General practice is to mount the projectors on masts no less than 13 m high.

Shipyards, constructional work. A comparatively low level of illumination is required over extensive areas and a high intensity at certain points.

Projectors should be mounted high to avoid glare and be located on the berth structure.

High mounting of a similar character should be employed for the lighting of buildings under construction. Skilfully planned floodlighting is a valuable aid to the modern building contractor in his race against time.

Industrial floodlighting design. The amount of light required for any of these industrial applications will depend upon the nature of the work carried out. The illuminance required for manufacturing operations is given in *Table 13.7* at the beginning of this section. The size of lamp required can be calculated by considering the area to be lighted, the illuminance required, depreciation factor, and the beam factor of the particular projector to be employed. In such cases it is not always the horizontal area that must be taken as the surface to be lighted.

Table 13.7 Recommended values of illuminance

The recommendations in this table are taken from the CIBSE Code 1994. For more detailed recommendations the reader is advised to consult the Code itself.

Task group and typical task or interior	Standard service illuminance (lux)
Storage areas and plant rooms with not continuous work	150
Casual work	200
Rough work	
Rough machining and assembly	300
Routine work	
Offices, control rooms, medium machining and assembly	500
Demanding work	
Deep-plan drawing or business machine offices	
Inspection of medium machinery	750
Fine work	
Colour discrimination, textile processing, fine machining and assembly	1000
Very fine work	
Hand engraving, inspection of fine machining and assembly	1500
Minute work	
Inspection of very fine assembly	3000

Decorative floodlighting. It is necessary in this application first to decide upon the average illumination required. This naturally depends on the reflection factor, which depends on the texture of the surface and whether it is clean or dirty. For clean Portland stone the factor is 60%, which is reduced to 20% if the surface is dirty. At the other end of the scale clean red brick has a reflection factor of 25%, dropping to 8% if dirty.

The question of arranging the projectors requires careful consideration. Lighting directly from the front is seldom satisfactory, while lighting from an angle can be most pleasing. Lighting of polished surfaces is seldom successful using floodlights, as they tend to act as plane mirrors, producing an image of the floodlight. One technique is to light objects in front of the polished surface so that it stands out in silhouette.

Football and other sports. Practice or casual training areas can be lighted from one side by low wattage projectors housing tungsten halogen lamps (e.g. Thorn 'Halines') mounted on 4–5 m poles, but tournament areas require a higher luminance and more rigorous glare control. Minimum mounting heights of 10 m are recommended and the area should be lighted from both sides.

Stadiums are lighted to higher standards, especially if matches are likely to be televised. The most popular method is to light the area by means of 2000 W CSI, MBIL or SON de-luxe lamps mounted on towers up to 20 m high at the corners of the stadium. Alternatively, the lamps may be mounted on the roofs of the stands or on stub towers mounted upon them.

Escape lighting and emergency lighting. In accordance with the Fire Precautions Act 1971 and the Health and Safety at Work Act 1974, BS EN 1838 : 1999 *Code of practice for the emergency lighting of premises* lays down minimum standards. Interpretations of these may vary from one authority to another. Installations and equipment are defined as follows.

Definitions of terms. The basic aim of legislation and supporting documents such as British Standards, Codes of Practice and local authority regulations is to encourage uniformity of application of emergency lighting on a national scale. A number of terms and their meanings associated with emergency lighting systems are indicated below.

Emergency lighting. Lighting from a separate source independent of the mains supply, which continues after failure of the normal lighting of premises. Such lighting may be for standby or escape purposes.

Standby lighting. Standby lighting is emergency lighting, fed from a separate source from the mains, which comes into operation during a power failure to enable essential activities to continue.

Escape lighting. Lighting fed from a separate source, which is brought into immediate operation in the event of a power failure, to enable a building to be evacuated quickly and safely.

Sustained luminaire. A lighting fitting containing at least two lamps, one of which is energized from the normal supply and the other from an emergency lighting source.

Self-contained luminaire or sign. A luminaire or sign in which all the associated control units are housed and which only requires connection to the normal supply.

Slave luminaire or sign. One operated from a central control system.

Maintained system. In a maintained emergency lighting system the same lamps remain alight regardless of whether the normal lighting supply has failed. The maintained system may be provided with power from a floating battery system, which means the battery charger and power rectifier are connected in parallel with the battery and load, *Figure 13.7.* The floating system is now rarely used as the high float voltage results in short lamp life and low end-of-discharge light output. Alternatively, an automatic transfer switch may be used to connect the battery to the load terminals at the instant of normal

286

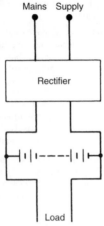

Figure 13.7 Circuit for a single floating battery system

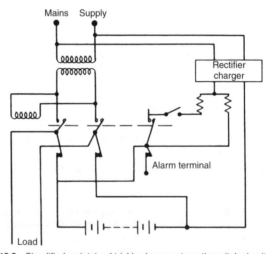

Figure 13.8 Simplified maintained trickle charge automatic switch circuit

supply failure, *Figure 13.8*. The taper-type charger is now replaced by the constant voltage type but is still to be found in use.

The floating battery maintained system provides d.c. to the load terminals. The transfer switch circuit is normally provided with power by a step-down isolating transformer, and provides low voltage a.c. until at the instant of supply failure the automatic transfer switch connects the battery to the load terminals thus providing d.c. during supply failure.

287

Non-maintained system. A system in which lamps are normally off but are automatically illuminated on failure of the mains supply, *Figure 13.9.* Sustained emergency lighting is a category of a non-maintained system whereby a lighting fitting contains lamps illuminated from normal supply and separate lamp(s) illuminated from the battery only when normal supply fails. In this case the luminance of the fitting may differ during normal and emergency periods.

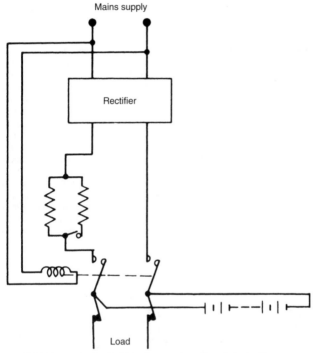

Figure 13.9 Non-maintained trickle charge automatic switch circuit

Central system. A number of lamps (luminaires) fed from a central secondary power supply is known as a central system.

Equipment available. Battery equipment available for emergency lighting systems may be classified as follows:

Category 1. Central storage batteries supplying tungsten filament lamps.

Category 2. Central storage batteries supplying fluorescent tubes, each fluorescent tube fitting being equipped with its own individual in-built static inverter to give high frequency a.c. to the tube. Various frequencies are used, 10, 20 and 40 kHz, for example.

Category 3.   Central storage batteries supplying bulk inverters which can produce several kilowatts of a.c. at normal mains frequency, i.e. 50 Hz.

Category 4.   Self-contained luminaires, each containing its own battery, usually of the sealed nickel cadmium type, complete with charging facility. Self-contained luminaires are available in a wide range of designs and are usually equipped with low power filament bulbs or transistor-inverter fluorescent tubes of 8 W and 13 W rating.

Chloride Keepalite systems. Operation of maintained and non-maintained Chloride Keepalite systems are described below and reference should be made to the accompanying electrical diagrams. *Figure 13.10* shows a Keepalite maintained system. Only one control switch is provided, this being of an on-off nature in the mains a.c. supply. After installation this is switched on and operation thereafter is entirely automatic. While the mains supply is healthy the maintained lights circuit is supplied with a.c. through the transformer, and the battery is kept in a fully charged condition by the constant voltage charger. On failure of the mains the maintained lights circuit and the non-maintained lights circuit are instantly and automatically connected to the battery by the changeover contactor. When the mains supply is restored the contactor takes up its normal position which disconnects the emergency non-maintained lights circuit and reconnects the maintained lights circuit to the transformer. The charger then takes over and automatically recharges the battery. When the fully charged condition is reached the battery voltage reaches the constant voltage setting, reducing the charge current to the low level needed to balance open-circuit cell losses.

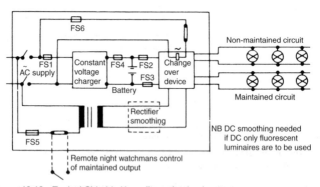

Figure 13.10   Typical Chloride Keepalite maintained system

Figure 13.11 shows a Keepalite non-maintained system. This also has only one switch in the mains supply circuit. This system energizes the emergency circuits only when the mains supply fails. While the mains is healthy the battery is kept in a fully charged condition by the charger. Following a mains failure the emergency lights are connected to the battery virtually

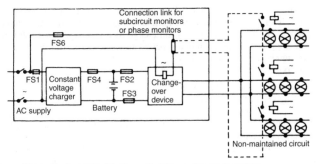

Figure 13.11 Typical Chloride Keepalite 'Non-maintained' system

instantaneously. Upon restoration of the mains supply the operation is similar to that described above for the maintained system.

Statutory requirements

A minimum illuminance of 0.2 lux (preferably 1 lux) on the centreline of an escape route without hazard, with a maximum diversity of 40 : 1. Since photometric data for spacing is based upon modern luminaire performance alone, the usual average illuminance is likely to be in the order of 1 lux. For open areas greater than 60 m² a minimum level of 0.5 lux applies, excluding the last 0.5 m to the wall.

Positioning of luminaires must be such as to indicate exit and change of direction signs and hazards such as steps or ramps on the exit route, which includes the area outside the final door. They must be placed so that they indicate the lines of escape.

Legal responsibility. Fire officers are responsible for the issue of fire certificates for most premises, except government offices, schools and local council property which come under the aegis of HM Factory Inspectorate. The inspector will demand a written guarantee from the contractor to the effect that the installation conforms to local standards and requirements, usually to BS 5266. The final legal responsibility rests with the owner or occupier.

Choice of luminaire. The choice of the type of luminaire will be governed partly by economic and partly by engineering considerations. Centrally controlled (slave) luminaires are less expensive than the self-contained type, but the necessity of providing separate wiring from the central control and power position to each fitting largely offsets this, especially as the provision of heat-resisting wiring is essential. In addition a plant room, housing the central batteries and control switchgear, is necessary. There is no certainty that in the event of a subcircuit failing the system will come into operation unless an expensive relay system is provided.

Self-contained luminaires require little or no special wiring as they can be operated from the same distribution boxes as the main system. They can easily be added to an existing installation and, since they automatically come into action when the supply to the part of the installation in which they occur is interrupted, can easily be tested by removing a fuse or operating an isolating switch. Each contains its own power supply, usually consisting of nickel cadmium cells located in the luminaire and operating miniature fluorescent tubes. They are maintained at full charge by a mains-powered solid-state module and operate for a minimum period of three hours from the interruption of supply. Consequently this type of emergency lighting system is rapidly gaining popularity compared to the centrally operated type.

Planning the installation. Escape routes should be marked on a floor plan. They should be as direct as possible but avoid congested areas or those where rapid spread of flame is likely. Danger points, doors, staircases and places where the route changes direction, should be specially emphasized and an emergency lighting fitting placed at each of them. The isolux diagram provided by the manufacturer should then be centred on each of these points and positions on the centreline of the escape route where the illuminance falls below 0.2 lux marked and extra luminaires positioned at them. It is essential to ensure that each luminaire is so placed as to be visible from the position of the preceding one, as in smoky conditions they may have to act as guides on the escape route.

Exit signs should be mounted between 2 and 2.5 m above floor level and positioned close to the points they indicate. Where no direct sightline to the exit exists subsidiary signs should be used to point the way to it. Exit signs should be of the 'maintained' type; i.e. they should be lighted all the time that the building is occupied.

Signs are to be kept in order and illuminated for as long as it takes to evacuate the building. Standards for lighting are following the pictogram route but without wording – that is a man running towards a solid door with a pointing arrow in between. These are gradually replacing the interim format (detailed in BS 5499) of the mixture of a running man inside a box, arrow and text. The old BS 5260 'Exit' word sign on its own is now obsolete. Note that it is not permitted to mix signs of different formats in an installation.

Maintenance and testing. Each self-contained sign should be energized once a month from its internal battery and every six months it should be left on for an hour (BS 5266 Part 1). All testing should take place at times of least risk and personnel operating in the area should be notified.

Recommendations for further reading

Chartered Institution of Building Services Publications and Codes, Delta House, 222 Balham High Road, London SW12 9BS, www.cibse.org

Lighting Industry Federation Factfinders, Swan House, 207 Balham High Road, London SW17 7BQ, www.lif.co.uk

ICEL: Emergency Lighting Design Guides. Tel: 020 8675 5432 email: info@icel.co.uk

This section edited by Thorn Lighting Ltd to whom acknowledgement is made.

14 Motors and control gear

D.C. Motors

D.C. motors are divided into three classes, as follows:

1. *The series-wound motor.* In this type (*Figure 14.1*) the field is in series
   with the armature. This type of motor is only used for direct coupling
   and other work where the load (or part of the load) is permanently
   coupled to the motor. This will be seen from the speed-torque characteristic, which shows that on no load or light load the speed will be
   very high and therefore dangerous.

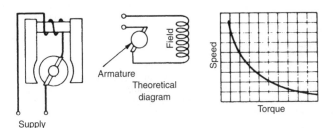

Figure 14.1   Series-wound motor

2. *The shunt-wound motor.* In this case the field is in parallel with the
   armature, as shown in *Figure 14.2*, and the shunt motor is the standard
   type of d.c. motor for ordinary purposes. Its speed is nearly constant,
   falling off as the load increases due to resistance drop and armature
   reaction.

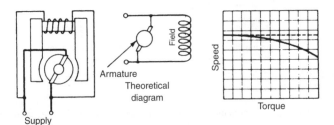

Figure 14.2   Shunt-wound motor

3. *The compound-wound motor.* This is a combination of the above two
   types. There is a field winding in series with the armature and a field
   winding in parallel with it (*Figure 14.3*). The relative proportions of

292

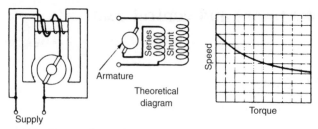

Figure 14.3 Compound-wound motor

the shunt and series winding can be varied in order to make the characteristics nearer those of the series motor or those of the shunt-wound motor. The typical speed-torque curve is shown in the diagram.

Compound-wound motors are used for cranes and other heavy duty applications where an overload may have to be carried and a heavy starting torque is required.

Speed control. Speed control is obtained as follows.

Series motors. By series resistance in parallel with the field winding of the motor. The resistance is then known as a diverter resistance. Another method used in traction consists of starting up two motors in series and then connecting them in parallel when a certain speed has been reached. Series resistances are used to limit the current in this case.

Shunt- and compound-wound motors. Speed regulation on shunt- and compound-wound motors is obtained by resistance in series with the shunt-field winding only. This is shown diagrammatically for a shunt motor in *Figure 14.4*.

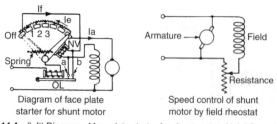

Diagram of face plate
starter for shunt motor

Speed control of shunt
motor by field rheostat

Figure 14.4 (left) Diagram of faceplate starter for shunt motor. (right) Speed control of shunt motor by field rheostat

Starting. The principle of starting a shunt motor will be seen from *Figure 14.4* which shows the faceplate-type starter, the starting resistance being in between the segments marked 1, 2, 3, etc. The starting handle is held in position by the no-volt coil, marked NV, which automatically allows the starter to return to the off position if the supply fails. Overload protection is obtained by means of the overload coil, marked OL, which on overload short-circuits the no-volt coil by means of the contacts marked *a* and *b*.

When starting a shunt-wound motor it is most important to see that the shunt rheostat (for speed control) is in the slow-speed position. This is because the starting torque is proportional to the field current and this field current must be at its maximum value for starting purposes. Many starters have the speed regulator interlocked with the starting handle so that the motor cannot be started with a weak field.

These methods of starting are not used much today but are left in because many installations still exist. Modern methods of control employ static devices described below.

Ward-Leonard control. One of the most important methods of speed control is that involving the Ward-Leonard principle which comprises a d.c. motor fed from its own motor-generator set. The diagram of connections is shown in *Figure 14.5*. The usual components are an a.c. induction or synchronous motor, driving a d.c. generator, and a constant voltage exciter; a shunt-wound d.c. driving motor and a field rheostat. The speed of the driving motor is controlled by varying the voltage applied to the armature, by means of the rheostat in the shunt winding circuit of the generator. The d.c. supply to the field windings of the generator and driving motor is obtained by means of an exciter driven from the generator shaft.

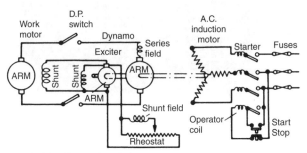

Figure 14.5 Ward-Leonard control

With the equipment it is possible to obtain 10 to 1 speed range by regulation of the generator shunt field and these sets have been used for outputs of 360 W and upwards. On the smaller sizes speed ranges up to 15 to 1 have been obtained, but for general purposes the safe limit can be taken as 10 to 1. Speed control obtained in this way is extremely stable and the speed regulation between no load and full load at any particular setting is from 7 to 10%, depending on the size and design of the equipment.

This type of drive has been used for a variety of industrial applications and has been particularly successful in the case of electric planers and certain types of lifts, with outputs varying from 15 kW to 112 kW, also in the case of grinders in outputs of 360 W, $3/4$ kW and $1 1/2$ kW with speed ranges from 6:1 to 10:1.

Thyristor regulators. The development of thyristors with high current-carrying capacity and reliability has enabled thyristor regulators to be designed to provide a d.c. variable drive system that can match and even better the many a.c. variable-speed drive systems on the market. This has meant a redesign of the d.c. motor to cater for the characteristics of the thyristor regulator. Machines have laminated poles and smaller machines may also have laminated yokes. This is to improve commutation by allowing the magnetic circuit to respond more quickly to flux changes caused by the thyristor regulators. Square frame designs of d.c. machines have also been developed with much improved power/weight ratios together with other advantages. A typical motor is shown in *Figure 14.6*.

An electrical schematic, *Figure 14.7*, shows a three-phase 6-pulse thyristor bridge, d.c. motor and feedback loop. It is usual for a thyristor-controlled d.c.

Figure 14.6 A Mawdsley's square frame industrial d.c. motor rated at 250 kW with forced ventilation

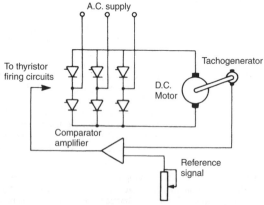

Figure 14.7 Three-phase six-pulse thyristor bridge, d.c. motor and feedback loop

motor to be fitted with a tachogenerator which provides a speed related signal for comparison with the reference signal. It is the error or difference between these two signals which advances or retards the firing angle of the thyristors in the regulator to correct the d.c. voltage and hence the speed of the motor. In addition to, or instead of using the motor speed, other parameters such as current, load sharing, web or strip tension and level may be employed.

Switched reluctance drives. A development in the d.c. variable-speed drive is that of the switched reluctance motor. The machine has salient poles on the stator and as in the case of other d.c. motors the number of poles does not determine the motor speed. The rotor is built up of shaped laminations but carries no winding at all.

Unidirectional current pulses are applied sequentially to the field poles at a rate determined by the required speed and rotor position. Control of the pulse timing is derived from a position transducer mounted on the motor shaft. By appropriate pulse timing either positive or negative torques can be achieved and full four quadrant control is available over a wide speed range and both constant torque or constant power characteristics can be provided.

A.C. Motors

Alternating current motors can be grouped as follows:

(a) Induction motors.
(b) Synchronous motors.
(c) Variable-speed commutator motors including the Schrage motor.
(d) Series motors.
(e) Single-phase repulsion, capacitor and shunt motors.
(f) Pole-changing and other special motors.

The first three are used in all sizes, and for all general purposes induction motors are employed on account of their simplicity, reliability and low first cost. Synchronous motors are generally installed where it is desirable to obtain power-factor improvement or where a constant speed is required. They are only economical in the case of larger loads, say of 75 kW and over, although there are instances where smaller machines are in use for special purposes.

The three-phase commutator motor was once the only a.c. motor for large outputs which gives full speed control, and although expensive it was widely used for duties where variable speed is required. Pole-changing and other special motors with speed control characteristics are now available and details of them are given later on in this chapter. Static inverters for use with cage induction motors are becoming very popular as variable-speed drives and are also described later.

Groups (d) and (e) represent the types used for small power motors, which also include induction motors. These small motors have been developed to a great extent because of the number of small machines incorporating individual drives. Normally small power motors include machines developing from 20 W to about $1\frac{1}{2}$ kW. The reason for inclusion of the $1\frac{1}{2}$ kW motor is that a different technique is used for manufacturing small motors which are turned out in large quantities by mass-production methods. Most of the manufacturers of these small motors can supply them with gearing incorporated giving final shaft speed of any value down to one revolution in 24 hours or even longer.

The induction motor, which can be termed the standard motor, is now made on mass-production lines, and as a result of the standardization of voltage and frequency, the cost of standard-sized new motors is exceptionally low. The absence of a commutator and, in the case of the cage motor, of any connection whatever to the rotor, combined with the simplicity of starting, make it the most reliable and cheapest form of power drive available.

There are a number of specialized motors which are used in a few unusual applications, but these will not be described as they are really rather of academic interest than of general use in industry. The linear induction motor falls within this category although this has been applied to overhead cranes and as a means of actuation where force without movement is required.

The use of synchronous motors for improvement of power factor is referred to in the section dealing with power factor correction, but it should be realized that the essential points of a synchronous motor are its constant speed (depending on the frequency) and the fact that the power factor at which it operates can be varied at will over a certain range – usually from 0.6 leading to 0.8 lagging – by varying the excitation current.

Induction Motors

The essential principle of an induction motor is that the current in the stator winding produces a rotating flux which because it cuts the rotor bars induces a current in the winding of the rotor. This current then produces its own field which reacts with the rotating stator flux thus producing the necessary starting and running torque.

The stator winding to produce this rotating flux is fairly simple in the case of a three-phase motor, being based on three symmetrical windings, as shown in *Figure 14.8*. In the case of single-phase and now vanishing two-phase it is not quite so easily understood.

Figure 14.8   Three-phase stator winding to produce rotating field

The stator field will revolve at synchronous speed and if no power whatever was required to turn the rotor it would catch up with the flux and would also revolve at synchronous speed. The condition for torque production would then have vanished. As, however, a certain amount of power is required to turn the rotor even if unconnected to any load, the speed is always slightly less than synchronous. As the load increases the speed falls in order to allow the additional rotor currents to be induced to give a larger torque.

The difference between the actual speed and synchronous speed is termed the slip, which is usually expressed as a percentage or a fraction of the synchronous speed. For standard machines the maximum slip at full load is usually about 4%.

Calculation of synchronous speed. Induction motors are made with any number of poles (in multiples of 2), but it is not usual to make motors with more than 10 poles, and for ordinary use 2, 4 and 6 poles are chosen, if possible, on account of the lower first cost and higher efficiency.

$$\text{Synchronous speed in rev/min} = \frac{\text{frequency} \times 60}{\text{number of pairs of poles}}$$

Thus a 2-pole motor on 50 Hz will have a synchronous speed of 3000 rev/min, a 4-pole 1500 rev/min, and 6-pole 1000 rev/min. The suitability of 1500 rev/min for many purposes has made the 4-pole motor the most common.

The actual rotor speed for 4% slip is given for various motors on 50 Hz in *Table 14.3* on page 312. Slip is calculated from

$$\text{Percentage slip} = \frac{(\text{syn.speed} - \text{rotor speed}) \times 100}{\text{syn. speed}}$$

and the rotor speed for any given slip will be

$$\text{Rotor speed} = \text{syn. speed} \frac{100 - \text{slip}}{100}$$

the slip being the percentage slip.

Variation of slip with torque. It can be shown that the torque of an induction motor is proportional to

$$T = \frac{kE_2 s R_2}{R_2^2 + (sX_2)^2}$$

where T = torque
k = constant
E_2 = rotor voltage
s = fractional slip
R_2 = rotor resistance
X_2 = rotor reactance.

The variation of slip with torque can therefore be calculated and typical torque-slip curves are given in *Figure 14.9*. These curves are for the same motor, and curve (*a*) is for the rotor short-circuited, whereas (*b*) and (*c*) are for cases where additional or added resistance has been put in the rotor circuit. It will be seen from these curves and also from the formula above that the torque at starting or low speeds is greatly increased by adding resistance in the

298

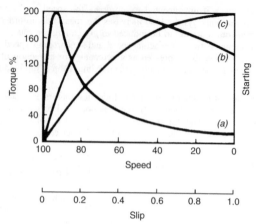

Figure 14.9 Torque-slip curves

(a) No added resistance in rotor $R = 1$.
(b) With added resistance $R = 4$
(c) With more added resistance $R = 20$

rotor circuit, and this principle is made use of in the wound-rotor induction motor which is used to start up against heavy loads.

It can also be shown that maximum torque occurs when

$$R_2 = sX_2 \text{ when the slip will be } s = R_2/X_2$$

Figure 14.10 shows the effect on the torque/speed characteristics of high reactance as an alternative to high resistance.

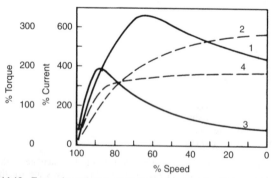

Figure 14.10 Torque/speed and current/speed curves for high resistance and high-reactance cage motors

Torque curves: 1. High resistance; 3. High reactance;
Current curves: 2. High reactance; 4. High resistance.

Wound-rotor or slip-ring motor. The slip-ring induction motor is used for duties where the motor has to start up against a fairly heavy load and the slip-rings are arranged for added resistance to be inserted in the rotor circuit for starting purposes. *Figure 14.11* shows how the various circuits are connected to the supply and to the variable rotor resistance.

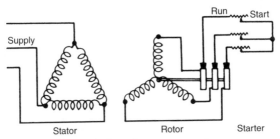

Figure 14.11 Starting wound-rotor motor

This type of motor is referred to as a wound-rotor motor, because for this purpose the rotor has to be wound with insulated conductors similar to those used for the stator. In the case of larger motors, a device is fitted to the rotor shaft enabling the slip-rings to be short-circuited and the brushes lifted off the slip-rings, thus reducing both electrical and friction losses. Starters for these motors are described in the subsection on 'Motor starters' later in this chapter.

With the external resistance in the rotor circuit, it is possible to obtain a large starting torque without drawing an excessive starting current from the line. Generally it will be found that full-load torque may be obtained when passing about $1\frac{1}{4}$ times full-load current; other values of torque, up to 2 to $2\frac{1}{2}$ times full load (depending upon the characteristics of the motor), may be obtained when passing from $2\frac{1}{2}$ to 3 times full-load current.

Cage motors. Cage motors should always be employed whenever possible because of their robustness, simplicity, lower cost and lower maintenance compared with slip-ring machines. They may be employed when the starting torque is sufficient to run the drive up to speed and when the starting current to do this is not too high for the conditions of the supply.

The rotor conductors of cage rotors normally consist of aluminium bars arranged round the periphery of the rotor and connected at each end by a ring of aluminium. Frequently both conductors and end-rings are die cast, although in larger sizes, of the order of 200 or 300 kW, the conductors and end-rings would be made of copper. The resistance of the rotor is thus low and very little starting torque is available in comparison with the current taken from the supply system. On the other hand, a very cheap and efficient motor is produced. *Figures 14.12(a)* to (*d*) show four different types of cage rotor machines.

Typical characteristics for an average purpose cage motor are shown in *Figure 14.13*. It will be seen that the torque developed at standstill by the motor on full voltage is 125% of full-load torque, the starting current being just over 600% full-load current, i.e. corresponding to the delta curves.

In view of the large starting current on full voltage, regulations imposed by supply companies protect the interests of neighbouring consumers by limiting the size of cage motors that may be started by switching direct on mains voltage. This rating of the machine that can be switched direct-on-line will vary depending on the electrical characteristics of the supply. In some cases the limit may be about 5 kW, in others over 75 kW on a 400 V supply.

In cases where the direct starting method cannot be employed because of the high initial peak current, the cage motor has to be started on reduced voltage. There are three methods of starting on reduced voltage, all of which involve a reduction in starting torque. This excludes static inverter starting which is discussed later. In such a method the voltage applied to the motor is also reduced. Another development in cage rotor design, also described later,

(a)

Figure 14.12a Drip-proof motor to BS 5000 Parts 10 and 99 and BS 4999 for general clean industrial conditions. It is available in ratings from 5.5 to 650 kW with foot or flange mounting. Enclosure is to IP22 (Brook Crompton Parkinson Motors)

(b)

Figure 14.12b Totally enclosed ventilated motor with IP54 enclosure in ratings from 0.18–425 kW foot or flange mounting. It complies with BS 5000 Part 10 (Brook Crompton Parkinson Motors)

(c)

Figure 14.12c Frame size 7–D280M totally enclosed fan-ventilated foot mounted motor to IP55. Ratings 18.5–425 kW (Brook Crompton Parkinson Motors)

(d)

Figure 14.12d Flameproof totally enclosed fan-ventilated motor for Group I applications in mines as certified by the Health & Safety Executive, and for Groups IIA and IIB Ex d industrial use, certified by BASEEFA. Available as foot or flange mounting with IP54 enclosure. Electrical performance to BS 5000 Part 99, 0.37 to 150 kW and outputs to BS 4683 Part 2 (Brook Crompton Parkinson Motors)

includes an integral eddy current inductor which produces a high torque at low current:

(a) Star-delta connections.
(b) The use of an auto-transformer.
(c) Primary resistance starting.

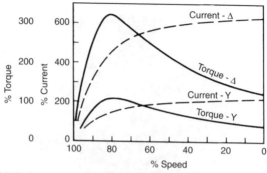

Figure 14.13 Torque/speed and current/speed curves for normal cage induction motor connected in star and delta

Starting on reduced voltage. Star-delta starting is the most usual method of starting on reduced voltage and consists of connecting the stator winding in star until a certain speed has been reached, when it is switched over to delta, the circuit connections being as shown in *Figure 14.14*. Six terminals must be available on the machine for these connections. By this means the voltage across each winding is reduced to 58% of the supply voltage. The star connection means that at starting the line current and torque values are one third of the full voltage figures. This method is simple and inexpensive, but can only be used where the starting load is small. Repeated starting can be obtained.

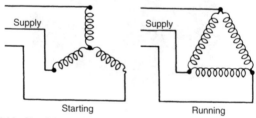

Figure 14.14 Star-delta starting for cage motors

Wauchope-type continuous torque star-delta starter. The Wauchope-type star-delta starter is an important advance in the technique of the starting of three-phase induction motors of cage design, and achieves a smooth acceleration approaching that of a slip-ring motor, although sustained excess torque during starting is not claimed.

Considering standard star-delta starting, if it were possible to change from star to delta instantaneously, the current would rise to a value approximately three times the current in star obtaining at the instant of changeover. But owing to the transition pause and consequent loss of speed, the delta peak current is very much increased, and may even approximate to the current which would be drawn from the line when the motor is connected in delta direct to the supply.

The Wauchope star-delta starter improves the starting characteristics associated with a standard star-delta starter by the insertion of resistance when changing over from star to delta, and incidentally, provides three steps of acceleration, instead of two steps as in the conventional starter.

Figure 14.15 shows the changes on connections in the Wauchope starter and the sequence of starting comprises in detail:

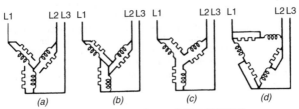

Figure 14.15 Wauchope star-delta starter – switching sequence

(a) The motor is connected in star and is permitted to accelerate to a stable speed in the normal manner.
(b) The three resistances are then connected in parallel with the motor windings. The line current is increased by the amount taken by the resistances, but this does not affect the running of the motor. This is simply a preparatory step and is in operation for a fraction of a second only.
(c) The star point of the windings is now opened. It will be seen from *Figure 14.16* that the resistances have been so connected that the motor windings are in delta, with a resistance in series with each winding. At this stage the voltage across the motor windings is increased, giving a corresponding increase in torque, and the motor accelerates to a higher steady speed.

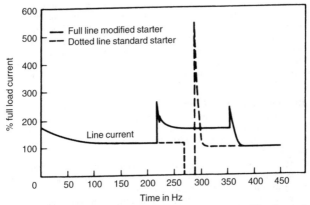

Figure 14.16 Variation of line current during star-delta starting Wauchope starter indicated by full line. Standard starter indicated by broken line (Simplex – GE Ltd)

(d) The resistances are then short-circuited and the motor windings connected in delta across full line voltage. The motor accelerates to full speed and the starting operation is complete.

Figure 14.16 shows the form of the resulting starting current for a machine utilizing a Wauchope star-delta starter.

Auto-transformer starting. The use of an auto-transformer is confined to those cases where a definite limit is required to the starting current, and the arrangement is shown in *Figure 14.17*, the transformer being disconnected from the supply in the running position. Transformer tappings enable the starting voltage to be selected to obtain the starting torque required for the load. Starting torque and current are each equal to the square of the transformer tapping (fraction) × direct switching value; e.g. 50% tapping gives $(1/2)^2$ or 25% of the direct switching value.

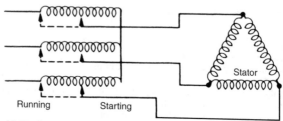

Figure 14.17 Auto-transformer starting for cage motors

Primary resistance starting. In this method of reduced voltage starting, the stator is connected through an adjustable three-phase series resistance. As the motor accelerates, the resistance is short-circuited in one or several steps. A heavier line current is required for a certain starting torque compared with other methods, e.g. 50% line voltage and current gives 25% torque and 80% line voltage and current gives 64% torque. Heavy peak currents are avoided. The number of starts is limited by the resistance rating. When high torque motors are employed, this method shows appreciable saving in first cost, and the simple yet robust construction will give very low maintenance costs. If it is not important to strictly limit the starting current, these advantages recommend its more extended use, and in an emergency or breakdown, the method allows considerable scope for ingenious improvisation.

Integral eddy current inductor. A further development in cage rotor design produces high torque at low current. It involves the inclusion of an eddy current inductor as part of the rotor, *Figure 14.18*. Manufactured in the output range 18.5 kW to 150 kW it is being used as an alternative to wound-rotor motors.

Known as the pipe cage motor, it has superior performance characteristics in terms of frequency of starting and driving high inertia loads while only requiring the cheaper direct-on-line starter.

A cage rotor is used having a core in alignment with that of the stator, but with copper rotor bars extended at the opposite drive end of the motor. This extension takes up the space normally occupied by the slip-rings of a

Figure 14.18 Complete pipe cage rotor for 30 kW 4-pole motor (Brook Crompton Parkinson Motors)

wound-rotor motor. Each extended rotor bar passes through a close fitting steel pipe before connection to the usual short-circuiting ring. The rotor bars are insulated in the slots and pipes by high temperature materials.

The pipe cage assembly acts as a series of individual short-circuited transformers with the rotor bars as their respective primary windings. The pipe forms both the core and secondary winding of these transformers which are short-circuited by the supporting discs. The pipes are made of steel and, therefore, exhibit a high electrical resistance which is further increased when the primary (i.e. rotor) current is at mains frequency. This is due to skin effect, thus making the effective pipe resistance dependent on rotor frequency and hence motor speed.

At start, with rotor current at mains frequency, a longitudinal current flows in the pipes and adds impedance to the rotor circuit. Both the resistance and inductive components of this impedance are dependent on rotor current and frequency and, therefore, decrease as the motor accelerates to full speed. It is this automatic adjustment of impedance that gives the pipe cage motor its torque, current, and speed characteristics. Starting current is decreased at start in comparison with a normal cage machine and the additional resistance increases the starting torque. *Figures 14.19* and *14.20* show the relationship of pipe cage motors to cage and slip-ring motors.

Cage versus slip-ring. A slip-ring motor may be chosen for a particular application in preference to a cage machine for one or more of several reasons:

1. To limit the starting current drawn from the line.
2. To obtain a high value of starting torque for a comparatively low current drawn from the line.
3. To obtain speed control.

If a cage motor can be used that will satisfy either (1) or (2) for a particular duty, it will be possible to dispense with the slip-ring machine on that particular score. Both the starting torque and starting current of cage motors depend on the characteristics of the rotor employed; and there is a wide choice of motor for different types of drive.

'High resistance' cage motor. As with the slip-ring (wound-rotor) motor, the starting torque of the motor can be increased by using a rotor with a higher

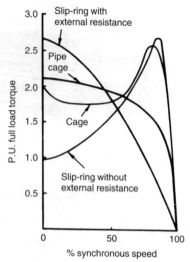

Figure 14.19 Performance of a pipe cage motor compared with cage and slip-ring motors – torque/speed curves

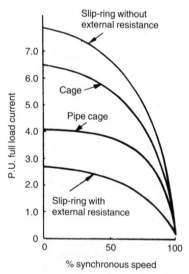

Figure 14.20 Performance of a pipe cage motor compared with cage and slip-ring motors – full load current/speed curves

resistance; but since this resistance has to remain permanently in circuit (as in the ordinary cage motor), it cannot be too high if the motor is to retain anything like normal efficiency. A 'high resistance' rotor has been used in applications such as forge hammers and power presses. The motor is allowed to slow down with increasing load so that the stored energy of the flywheel and other moving parts of the system may be utilized to relieve the motor of very heavy overloads. By this means also the supply line is relieved of very heavy peak currents, since the additional peak torque is provided by the flywheel and not by the motor. With such machines, therefore, since the load is of an intermittent nature, efficiency is not of so great importance as overload capacity.

High resistance rotors have been frequently used on cranes and hoists where a high starting torque is required, but generally it may be said that, unless there is some definite advantage to be gained, the inefficiency will be such as to preclude their use in any situation where energy consumption is of importance.

Double cage high torque motor. In order to overcome the disadvantages of the cage motor, and avoid having to use the more expensive slip-ring motor and its associated gear, an alternative is the use of the double cage rotor in which the resistance of the cage rotor is increased temporarily while starting.

Table 14.1 Efficiency and power factor of 4-pole induction motors on 50 Hz–full load

The following are average values for standard motors running at 1440 to 1470 rev/min.

kW	Efficiency				Power factor			
	Single phase				Single phase			
	Split phase	Capacitor	2 phase	3 phase	Split phase	Capacitor	2 phase	3 phase
	Per cent	Per cent	Per cent	Per cent				
$3/4$	65	70	73	74	0.80	0.90	0.79	0.81
$1\frac{1}{8}$	69	74	76	78	0.81	0.90	0.79	0.81
$1\frac{1}{2}$	72	76	78	80	0.82	0.91	0.82	0.84
$2\frac{1}{4}$	74	78	82	83	0.82	0.92	0.82	0.84
3	76	80	83	84	0.82	0.93	0.82	0.84
$3\frac{3}{4}$		82	84	85		0.93	0.84	0.86
$5\frac{5}{8}$		83	85	86		0.94	0.85	0.87
$7\frac{1}{2}$		84	87	88		0.94	0.86	0.88
$9\frac{3}{8}$		84	87	88		0.90	0.86	0.88
$11\frac{1}{4}$		85	88	88		0.90	0.87	0.89
15		86	88	90		0.90	0.88	0.90
$22\frac{1}{2}$			89	90			0.88	0.90
30			90	90			0.89	0.90
$37\frac{1}{2}$			91	91			0.90	0.91
$56\frac{1}{4}$			91	91			0.90	0.91
75			92	92			0.91	0.92

Table 14.2 Full load currents (amperes) of alternating current motors

The values below may vary slightly with different types of motors but can be accepted as reasonably accurate

kW	Single phase										Two phase		Three phase					
	Split phase (Volts)					Capacitor (Volts)					Volts		Volts					
	100	200	230	400	480	100	200	230	400	480	200	400	200	220	350	400	440	500
3/4	12.9	6.5	5.6	3.2	2.7	11.6	5.9	5.1	2.9	2.4	2.9	1.4	3.4	3.1	2	1.7	1.5	1.4
1 1/2	24	12	10.4	6	5	22	11	9.5	5.4	4.5	5.7	2.9	6.6	6	3.8	3.3	3	2.6
2 1/4	35	17	15	9	7	31	15	13	7.8	6.5	8.4	4.2	9.8	8.9	5.7	4.9	4.5	3.9
3	45	23	20	11	9.4	40	20	18	10	8.5	11	5.6	13	12	7.5	6.5	6	5.2
3 3/4	56	28	25	14	12	51	25	22	13	11	14	6.8	16	14	9.1	7.9	7	6.3
4 1/2	66	33	29	17	14	60	29	26	15	13	16	8.1	19	17	11	9.4	8.2	7.5
5 5/8	82	41	36	21	17	74	37	33	19	15	20	10	24	21	14	12	11.5	9.4
6	87	43	38	22	18	78	39	34	20	16	21	11	25	22	15	13	12	9.8
7 1/2	109	54	47	27	23	98	48	42	24	20	26	13	30	27	17	16	14	12
11 1/4	159	79	69	40	33	147	73	64	37	31	38	19	44	40	26	22	20	18
15	209	105	91	52	44	193	97	84	48	41	50	25	58	53	34	29	26	23
16 3/4	256	128	111	64	53	237	118	103	59	49	62	31	72	66	42	36	33	29
22 1/2	306	152	134	77	64	283	142	124	71	59	74	37	86	78	50	43	39	34
30	400	200	174	100	83	370	185	161	93	77	96	48	111	101	64	56	51	45

37 1/2	487	244	212	122	102	450	226	196	113	94	118	59	137	124	79	68	62	55
45	586	292	254	147	122	542	270	235	136	113	140	70	162	147	94	81	74	65
56 1/4	715	358	310	179	150	662	330	287	166	139	171	86	198	180	114	99	90	79
75	941	471	410	235	196	870	435	380	217	182	228	114	263	239	152	132	120	105
102 1/2	–	–	–	–	–	–	–	–	–	–	336	168	388	356	225	194	176	155
150	–	–	–	–	–	–	–	–	–	–	446	223	517	468	299	258	235	207

The current required for any alternating current motor can be obtained from the following equations. The power factor and efficiency can be obtained from Table 14.1.

Single phase

$$\text{Current} = \frac{kW \times 1000}{\text{Voltage} \times \text{power factor} \times \text{efficiency}}$$

Three phase

$$\text{Current} = \frac{kW \times 1000}{1.732 \times \text{line voltage} \times \text{power factor} \times \text{efficiency}}$$

Two phase, four-wire supply

$$\text{Current} = \frac{kW \times 1000}{2 \times \text{line voltage} \times \text{power factor} \times \text{efficiency}}$$

Two phase, three-wire supply

Current = In outers, as above

In common, 1.414 × outer value

The double cage rotor in its simple form consists of two separate cages. The outer or starting cage is made of high resistance material and is arranged to have the smallest possible reactance. The inner cage is of the ordinary low resistance type, and since it is sunk deep into the iron, has a high reactance. The four qualities – reactance and resistance of inner and outer cage – can be varied in an infinite number of combinations and many different shapes of speed-torque curve can be obtained. At starting, the frequency of the currents in the rotor conductors is the same as the supply frequency, thus the high reactance of the inner cage produces a limiting effect and reduces the current flowing in this winding. The outer cage, being of high resistance, develops a high starting torque depending largely on the value of its resistance. As the rotor accelerates and approaches synchronism, the frequency of the e.m.f.s in its conductors falls and the effective reactance in the inner cage is reduced; the inner cage now carries practically all the current until finally, when near synchronism, the rotor operates with the characteristics of an ordinary low resistance rotor. The general result is to produce a machine having a high start torque and a high running efficiency, with reasonably small values of starting current.

Often the higher starting torque allows the motor to be started on reduced voltage provided the load against which it is required to accelerate is not too great.

By altering the relative values of resistance and reactance, a wide variety of torque-speed characteristics can be obtained (*Figure 14.21*). For example, one make of standard high torque motors available up to about 55 kW for direct switching develops at least twice full-load torque at standstill and takes a starting current of about 350% full-load current. In another range of standard machines for star-delta starting, 100% of full-load torque is developed at starting in the star connection, the starting current being 150–175% full-load current.

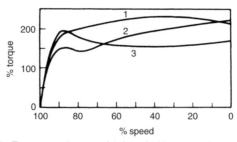

Figure 14.21 Torque-speed curves of various double cage motors

It should be noted that the cost of a double cage machine and its starting gear is considerably less than an equivalent slip-ring motor with resistance starter, and performance is obtained without sacrifice of running efficiency.

Extra high-speed induction motors. Certain types of wood- and metal-working machines and portable tools require very high speeds of rotation (e.g. up to 27 000 rev/min). Such speeds cannot be obtained with a direct drive from an ordinary induction motor supplied at standard frequency (50 Hz), as the maximum speed is 3000 rev/mm (corresponding to two poles). In these special cases, where a direct drive is particularly advantageous, the solution to the problem is to raise the frequency of the supply to the motor. This can be accomplished by means of a frequency changer set, or frequency booster.

A frequency booster is a slip-ring induction motor. If this is excited from the a.c. mains and driven in the opposite direction to that of the rotating magnetic field of its primary winding, a high frequency supply can be drawn from the secondary winding.

Either the stator or rotor can be made to act as primary, selection being according to whichever best suits the voltage and current requirements.

The frequency booster may be driven by a motor (direct or belt coupled) or by any other source of mechanical power.

With different driving-motor speeds and excitation windings, different frequencies can be generated as indicated by the following formula:

High frequency output

$$= \left(\text{Pairs of poles in booster} \times \frac{\text{rev/min of motor}}{60} \right)$$

$+$ Excitation frequency

Special precautions are necessary in the arrangement of the switchgear. The driving motor starting switches must be interlocked with the switches controlling the power supply of the induction machine's stator and rotor, so that it is impossible to connect either the input or the output of the high frequency unit before the machines are run up to speed. Similarly, it must be arranged so that in the event of the motor circuit being tripped the input and output circuits of the booster are tripped at the same time.

The standard high frequency portable electric tool is designed to operate at 200 Hz. Compared with the 'universal' type of motor, the speed does not drop in accordance with the load applied, and there are no commutators or brushes to wear or to be replaced.

Synchronous Motors

The synchronous motor is essentially a reversed alternator and is often used for power-factor correction. As its name implies, it has a constant speed (running at synchronous speed at all loads), and its power factor can be controlled by varying the exciting current. It can thus be made to take a leading current for power-factor improvement purposes. The synchronous motor itself is not self-starting and it must be synchronized on to the supply when it has been run up to speed by a special starting motor or by some other means.

Self-contained motor. Refer to *Figure 14.22* which shows the arrangement of a self-contained synchronous motor suitable for driving a steady load but does not show any method of starting. The three-phase supply is taken direct to the stator and a d.c. supply is necessary for excitation. This can either be obtained from a separate d.c. system (sometimes used where there are several motors in use) or from the individual exciter mounted direct to the motor as shown. Power-factor control is obtained by varying the excitation – this

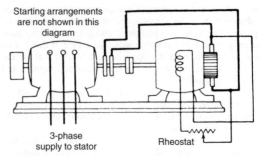

Figure 14.22 Diagrammatic arrangement of motor with exciter. Power factor is controlled by rheostat in exciting circuit of exciter

being regulated in large motors by means of a rheostat in the field circuit of the exciter.

This type of motor in its simple form has no starting torque and will not therefore start up under load. Also if the overload capacity is exceeded the motor will fall 'out of step' and will shut down. It must then be started up and synchronized in the usual manner.

Synchronous-induction motor. The diagram for a typical self-starting synchronous-induction motor is shown in *Figure 14.23*. It will be seen that by means of a starting resistance the machine will start up as an induction motor. As full speed is reached the motor will pull into synchronism (against full load if required) and the starting resistances are then short-circuited.

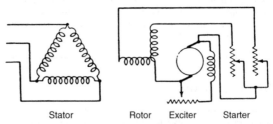

Figure 14.23 Synchronous induction motor. A two-phase rotor is used. Diagram shows method of starting as an induction motor

Table 14.3 Synchronous speed and rotor speed at 4% slip

No. of poles on stator	Synchronous speed on 50 Hz, rev/min	Rotor speed at 4% slip
2	3000	2880
4	1500	1440
6	1000	960
8	750	720
10	600	576

A two-phase winding is used on the rotor and arranged so that the neutral point is used as one connection for the excitation circuit.

Hunting. One of the features of a synchronous motor is that on a fluctuating load it may hunt. In modern industrial motors this is prevented by means of a damping winding in which eddy currents are induced by the variations in speed should hunting occur.

Hunting is more likely to occur with weak excitation than with strong. Temporary hunting can therefore often be cured by strengthening the field.

Single-Phase Motors

Where a three-phase supply is not available most small power applications can be met by using single-phase motors. In some circumstances even if there is a three-phase supply there may be economic benefit to be gained from using a single-phase machine in the way of simpler wiring and control gear.

In order to be self-starting an electric motor must have a rotating magnetic field. The phase displacement in a three-phase supply produces this, but a single-phase motor requires an auxiliary (starting) winding designed to give a displacement similar to a two-phase supply before this effect is produced. It can be achieved in various ways, each of which produces a motor with a different set of characteristics. Most of the motors described below are of the induction type, only the series and repulsion motors having wound rotors.

Split-phase motor. The starting winding of a split-phase motor, *Figure 14.24*, uses fine wire and thus has a high resistance. It is also arranged to have a low reactance. The current in the start winding thus leads that in the main winding and a rotating magnetic field is set up similar to that of a two-phase motor. The starting winding works at a high current density and it must be switched out as soon as possible when the machine reaches about 75% of full-load speed. This type of motor is suitable for low inertia loads and infrequent starting. Starting current is relatively large and account should be taken of this when installing, to avoid excessive voltage drop.

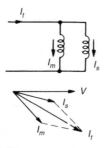

Figure 14.24 Split-phase motor

Capacitor start motor. A capacitor is inserted in series with the starting winding to reduce the inductive reactance to a low or even a negative value,

314

Figure 14.25. The starting winding current therefore leads the main winding current by almost 90°. A large a.c. electrolytic capacitor is used and, since this is short-time rated, it must be switched out as soon as the motor runs up to about 75% of full-load speed. The motor is suitable for loads of higher inertia or more frequent starting than is the split-phase motor; starting torque is improved and starting current reduced.

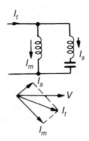

Figure 14.25 Capacitor start induction run motor

Capacitor start and run motor. A paper capacitor is permanently connected in series with the starting winding, *Figure 14.26*. Starting torque is low but running performance approaches that of a two-phase machine. Generally quieter than split-phase or capacitor start motors; efficiency and power factor are improved.

Figure 14.26 Capacitor start and run motor. Phasor diagram is shown in Figure 14.25

Capacitor start, capacitor run motor. For this machine a large electrolytic capacitor is used for starting but this is switched out when the motor runs up to speed and a smaller paper capacitor is left in circuit while the machine continues to operate. *Figure 14.27*. Thus the good starting performance of the capacitor start motor is combined with the good running performance of the capacitor start and run machine.

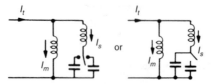

Figure 14.27 Capacitor start, capacitor run motor. Phasor diagram is shown in Figure 14.25

Shaded-pole motor. A short-circuited copper ring is placed round a portion of each pole, and this ring has currents induced in it by transformer action, these cause the flux in the shaded portion to lag the flux in the main pole and so a rotating field is set up. Starting torque is small and efficiency is poor since losses occur continuously in the shading ring (*Figure 14.28*).

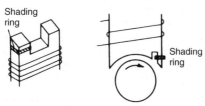

Figure 14.28 Shaded-pole motor

Typical performance details. *Table 14.4* gives performance details of the various motors discussed and compares them with a three-phase machine of a similar size. Small power motors only are considered.

Table 14.4 Performance details of single-phase machines compared with similar sized three-phase motors

Item	Three phase	Split phase	Capacitor start	Capacitor start and run	Shaded pole
Full load efficiency %	60/80	50/60	50/60	50/60	30
Full load power factor	0.6/0.8	0.6/0.7	0.6/0.7	0.9	0.5/0.7
Start current × full load current	6	8–11	5	4	2
Start torque %	200	180	280	40	30
Run-up torque %	200	165	190	35	25
Pull-out torque %	250	200	200	150–250	120
Speed variation	Some*	None	None	Some*	Some*
Starting cycles/hour	60	10–$\frac{1}{2}$ sec	15–3 sec	60	60
Limitations	–	High start curr.	Short time rating of caps.	Low start torque	Low start torque and low efficiency
Typical applications	–	For general use when high start torque or freq. start not needed	High start torque; low start current. High inertia drives, pumps, compressors	Generally confined to fan drives or where low starting torque acceptable	

*Some speed variation possible with fan drives by reducing the voltage, see text

Open-circuiting devices. Reference has been made to the open-circuiting devices in split-phase, capacitor-start, induction-run and capacitor-start, capacitor-run motors. These machines are fitted with a device to open-circuit the starting winding once the load has been accelerated to a certain speed. In some cases the device is a relay, the operating coil of which is connected in series with the running winding or in parallel with the starting winding to contacts which open the starting winding. In most cases, however, a speed-sensitive arrangement is used designed to operate as near as possible to the cross-over point between start and run winding speed-torque curves, varying for motors of different polarities or supply frequencies.

Speed variation. The induction motor is a constant speed machine but, in conjunction with fan drives, capacitor start and run and shaded-pole motors, can be controlled to give a limited speed variation. See later how the static inverter enables a cage machine to operate more economically at variable speed.

Under constant load conditions a motor will stall or fail to start when the voltage is lowered below a certain value. However, the load presented by a fan varies as the square of the speed, and stable running conditions continue at lower voltages despite loss of torque, see *Figure 14.29*.

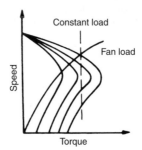

Figure 14.29   Speed control of fan load

Voltage variations may be achieved in a number of ways.

1. *Resistors.* Heat losses occur in resistors, and voltage drops occur at starting.
2. *Tapped windings.* Careful matching of fan and motor is essential. Voltage supply variations cause changes in speed which cannot be corrected. Manufacturing problems are created.
3. *Auto-transformer or Variac.* These devices are generally too expensive and automatic control is difficult.
4. *Thyristors.* Thyristors can be costly but as their use increases the costs fall. They are economic in multiple control systems. Negligible losses are incurred. They can be easily controlled through external circuits such as thermostats to give automatic regulation. Some problems exist with harmonic distortion of the public supply. When using thyristors special attention must be paid to peak currents at about two thirds full speed, which might result in excessive heating. Thyristor control is possible down to 10% of full-load speed, but between 10% and 40% full-load speed noise may be a problem. The motor is usually derated to 80–90% of full output.

Single-phase series commutator motor. This type of motor is similar to a d.c. series machine and it will run satisfactorily on a single-phase a.c. supply,

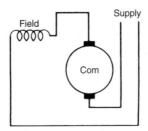

Figure 14.30 Simple series motor

Figure 14.30. The torque-speed characteristic is similar to its d.c. counterpart. Such motors are commonly used on domestic appliances like vacuum cleaners, power tools, etc.

Repulsion motors. There are many forms of repulsion motors, but the main principle is that a stator winding similar to a series motor is used with a wound rotor having a commutator which is short-circuited. The brushes are set at an angle (about 70°) to the main field and by means of transformer action the field and armature fluxes are such that they repel each other and the rotor produces a torque, *Figure 14.31.*

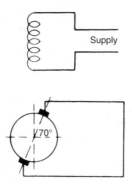

Figure 14.31 Simple repulsion motor

Repulsion motors can be started either by a variable series resistance or by auto-transformer, and a fair starting torque can be obtained. On this account the principle is used for starting in the repulsion-start, single-phase induction motor. In this motor the rotor is as used for a repulsion motor, but after starting the two brushes are lifted and the commutator short-circuited all round by means of a copper ring. The motor then runs as an induction motor.

The speed of a repulsion motor driving a constant-torque load may be controlled either by movement of the brushes or by variation of voltage applied to the motor. The former method is usually adopted because of its simplicity and the avoidance of additional control gear.

The performance of a repulsion motor start induction motor is generally similar to that of a high torque capacitor start machine. These motors are now available from only a small number of specialist manufacturers.

British standards. Most single-phase motors are in the small power range. Performance is covered by BS 5000 Part 11, Part 99 (IEC 60034-1) and dimensions by BS 2048.

Speed Variation of A.C. Motors

Standard types of a.c. motors do not permit any real speed variation as their speeds are fixed by the frequency of the supply on which they operate. The synchronous motor has a definitely constant speed and the induction motor can be assumed as constant speed as the maximum slip is not usually more than 5% except when supplied at variable frequency by a static inverter. A limited speed variation can be obtained by rotor resistances.

Rotor resistance control. The characteristic of the slip-ring machine with rotor resistance control approaches very closely to that of the series d.c. motor in that the speed rises as the load falls off and, therefore, it can only be satisfactorily employed for speed regulation where the load is fairly constant. Speed reduction by rotor resistance control is wasteful because of the power dissipated in the resistance, resulting in a low overall efficiency. In general, accurate control of speed cannot be obtained with any degree of satisfaction below 30–40% of full speed due to the fact that slight variation in the load causes wide fluctuations in the speed. In addition, difficulty may be experienced in maintaining constant speed due to the change of resistance with temperature of the external resistances, the speed tending to fall as the temperature of the controlling resistance increases. With a suitable external resistance of the liquid type, stepless variation between limits may be obtained, but with the metallic resistance grids and a drum controller, the number of steps of speed is, of course, limited to the number of settings in the controller, the resistances being graded to suit these steppings.

Pole-changing motors. As the speed of an induction motor depends on the number of poles, two, three or even four different speeds can be obtained by arranging the stator winding so that the number of poles may be changed. Speed ratios of 2 : 1 may be obtained from a single tapped winding and other ratios are available from two separate windings. The two systems may be combined to give, on a 50 Hz supply, speeds of, say, 3000, 1500, 1000 and 500 rev/mm. No intermediate speeds can be obtained by this method. These motors are sometimes used for machine tools, for example drilling machines, and this method gives a very convenient speed change.

P.A.M. motors. Pole amplitude modulation is a system of winding that offers various speed ratios by pole-changing without the need of separate windings. The system is patented by Prof. Rawcliffe of Bristol University and is a development of the original single-tapped two-to-one ratio Dahlander winding of 1897.

A single tapped winding is used and by reversing one half of each of the phases the flux distribution in the machine is changed, thereby producing a resultant field of different polarity. The main advantage of this system is the better utilization of the active material in the motor, resulting in a smaller frame size for a particular speed combination than would obtain using the two-winding method. Both the efficiency and power factor are higher with this method than for the equivalent two-winding machine.

Change in speed both up and down can be made without imposing undue transients on the supply system. Speed ratios of 3 : 1 or greater are not practical. The p.a.m. principle can be applied to slip-ring as well as cage machines.

Cascade induction motors. Induction motors can be arranged in cascade form to give intermediate speeds. In this arrangement two motors are arranged so that the rotor of one motor is connected in series with the stator of the second, *Figure 14.32*. Only one stator is fed from the supply.

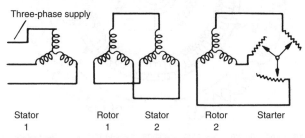

Three-phase supply

| Stator | Rotor | Stator | Rotor | Starter |
| 1 | 1 | 2 | 2 | |

Figure 14.32 Three-phase cascade connections

The speed of the common shaft will be equal to that of a motor having a number of poles equal to the sum of those of the two motors. The speed of a cascade arrangement is thus a low value – usually an advantage for driving heavy machinery. For speed variations they can be arranged so that either the main motor can be used separately or in cascade. For instance, a combination of a 4-pole and 6-pole motor will give either 1500 or 1000 rev/min separately, or combined the speed will be

$$\frac{50 \times 60}{2 + 3} = 600 \text{ rev/min}$$

The three-phase commutator motor (Schrage). A fully variable speed motor for use on three-phase is the commutator motor, one type of which is that due to Schrage, *Figure 14.33*. The primary winding is situated on the rotor and is fed by means of slip-rings. The rotor also carries a secondary winding which is connected in the usual way to the commutator and through the brushes to another secondary winding on the stator.

Three pairs of brushes are required, each pair feeding one phase of the stator winding, as shown in the diagram. Speed variation is obtained by moving each pair of brushes relative to each other, this being done by hand or automatic control through suitable wormgear.

The speed range is roughly 3 to 1 for normal load – this ranging from 40% above to 60% below synchronism. The speed varies from 5 to 20% with the load, but this can of course be counteracted by further movement

320

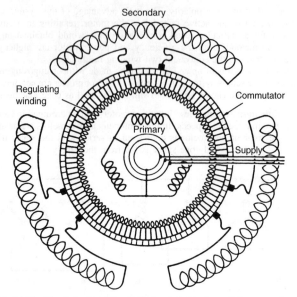

Figure 14.33 Diagram showing windings of three-phase Schrage commutator motor

of the brushes. Motors are usually started by placing brushes in lowest speed position, giving a starting torque up to $1\frac{1}{2}$ times full-load torque.

These motors are expensive in first cost but have proved very satisfactory for driving machinery requiring speed control, such as printing machines, textile mills, etc.

The higher initial cost over the slip-ring motor of corresponding rating is soon recovered because of the elimination of rheostat losses.

Motors for a speed range of $3:1$ are available with ratings from $21\frac{1}{4}$ W to 187 kW. Larger motors up to about 300 kW can be built for a smaller speed range. Larger ranges of speed up to $15:1$ are possible, but they involve a more costly motor, as the frame size for a given torque is governed by the speed range.

Variable-speed stator-fed commutator motor. Many applications requiring a variable-speed a.c. motor may be met by the stator-fed commutator type, *Figure 14.34*.

This motor is similar to a slip-ring induction motor, with the difference that the rotor winding is connected to a commutator instead of to slip-rings. Speed regulation is obtained by means of a separate induction regulator connected between the mains and the brushes on the commutator.

An auxiliary winding is often employed on the stator connected in series with the brushgear and the regulator, the purpose being to increase the power factor of the motor throughout its range of speeds.

The regulator acts as a variable-ratio transformer, supplying a variable voltage at supply frequency to the brushes. As the regulator secondary voltage

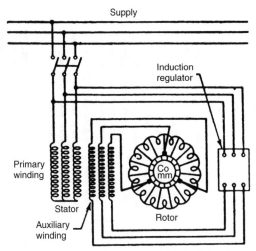

Figure 14.34 Internal and external connections of stator-fed variable-speed commutator motor

is reduced, the speed of the motor rises; the motor speed will rise above synchronous speed as the voltage is increased in the opposite direction.

In the majority of cases, operation of the regulator by a handwheel for speed adjustment is all that is necessary. As the regulator needs only a cable connection to the motor, remote control of the motor is very simple.

The speed of a stator-fed commutator motor can be varied without steps within the limits of the speed range. Normal ranges of speed regulation are 1:1.5, 1:2, 1:3, 1:4, but machines can be designed for ranges of 1:10, or greater. The machines for 1:3 speed range are usually designed so that the maximum speed is about 35% above synchronism and the minimum about 55% below.

The speed of the motor, like that of any shunt motor, is higher at no load than at full load. If very close speed regulation is desired, it may be obtained either by electrical compensation in the motor, or by the provision of automatic control of the regulator.

For most forms of drives, stator-fed motors develop sufficient starting torque when directly switched on to the line with the induction regulator in the lowest speed setting. Starting with the regulator in the other positions increases both starting torque and current.

Motors are available with ratings from $3/4$ kW up to about 375 kW for a speed range of 3:1. The size is a little larger than a slip-ring induction motor for the same duty. The motor is readily built with totally enclosed or totally enclosed fan-cooled enclosure because of the fixed brushgear and absence of slip-rings. Since the motor is stator-fed it may be designed for high voltage supplies.

'Tandem' motors. For lift and crane operation it is sometimes desirable that the power unit should be capable of giving a stable creeping speed in

Table 14.5 Characteristics of a.c. motors

Type	Speed variations on load	Possible speed control	Starting torque % Full Load	Starting current % Full Load	Notes
Polyphase induction cage	Falls up to 5 to 6% from no load to full load	No variation or control possible*	120–150 33–55 20–80	500–600 150–250 100–400	Direct starting Star-delta starting Auto-transformer Standard industrial motor. Speed assumed constant for general use.
Polyphase induction high torque	As above	None	180–300 60–100	390–525 130–175	Direct starting Star-delta Several makers supply specially designed motors with high starting torque with moderate starting current.
Polyphase induction slip-ring wound-rotor	As for cage	Can be varied by rotor resistance but variation depends on load. Up to 30% approx.	100–150	150–250	For general use where motor must start up on heavy load. Otherwise performance similar to cage.
Polyphase commutator (Schrage)	Varies up to 20% for one brush position	Variation 3 to 1 by moving brushes	150–200	150–200	A.C. variable speed motor for moderate or large sizes. Expensive in first cost. Speed control is ideal.

Polyphase commutator (stator fed)	Varies about 20%	Variation 3 to 1 by induction regulator	175 % Full Load	125 % Full Load	Starting characteristics at minimum speed position. Motor ratings from 1 to 375 kW May be designed for high-voltage operation.
Synchronous	Constant [power factor can be varied.]	None	None	–	For large continuously running drives. The auto-synchronous starts as an induction motor giving F.L. torque.
Single-phase induction split-phase	Falls slightly as load increases up to 6%	None	200–220 % Full Load small powers only	650–750 % Full Load	Used for small powers only on account of high starting current. Speed assumed constant

*Except when a variable frequency supply is available

addition to the normal hoisting or lowering speed. To meet these conditions the 'Tandem' motor has been developed for use where the electrical supply is alternating current.

The 'Tandem' motor consists of two component motors, built as one complete unit. The slow-speed motor is of the cage type and forms the main supporting part of the set. A slip-ring motor forms the high-speed unit.

Where the available power supply is 50 Hz, the two speeds usually selected are 960 rev/min for the slip-ring machine and 150 rev/min for the cage motor, but other speeds can be provided if desired. Speed changing can be effected while the motor is running and is carried out by transferring the supply from one stator winding to the other.

On changing down from the high speed to the low speed, the latter is reached by powerful regenerative braking during which the total reverse torque may amount to 400% of full-load torque. Under normal conditions the fairly high inertia of the rotor absorbs much of this torque, thereby preventing too rapid deceleration and excessive stress in the gears and other mechanism of the drive.

Further adjustment of the braking may be effected by inserting a choke or buffer resistance in series with the stator winding of the cage low-speed motor.

Slip energy recovery schemes. If instead of wasting the energy in the resistances associated with the speed control of slip-ring machines the power can be fed back into the supply then an improvement in the efficiency of the drive at reduced speeds can be obtained. A direct connection between the slip-rings and supply is not possible because both the slip-ring voltage and frequency vary with motor speed.

In the modified Kramer scheme a thyristor bridge is connected in the rotor circuit as shown in *Figure 14.35*. This converts slip-ring frequency power to d.c. which is fed to a separate d.c. motor-driven asynchronous generator which returns the slip power to the supply. A completely static system is available in which the motor generator is replaced by a static inverter. In another system the d.c. output from the thyristor bridge may be fed to a d.c. motor coupled to the main slip-ring motor so that the combined torques of both machines provide a constant power output to the drive over the speed range.

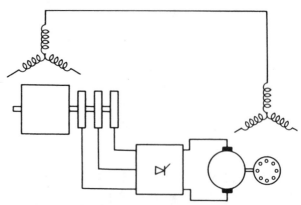

Figure 14.35 Modified Kramer slip recovery system

Variable frequency drives. The speed of an induction motor can be altered by varying the supply frequency. This variable frequency can be provided in a number of ways.

Until the advent of the thyristor the most common method of obtaining a variable frequency was by a motor generator set as shown in *Figure 14.36*. As such machines are expensive it is usual to use one set to supply a number of motors as indicated. Three full power electromechanical power conversions are involved in the use of a motor generator set and it is therefore advantageous to use systems in which the frequency conversion is accomplished electrically within the converter. Output frequency is still altered by changes in converter speed, but this is accomplished by using only a small driving motor. The converter itself is an unwound stator within which rotates a rotor whose windings are connected to both slip-rings and a commutator, *Figure 14.37*.

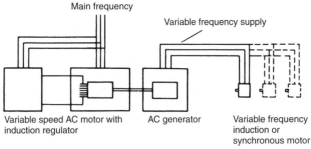

Figure 14.36 Motor generator set supplying variable frequency motors

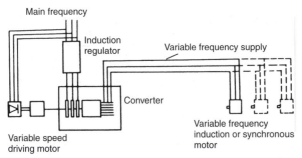

Figure 14.37 Parry motor-driven frequency converter supplying variable frequency motors

Self-driven converters are also available which are basically stator-fed commutator motors in which the output is taken not as mechanical power at the shaft, but as slip frequency power from a set of slip-rings connected to the rotor windings, *Figure 14.38*. The induction regulator in such an equipment controls the converter speed and therefore the frequency, the output voltage

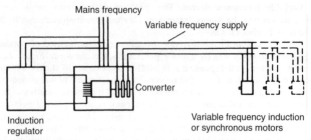

Figure 14.38 Self-driven frequency converter supplying variable frequency motors

inherently varying with the frequency to maintain a constant ratio between voltage and frequency.

The availability of thyristors makes it possible to change from a fixed frequency to a varying frequency by purely static means. Where the required frequency range is well below the mains frequency the cyclo-converter system can be employed. The required frequency can be synthesized by appropriate sequential switching of the motor supply terminals to successive supply phases. Two sets of thyristors are needed, supplying respectively the positive and negative half cycles of the a.c. supply to the motor. Regenerative operation of the motor is possible by allowing power flow back to the supply from the kinetic energy of a high inertia drive during rapid deceleration. *Figure 14.39* shows the connections for a cyclo-converter drive.

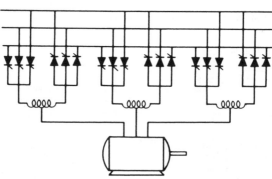

Figure 14.39 Arrangements of thyristors for a cyclo-converter to supply a variable frequency motor

Static frequency converter drive. Essentially a cage induction motor is a fixed-speed machine, the speed being related to the supply frequency and number of poles. The advent of variable frequency static control systems enables a standard 50 Hz cage machine to be driven at any speed corresponding to between 0.5 and 100 Hz, i.e. twice rated speed. One such system is the SAMI made by ABB and widely used in industries such as steel, petrochemical, paper, water and heating and ventilating.

Wherever variable-speed drives are required the static frequency converter and cage motor can usually be employed. Energy saving is a feature of such installation because the optimum speed for a given output can be attained.

Essentially a SAMI frequency converter consists of a 6-pulse diode bridge rectifier connected to the a.c. supply network and a p.w.m. inverter fed with d.c. from the rectifier, to provide the variable frequency to the motor, *Figure 14.40*. The d.c. link between the two static components includes an LC filter to smooth out any remaining a.c. from the rectifier and also to act as an energy store.

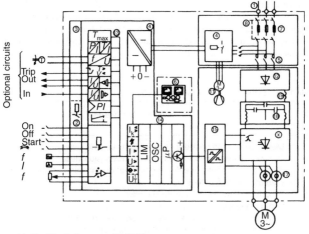

Figure 14.40 Block diagram of SAMI frequency converter

Motor Dimensions

Dimensions and outputs of electric motors for use with a cooling medium with a temperature not exceeding 40°C are covered by BS 4999 Part 141 and BS EN 50347. BS 4999 gives general information in which is included the nomenclature for identifying frame sizes and lays down standard dimensions and symbols that are mandatory and non-mandatory. BS EN 50347 specifies outputs in kilowatts and the associated shaft dimensions allocated to the frame sizes defined in BS 4999, output is related to speed, type of enclosures, class of rating, supply conditions and type of rotor. Part 141 of BS 4999 also indicates slide rail dimensions.

System of frame nomenclature. This is laid down in BS 4999 Part 103. For frame sizes up to and including 400 which are primarily intended for low voltage induction-type motors the nomenclature is as follows. First a letter indicates the type of enclosure: C is enclosed ventilated; D is totally enclosed except flameproof; and E is for all types of flameproof enclosures. Second a number of two or three digits indicates the height in millimetres of the shaft centre above the feet on a foot-mounted frame. Third a letter, S, M or L, characterizes the longitudinal dimensions where more than one length is used. Finally, for other than foot-mounted types, a letter is employed to indicate the type of mounting: D for flange mounting; V for skirt mounting; C for face flange mounting; P for pad mounting; and R for rod mounting. For example, a t.e.f.c. motor of frame size 160 M, suitable for flange mounting, is designated D160 MD.

Single letter symbols for dimensions are specified in *Table 14.6* Double letter symbols for dimensions are shown in *Table 14.7*. They are all shown on *Figure 14.41*. It is not possible to reproduce all the dimensions relating to the different types of induction motors covered by the standard but *Table 14.8* shows the fixing dimensions of foot-mounted frames (all enclosures) for low voltage a.c. induction motors. The stator terminal box is positioned on the left-hand side of the motor, looking at the non-driving end, with its centre on or about the centreline of the motor.

Table 14.6 Symbols for dimensions (BS 4999 Part 103)

Letter symbol	Dimension description
A	Distance between centre lines of fixing holes (end view)
B	Distance between centre lines of fixing holes (side view)
C	Distance from centre line of fixing holes at driving end to shaft shoulder
D	Diameter of shaft extension
E	Length of shaft extension from shoulder
F	Width of keyway
G	Distance from bottom of keyway to opposite side of shaft
H	Distance from centre line of shaft to bottom of feet
H^1	Distance from the centre line of shaft to mounting surface, e.g. bottom of feet for machines with low shaft centres
J	Radius of circle to which mounting pads (or faces for rods) are tangent
K	Diameter of holes in the feet or mounting pads (or faces)
L	Overall length
M	Pitch circle diameter of fixing holes
N	Diameter of spigot
P	Outside diameter of flange
R	Distance from surface of mounting flange to shaft shoulder
S	Diameter of fixing holes in flange
T	Depth of spigot

Outputs and shaft numbers for single-speed, t.e.f.c. cage rotor, MCR motors with Class E or Class B insulation are shown in *Table 14.9*. They are suitable for a three-phase 50 Hz supply not exceeding 660 V. Other tables in the standards give similar information for enclosed ventilated, slip-ring, enclosed ventilated slip-ring and flameproof t.e.f.c. motors. A further table provides the same information relating to t.e. frame surface cage rotor MCR motors.

Table 14.7 Symbols for dimensions (BS 4999 Part 103)

Letter Symbol	Dimension description
AA	Width of end of foot or pad (end view)
AB	Overall dimension across feet (end view)
AC	Overall diameter
AD	Distance from centre line to extreme outside of terminal box (end view) or other most salient object mounted on side of machine
AL	Overall length of slide rail excluding adjusting screw
AT	Thickness of slide rail foot
AU	Size of mounting holes in slide rail
AX	Height of slide rail
AY (max.)	Maximum extension of adjusting screw of slide rail
AZ	Width of slide rail at base
BA	Length of foot (side view)
BB	Overall dimension across feet or pad (side view)
BT (min.)	Horizontal travel on slide rail
DH	Designation of tapped hole or holes (if any) in shaft extension
DJ	Centre distance of the two tapped holes in shaft extension (if appropriate)
EB	Distance from end of bearing housing to shaft shoulder
EC	Length of shaft from end of bearing housing to end of shaft
ED	Minimum length of keyway
HA	Thickness of feet
HB	Distance from centre line to top of lifting device, terminal box or other most salient object mounted on top of the machine
HC	Distance from top of machine to bottom of feet
HD	Distance from bottom of feet to top of lifting device, terminal box or other most salient object mounted on top of the machine
KA	Usable tapped depth of hole in pad (or facing) of pad-mounted or rod-mounted machine
KK	Diameter of holes in terminal box for cable entry
LA	Thickness of flange
LB	Distance from fixing face of flange to non-drive end
LD	Centre line of terminal box to fixing face of flange
LE	Distance from extreme point of non-drive end to centre line of fixing holes in nearest foot
LF	Distance from end of shaft to fixing face of skirt-mounting flange
LG	Overall length including pulley
LH	Distance from centre line of pulley to centre line of fixing holes in nearest foot
LK	Distance from end of bearing housing (drive end) to centre line of fixing holes in nearest foot
LL	Distance from centre line of terminal box to centre line of fixing holes in foot nearest to shaft extension
XA (max.)	Centre line of bolt at adjusting screw end of slide rail to the beginning of the platform
XB	Width of slide rail at top

(continued overleaf)

Table 14.7 (continued)

Letter Symbol	Dimension description
XC	Bolt diameter for which clearance is provided in the slot of the slide rail
XD	Height of adjusting screw centre line above the platform
XE	Distance between the centre lines of the mounting-bolt holes (side view)
XF	Distance between the centre line of the mounting-bolt hole at the adjusting screw end and the adjacent end of the slide rail

Table 14.8 Fixing dimensions of foot-mounted frames, all enclosures (BS 4999 Part 141) (Primarily l.v. induction motors) *All dimensions are in millimetres*

Frame number	H Nominal	Negative tolerance*	A	B	C	K†	Fixing bolt or screw
56	56	0.5	90	71	36	5.8	M5
63	63	0.5	100	80	40	7	M6
71	71	0.5	112	90	45	7	M6
80	80	0.5	125	100	50	10	M8
90 S	90	0.5	140	100	56	10	M8
90 L	90	0.5	140	125	56	10	M8
100 S	100	0.5	160	112	63	12	M10
100 L	100	0.5	160	140	63	12	M10
112 S	112	0.5	190	114	70	12	M10
112 M	112	0.5	190	140	70	12	M10
132 S	132	0.5	216	140	89	12	M10
132 M	132	0.5	216	178	89	12	M10
160 M	160	0.5	254	210	108	15	M12
160 L	160	0.5	254	254	108	15	M12
180 M	180	0.5	279	241	121	15	M12
180 L	180	0.5	279	279	121	15	M12
200 M	200	0.5	318	267	133	19	M16
200 L	200	0.5	318	305	133	19	M16
225 S	225	0.5	356	286	149	19	M16
225 M	225	0.5	356	311	149	19	M16
250 S	250	0.5	406	311	168	24	M20
250 M	250	0.5	406	349	168	24	M20
280 S	280	1.0	457	368	190	24	M20
280 M	280	1.0	457	419	190	24	M20
315 S	315	1.0	508	406	216	28	M24
315 M	315	1.0	508	457	216	28	M24
315 L	315	1.0	508	508	216	28	M24
355 S	355	1.0	610	500	254‡	28	M24
355 L	355	1.0	610	630	254‡	28	M24
400 S	400	1.0	686	560	280‡	35	M30
400 L	400	1.0	686	710	280‡	35	M30

*There are no positive tolerances.
†The K dimensions are selected from the coarse series in BS 4186.
‡Dimension C assumes ball and roller bearings.

331

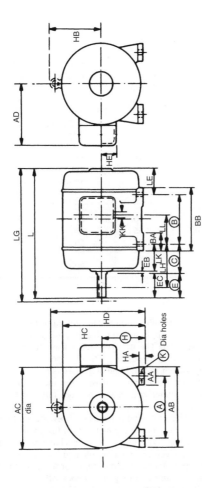

Figure 14.41 Symbols for foot-mounted frames

Notes: The ringed symbols are defined in *Table 14.6*. Unringed symbols are defined in *Table 14.7*. The above diagram is intended to be purely illustrative of the dimensions defined in *Table 14.6*

Table 14.9 Single-speed TEFC cage motor (BS EN 50347) suitable for a three-phase, 50 Hz supply not exceeding 415 V

Frame number	Output, kW				Shaft number	
	Synchronous speed, rev/min					
	3000	1500	1000	750	3000	1500 or less
D80	1.1	0.75	0.55	–	19	19
D90 S	1.5	1.1	0.75	0.37	24	24
D90 L	2.2	1.5	1.1	0.55	24	24
D100 L	3.0	2.2 and 3.0	2.2	0.75 and 1.1	28	28
D112 M	4.0	4.0	2.2	1.5	28	28
D132 S	5.5 and 7.5	5.5	3.0	2.2	38	38
D132 M	–	7.5	4.0 and 5.5	3.0	38	38
D160 M	11 and 15	11	7.5	4.0 and 5.5	42	42
D160 L	18.5	15	11	7.5	42	42
D180 M	22	18.5	–	–	48	48
D180 L	–	22	15	11	48	48
D200 L	30 and 37	30	18.5 and 22	15	55	55
D225 S	–	37	–	18.5	55	60
D225 M	45	45	30	22	55	60
D250 S	55	55	37	30	60	70
D250 M	75	75	45	37	60	70
D280 S	90	90	55	45	65	80
D280 M	110	110	75	55	65	80
D315 S	132	132	90	75	65	85
D315 M	150	150	110	90	65	85

Motor Control Gear

For safety, all motor starters should automatically return to the 'off' position in the event of failure of the supply and for this purpose an undervoltage or 'no-volt' release must be fitted. The undervoltage release forms an inherent part of the starting switch in all electromagnetically operated starters, but is additional to the starting-switch in hand-operated starters.

In faceplate d.c. starters the no-volt release takes the form of an electromagnet which holds the starter arm in the 'full on' position by magnetic attraction (see *Figure 14.4*).

With drum-type a.c. starters having a manually operated starting handle, the starting switch is fitted with a spring which biases the switch to the 'off' position, but is retained in the 'on' position by a mechanical latch. Fitted in the starter is a shunt-wound electromagnet, or solenoid, excited by the supply voltage. In the event of supply failure, the plunger or armature of the solenoid or electromagnet is released, and is arranged to knock off the hold-on catch and so return the switch to the 'off' position. A pushbutton for stopping the motor can be connected in series with the no-volt coil (see *Figure 14.42*).

Contactor control gear. In contactor starters, the no-volt feature is inherent as shown in *Figure 14.43*. Pressing the starter button energizes the operating coil which closes the contactor. In order to keep the contactor closed

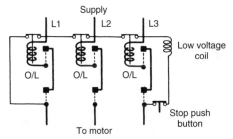

Figure 14.42 Diagram of drum type hand-operated direct-on-line a.c. starter

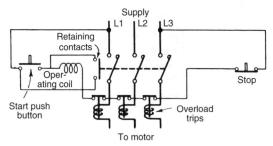

Figure 14.43 Diagram of direct-on-line contactor starter

when the 'Start' pushbutton is released, retaining contacts are required. These are closed by the moving portion of the contactor itself and thus maintain the operating-coil circuit, once it has been made and the contactor closed, irrespective of the position of the 'Start' pushbutton. Depression of the 'Stop' pushbutton, or failure of the supply to the operating coil, immediately causes the contactor to open. By energizing the operating coil from the same circuit that supplies the motor, such an arrangement is equivalent to a no-volt release. A similar contactor for control of a d.c. motor is shown in Figure 14.44.

When starting and stopping is automatically provided by means of a float, pressure or thermostatic switch, as in the case of motor-driven pumps, compressors, refrigerators and the like, two-wire control is used as in *Figure 14.45*. In these cases, it is desirable that some form of hand reset device be incorporated with the overload release, to prevent the starter automatically reclosing after tripping on overload or fault until the reset button has been pressed after clearance of the fault.

Overload protection. An important function of motor control gear is to prevent damage due to excessive current. This may be due to either mechanical overload on the motor or to a defect in the motor itself. In either case it is essential that the supply be disconnected before any damage is done to the motor. An overload device thus usually operates by releasing the latching-in device by disconnecting the supply to the no-volt coil, or, in contactor starters, by opening the operating coil circuit.

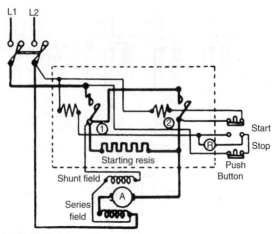

Figure 14.44 Typical d.c. contactor starter

Contactor No. 1 is the line contactor
Contactor No. 2 is the running contactor

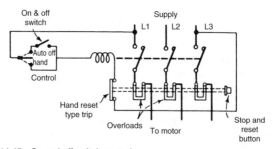

Figure 14.45 On and off switch control

Time-lag considerations. It is essential to guard against unnecessary operation from temporary overloads due to the normal operation of the machines which are being driven. The detrimental effect of an overload on a motor is a matter of time – a slight overload taking considerable time to develop sufficient heat to do any damage, whereas a heavy overload must be removed much more quickly.

Overload devices therefore usually have a time-lag feature giving a curve of which *Figure 14.46* is typical. Both the overload value should be adjustable and also the time feature for important motors. For d.c. or single-phase one overload device is usually sufficient, while for three-phase at least two of the lines must be protected, and it is considered advisable by many engineers to have an overload device in each phase. A three-phase motor will often run as a single-phase machine if one line gives trouble, and this is often avoided by

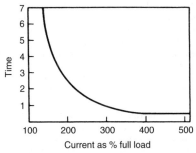

Figure 14.46 Curve showing variation in speed of operation with load for overload time lag

protecting all three lines with an overload relay which incorporates a single-phase tripping device. If the current in any one phase fails the relay trips out.

Overload devices. These are generally of two types – electromagnetic and thermal. The electromagnetic type consists of a coil or solenoid carrying the line current (or a proportion of it) with an armature which when attracted sufficiently operates the release circuit or latch. A time-lag feature is obtained by means of a dashpot or similar arrangement as otherwise the action would be practically instantaneous.

Thermal overload devices have been developed to a considerable extent due to their low cost. They may be bi-metal strips or solder pot elements, and in either case, as the action is due to their heating up, a time element is always present.

The action of a bi-metal strip overload release depends on the movement resulting from the different rates of expansion of the two metals forming the combined strip when heated. The bi-metal strip may be directly heated by the current, or indirectly heated by a coil of resistance wire which carries the current. In the 'solder pot' form of release, a spindle carrying a ratchet wheel is embedded in a low-melting-point fusible alloy. This alloy is heated and melted by excess currents and, when molten, permits the ratchet wheel to rotate and so trips the starter. This type of overload device has never enjoyed much popularity in the UK but has been used in the USA successfully for a number of years. Thermal trips are usually of the hand reset type; the reset feature may be combined with the stop pushbutton.

Thermal overloads are usually confined to the smaller control units up to 15–22 kW, but it will be realized that, with modern industrial tendencies, small motors represent a very large proportion of the machines now being installed.

Thermistor protection. Thermistors are semiconductors which exhibit significant changes in resistance with change in temperature, i.e. they are thermally sensitive devices. They are based on barium titanate and are formulated to produce a negligible increase in resistance with change in temperature until the Curie point is reached. At this point any further increase in temperature causes the resistance to change rapidly. This characteristic is used to operate as a protective device.

The thermistor is housed in the stator windings of the motor at that point which is considered to be the 'hot-spot', i.e. the highest temperature point for a given overload current. The associated control gear is generally housed in the starter and takes the form of an amplifier and possibly a relay. Thermistors handling large currents are available and these can operate directly via a small interposing relay. Thermistors are claimed to offer better protection than any other single system because they measure directly the temperature of the motor winding. But the disadvantage is that since they have to be embedded in the winding they introduce a weakness into the insulation system. This is one of the reasons that they do not find application in h.v. motors being restricted in the main to 415 V machines.

Multi-pushbutton control. In the case of electrically operated contactor gear both starting and stopping can be controlled from any number of points by connecting additional pushbuttons either in series or parallel, as required. This has a definite advantage when it is inconvenient to have the control gear mounted close to the place where the operator is situated. For emergency stopping any number of 'Stop' buttons may be used to save time in cases of emergency.

The same principles are used in connection with interlocking for lifts and similar machinery where it is necessary for certain items to be in position before the machine can be started up.

Multi-stage starters. Direct-on starting is allowed for motors up to a certain size, the limit being fixed by the supply system. For 'strong' systems l.v. motors of up to 100 kW or more can be started and at 3.3 kW the rating can rise to 300 kW or greater. For 'normal' systems the limit at 415 V may be anywhere between 5 kW and 30 kW. The supply authority will advise on this matter. Above these limits it is necessary to employ some method of starting which restricts the starting current to between two and three times full-load current. For cage motors using either star-delta or auto-transformer starting, the changeover from 'start' to 'run' is effected in contactor starters by employing two contactors with either a series or time relay for timing the moment of the changeover.

Wound-rotor starters. Where slip-ring motors have to start up against severe load conditions liquid resistance starters are found very satisfactory. By varying the electrolyte the added resistance is under control, and liquid starters have the advantage that the resistance is reduced continuously and smoothly instead of by steps as in other systems.

For wound-rotor motors hand-controlled starters are in more general use, but contactors can be used – one main contactor for the stator and two to five stages for cutting out the rotor resistance.

The changeover from one stage to another may be controlled by current or time. For current control a current or series relay is used to operate the changeover from one contactor to the next when the current has fallen to a certain value by virtue of the motor speeding up. This method ensures more correct starting but requires fairly accurate adjustment.

The time delay control is by means of a dashpot time-lag or similar device, and as soon as the first contactor is operated the relay comes into circuit. After the set time has expired the changeover takes place and the process is repeated for subsequent stages.

Figure 14.47 shows a scheme for control of a reversing heavy-duty d.c. motor.

Air-break limits. Manually operated air-break switchgear and starters are only satisfactory up to a certain size. For d.c., however, air-break gear

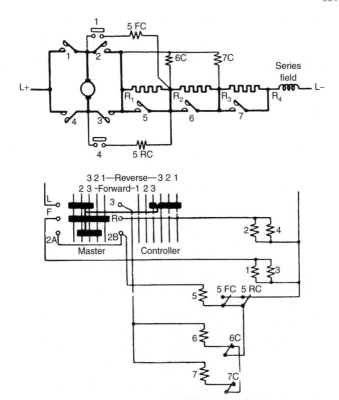

Figure 14.47 Typical d.c. contactor equipment for reversing heavy-duty motor

Relay 5RC controls contactor 5 in reverse direction
Relay 5FC controls contactor 5 in forward direction
Relay 6C controls contactor 6 in either direction
Relay 7C controls contactor 7 in either direction

is used up to large power outputs, but it is necessary to renew the contacts fairly frequently.

For controlling a.c. motors manually operated air-break gear other than contactor gear is satisfactory up to about $1\frac{1}{2}$ kW, but for larger outputs vacuum, SF_6 or air-break circuit-breakers are generally specified.

Fuses. Overload devices in starters are designed to interrupt the circuit in the event of the current rising above a predetermined value due to the mechanical overloading of the motor, and are not usually designed to clear short-circuits. It is very desirable to include a circuit-breaker, or fuses, of sufficient breaking capacity to deal with any possible short-circuit that may occur. The required ratings of high-breaking capacity cartridge fuses for motor circuits can be obtained from fuse manufacturers' lists.

Quick-stopping of motors. The application of electric braking, that is, braking by causing the motor to develop a retarding torque, to certain classes of industrial drive, such as rolling mills, electric cranes and hoists, is normal practice. It is being used on many other industrial drive systems and there are two fields of application:

1. Drives where a large amount of energy is stored in the rotating parts of the driven machine.
2. Drives requiring rapid and controlled deceleration in the event of an accident or emergency.

There are two ways in which a polyphase induction motor may be made to develop a braking torque and stop quickly:

1. By causing it to cease to act as a motor and to operate as a generator.
2. By reversing it, so that it is, in effect, running backwards until its load is brought to rest. The first method is exemplified by the method of d.c. injection, and the second by 'plugging'.

Braking by d.c. injection. In this arrangement, when the motor is to be braked, it is disconnected from the a.c. supply and direct current is fed to the stator winding. The effect of this is to build up a static magnetic field in the space in which the rotor revolves and the current thereby generated in the rotor winding produces a powerful braking torque in exactly the same way as is done in a shunt-wound d.c. motor.

In a workshop where a d.c. supply is available, braking is particularly easy, as little additional apparatus is required. In all other cases, the necessary d.c. can be provided by a rectifier built in or adjacent to the control box. No mechanical connections are required, all the braking control being included in the controller cabinet. When the stop button is pressed, the a.c. contactor (see *Figure 14.48*) opens, causing the d.c. contactor to close, and connect the stator of the motor to the rectifier. The motor is rapidly brought to rest. By means of a timing relay, which is arranged to give varying periods of application of the d.c. to suit the loading of the motor, the d.c. contactor is then tripped out.

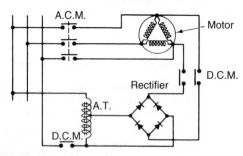

Figure 14.48   D.C. injection

A.T. Auto transformer
A.C.M. A.C. main contactor
D.C.M. D.C. main contactor

The method is applicable both to cage and slip-ring motors, although more usually applied to cage. When applied to slip-ring motors, the rotor circuit is made through the rotor resistance, a tapping on this being selected to give the best results. The d.c. may be arranged for injection on either the stator or rotor winding, or on both.

Braking by reversal (plugging). Rapid braking of a.c. (and d.c. motors) may be obtained by reconnecting the windings to the supply in the order corresponding to reversed direction of rotation. The changeover is usually made quickly, as by the operation of contactors, and the method is widely employed with automatic control systems where particularly rapid stopping is desired. When deceleration is completed and the motor stops, it will of course reverse unless the windings are disconnected.

Generally speaking, the modifications necessary to the normal starting gear required for an induction motor are not great in order to make it suitable for plugging the motor.

Plugging with direct-on starter. All that is necessary is to replace the usual starter switch by a changeover reversing switch, which may be hand-operated, or automatic, as demanded by the needs of the drive. *Figure 14.49* shows a direct-on pushbutton starter adapted to carry out the plugging operation, the only alterations being the addition of the reversing switch, the use of a special stop button, and usually a variable limiting resistance fitted in each phase feed of the plugging circuit. The resistances allow a variable time factor, and control the value of the braking torque and current so as to avoid damage to any part of the motor or driven machine.

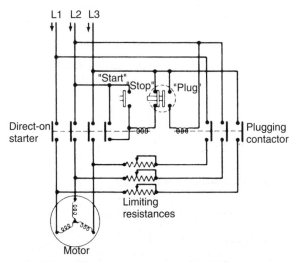

Figure 14.49 Wiring of direct-on contactor starter for plugging

Plugging with auto-transformer or star-delta starter. The position is rather more difficult, because means must be provided to make sure that the motor

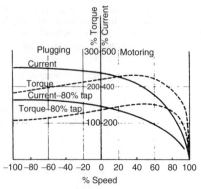

Figure 14.50 Torque/speed and current/speed curves for double cage motor on full and reduced voltage when motoring and plugging

is connected to the starting position of the starter before it is plugged. This can be done by interlocking the starter and reversing switch, or by providing a changeover switch for the reversal that also reduces the voltage to the motor.

Figure 14.50 shows the curves for a cage motor with a high resistance or double cage rotor when plugging on full voltage, and on reduced voltage such as would be obtained with an auto-transformer.

Plugging slip-ring motors. The switch for the stator must be a changeover reversing switch and some form of interlock is necessary to ensure that all rotor resistance is in circuit before the motor is plugged. In certain cases an extra plugging step may be necessary, although the starter will be simplified if the starting resistance on the first step is made sufficiently great to limit the braking current to the desired value. It should be remembered that the current passed when plugging will be almost twice that passed by the same resistance at standstill. In *Figure 14.51* curves are given for a slip-ring motor, with and

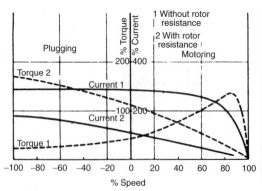

Figure 14.51 Torque/speed and current/speed curves for slip-ring motor with and without rotor resistance when motoring and plugging

without rotor resistance. It will be seen that if a large step of resistance is used, the resulting curves differ somewhat from those normal for the machine and that current and torque tend to increase above their standstill values. It is usual to determine the value of a star-connected resistance necessary to limit the braking current to any desired value from the formula:

$$R = 1.8 \times E/1.73 \times I$$

where E is the open circuit rotor voltage at standstill, and I the permissible braking current.

Plugging switch. The disadvantage of plugging is that the motor may carry on and start to accelerate in the reverse direction. On drives where this would be objectionable, however, it can be prevented by including a plugging switch on the motor shaft. This is a simple switch of the centrifugal type which opens when the motor reaches zero speed, and so trips the main stator switch of the motor by interrupting the no-volt or hold-on coil circuit. Another method of controlling the time the reverse contactor remains closed on plugging is by means of a timing relay set to drop out the reverse contactor just before the motor reaches zero speed.

With both d.c. injection and plugging the load is not held when the equipment comes to rest. If this is necessary, braking and stopping by means of an electromechanical brake must be used.

15 Switchgear and protection

Switchgear

The choice of suitable switchgear depends to larger extent on the actual duty than any other type of electrical plant. In addition to switching on or off any section of an electrical installation, the switchgear generally includes the protective devices which are necessary in order that the particular section may be automatically isolated under fault conditions.

The term 'switchgear' covers any device designed for controlling, that is interrupting, an electrical circuit and may include switches, circuit-breakers and isolators. Fuses also fall into this category but are generally used in conjunction with some other device, generally a switch or a circuit-breaker. The broad definitions of these devices are as follows:

> *A switch* – is a device which is designed to interrupt the normal rated current of the circuit in which it is used. It may be required to make on to a fault but is not required to interrupt fault current.
>
> *A circuit-breaker* – is designed to interrupt fault current and thereby clear a fault from the system. There are occasions when circuit-breakers may be backed up by fuses when they are installed in circuits having higher fault levels than their rated breaking capacity.
>
> *An isolator* – is designed to enable a circuit to be electrically isolated from its source of supply. It will not normally be used for switching duty although in some situations it could be designed to make on to a fault.
>
> *A fuse* – is a one-off operating device which is designed to interrupt a circuit to clear a fault.

Originally all switchgear consisted of open knife switches mounted on a slate or composition panel and operated by hand. The protective device consisted of a fuse which was generally mounted close to the switch. The use of high voltage a.c. and the great increase in total power in a system has necessitated the use of oil-break, air-break, vacuum, SF_6 or air blast switchgear.

For low voltages (up to 1000 V) knife-type switches are still used, and occasionally open-type boards may be encountered, but generally most switchgear today is enclosed. Metal-clad switch or combined switch and fuse units are used either singly or grouped to form a switchboard. For the lower current ratings insulated cases are obtainable in place of the metal type, these being particularly popular for domestic installations.

The knife switches are usually spring controlled, giving a quick make and break with a free handle action which makes the operation of the switch independent of the speed at which the handle is moved. In all cases it is impossible to open the cover with the switch in the 'on' position. The normal limit for this type of switch is from 300 to 400 A, but larger units can be made specially.

Miniature circuit-breakers or moulded-case circuit-breakers are used widely as protective devices in consumer premises and for group switching and protection of fluorescent lights in commercial and industrial buildings.

Moulded-case circuit-breakers with ratings up to 3000 A and capable of interrupting currents up to 200 kA (for the larger ratings) are becoming popular for control of l.v. networks.

When any breaking capacity rating is required, together with automatic operation, it is necessary to use special control devices to interrupt the fault current. Low voltage three-phase systems of up to 415 V are usually controlled by air circuit-breakers with or without series fuses. For higher voltage systems up to 36 kV, oil, air-break, vacuum and SF$_6$ breakers are available. For higher voltages, SF$_6$ circuit-breakers take over and are used up to 420 kV and beyond. For l.v. systems the air-break circuit-breaker is usually a moulded-case unit. The air-break circuit-breaker for 3.3 to 11 kV has an arc control device which is suitable for motor switching and is used mainly in the power stations built in the 1960s and 1970s. Its cost makes general use in industry and distribution systems unusual. These circuit-breakers predated the introduction of vacuum and SF$_6$ devices but had the great benefit of eliminating oil.

Oil circuit-breakers are still popular for h.v. distribution systems despite the perceived fire risk. These consist of an oil enclosure in which contacts and an arc control device are mounted. The arc is struck within the control device and the resultant gas pressure sweeps the arc through cooling vents in the side of the pot. High reliability and simple, though relatively frequent, maintenance are available from these devices.

Figure 15.1 shows a typical truck-mounted oil circuit-breaker with its cubicle for 12 kV. Vacuum and SF$_6$ circuit-breakers are also available in a similar form of truck mounting with similar cubicles. Vacuum circuit-breakers were the first type of oil-less circuit breaker to be available and have been used in industrial situations since the later 1960s. Vacuum interrupters are sealed-for-life ceramic 'bottles' containing movable contacts in a high vacuum. The circuit breaking performance of this design is very high and a large number of short-circuit operations can be achieved before any replacement is necessary. In fact, in most cases, this will never be required. Some maintenance is, however, normally necessary to check the small operating movement of the contacts and where necessary reset this.

SF$_6$ circuit-breakers come in a number of forms, all utilizing the good dielectric and arc extinguishing properties of this gas to provide another type of oil-less circuit-breaker. While the life of the contacts is not as great as those in a vacuum, the SF$_6$ circuit-breaker has other advantages that make it equally as acceptable for industrial and distribution use. Special designs of oil-less units are available such as shown in *Figure 15.2* which illustrates a double tier unit of vacuum circuit-breakers which allows a compact arrangement for speedy assembly on site.

All these new units are designed to minimize the maintenance that is required to the interrupting unit although regular checks on the mechanical operation and the cleanliness of exposed insulation is always advisable.

All h.v. circuit-breaker systems up to 36 kV are three-phase units, but for higher voltages, three separate single-phase breakers coupled together are generally used to facilitate the design of the insulation system for phase-to-phase voltage.

A major part of circuit-breaker cubicle cost is the protection and instrumentation systems that are associated with the particular circuit protection and its interlocking with adjacent circuits. Protection from overload is obtained by means of a device which releases the mechanism and opens the breaker. For lower current rated, lower voltage breakers, the protection may be provided by

Figure 15.1 11 kV switchgear with oil circuit-breaker withdrawn from its housing (Reyrolle Distribution Switchgear)

overload coils or thermal releases inside the unit itself. For large units which are protected by complex relay systems, the operation of one of the relays of the protection system releases the tripping mechanism in a similar manner.

Some essential features of all switchgear are:

(a) Isolation of the internal mechanism for inspection. This is important and full interlocks are always provided to prevent the opening of any part of the enclosure unless access to the higher voltage supply is prevented.

Figure 15.2 Double tier 12 kV vacuum circuit-breakers permitting speedy assembly on site (Reyolle Distribution Switchgear)

(b) Insulation from breaker contacts to the side of the enclosure must be adequate for the voltage and maximum load which the breaker will be called upon to deal with.

(c) Provision for manual operation for testing purposes and in case the electrical control (if provided) fails to operate.

(d) Provision for instruments which may be required. These may be in the form of either an ammeter or voltmeter, or both, on the unit itself or the necessary current and voltage transformers for connecting to the main switchboard or a separate instrument panel.

For switchgear up to 11 kV and most circuit-breakers up to 33 kV, isolation is effected in the following ways:

(a) By isolating links in or near the busbar chamber.
(b) Draw-out type of gear in which the whole of the circuit-breaker is withdrawn vertically or horizontally from the busbar chamber before it can be opened up.
(c) Truck-type, in which the circuit-breaker with its secondary connections is isolated generally in a horizontal manner before inspection or adjustment.

It should be noted that in certain cases, double isolating devices are necessary, i.e. both on the incoming and outgoing side.

Isolation is, of course, always required on the incoming side, but it is also necessary on the outgoing side if that part of the network can be made alive through any other control gear or alternative supply. With conventional switchgear, isolation of both sides takes place automatically.

A circuit-breaker is usually classified according to the voltage of the circuit on which it is to be installed; the normal current which it is designed to carry continuously in order to limit the temperature rise to a safe value; the frequency of the supply; its interrupting capacity in kA; its making capacity in kA(peak), i.e. the instantaneous peak current; and the greatest r.m.s. current which it will carry without damage for a specified length of time, usually 1 or 3 seconds.

BS EN 60947-2 covers circuit-breakers of rated voltage up to and including 1000 V a.c. and 1200 V d.c. It supersedes BS 4752 and BS 862 : 1939 covering air-break circuit-breakers for systems up to 600 V and BS 936 : 1960 *Oil circuit-breakers for m.v. alternating-current systems*. It also supersedes those parts of BS 116 : 1952, *Oil circuit-breakers for a.c. systems*, and BS 3659 : 1963, *Heavy duty air-break circuit-breakers for a.c. systems*, which relate to breakers having rated voltages up to and including 1000 V.

All circuit-breakers must be capable of carrying out a given number of mechanical and electrical operating cycles, each one consisting of a closing operation followed by an opening operation (mechanical endurance test) or a making operation followed by a breaking operation (electrical endurance test). The numbers of cycles for the mechanical endurance test are shown in *Table 15.1*.

Overload and Fault Protection

Electrical faults may be broadly divided into *short-circuits* and *overloads* and it is important to differentiate between the two. A short-circuit is generally characterized by a breakdown of insulation, either to earth or between poles, whereas an overload is usually brought about by the device fed from the switchgear being subjected to too high a demand, for example the stalling or partial stalling of a motor drive.

Protection against electrical faults is generally provided either by fuses or circuit-breakers. In some instances fuses are used in conjunction with circuit-breakers to take over the interruption of higher short-circuit currents. This is so with miniature or lower rated moulded-case circuit-breakers.

Table 15.1 Number of operating cycles for the mechanical endurance test (from BS 4752)

1	2	3	4	5	6	7
Rated thermal current in amperes	Number of operating cycles per hour*	Number of operating cycles				
		All circuit-breakers	Circuit-breakers designed to be maintained‡		Circuit-breakers designed not to be maintained	
		With current without maintenance† n	Without current n′	Total n + n′	Without current n″	Total n + n″
$I_{th} \leq$ 100	240	4000	16000	20000	4000	8000
$100 < I_{th} \leq$ 315	120	2000	18000	20000	6000	8000
$315 < I_{th} \leq$ 630	60	1000	9000	10000	4000	5000
$630 < I_{th} \leq 1250$	30	500	4500	2500	3000	
$1250 < I_{th} \leq 2500$	20	100	1900	2000	900	1000
$2500 < I_{th}$	10	(Subject to agreement between manufacturer and user)				

*If the actual number of operating cycles per hour does not correspond to values in column 2, this shall be stated in the test report.

†During each operating cycle, the circuit-breaker shall remain closed for a maximum of 2 seconds.

‡The manufacturer shall supply detailed instructions on the adjustments or maintenance required to enable the circuit-breaker to perform the number of operating cycles in column 5.

I_{th} is the rated thermal current of the circuit-breaker.

In h.v. distribution circuits another arrangement is that of fuses fitted in series with an oil switch; upon the blowing of a fuse link a striker pin is ejected which trips the oil switch.

Types of fuses. The simplest and cheapest form of protection against excess current is the fuse. Two types of fuse are in use:

1. The semi-enclosed type, comprising removable plastics or porcelain holder with handle through which the tinned copper fuse-wire passes. As the fusing factor of the semi-enclosed types is higher than the totally enclosed type and they are subjected to high prospective fault currents their use is no longer recommended although it will be many years before they finally disappear.
2. The totally enclosed or cartridge type fuse in which the fuse itself is enclosed by a cylinder of hard, non-combustible material having metal capped ends, and is filled with non-flammable powder, or other special material. Regulation 533-01-04 says fuses should preferably be of the cartridge type.

Rating of fuses. Fuse ratings for given circuits should be selected with care. In low voltage general purpose circuits, with non-inductive loads such as heating and lighting, the selected fuse should have a current rating which exceeds the full-load current of the circuit, but which is less than the cable current rating. The cable is then fully protected against both overload and short-circuit faults.

On the other hand, in circuits which contain inductance or capacitance, the resulting inrush currents dictate the choice of fuse current rating. Thus in a motor circuit, the starting current and its duration have to be related to the fuse time/current curves, and a fuse chosen to withstand this surge will have a current rating up to three times the motor full-load current. In such a circuit, the fuse provides short-circuit protection only, and overload protection is provided by other means (see section on fuse selection).

Cartridge fuses. Cartridge fuses are available for systems up to 660 V in current ratings from 2 to 1600 A and breaking capacity ratings in excess of 50 kA. Some fuses can interrupt currents up to 200 kA. Higher kA ratings can be obtained but they are not standard. *Figure 15.3* shows the cross-section of an l.v. fuse.

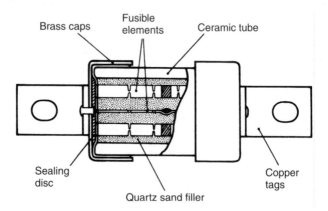

Figure 15.3 Construction of a modern cartridge fuse (GEC Installation Equipment Ltd)

The breaking capacity rating at 11 kV is not less than 13.1 kA and current ratings are available up to 350 A.

Other advantages of well-designed cartridge fuses are:

Discrimination. The fuse nearest the fault will operate, thus ensuring that only the faulty circuit is isolated and healthy circuits are unaffected. This discriminating property is inherent in cartridge-type fuses, as a glance at the time/current curves for different sizes of fuse in *Figure 15.4* will show. It will be seen that the speed of operation for any particular value of overload or fault current increases as the fuse gets smaller.

High speed of operation on short-circuit. This property enables fuses to be used for the back-up protection of motor starters and low breaking capacity circuit-breakers. For such purposes high breaking capacity in itself is not sufficient. The speed of operation of the fuses used for back-up protection

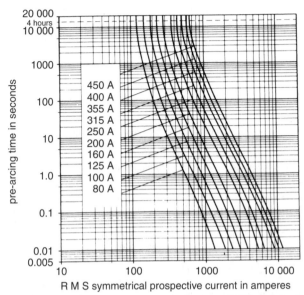

20 000
4 hours
10 000

1000

100

10

1.0

0.1

0.01
0.005

pre-arcing time in seconds

450 A
400 A
355 A
315 A
250 A
200 A
160 A
125 A
100 A
80 A

10   100   1000   10 000

R M S symmetrical prospective current in amperes

Figure 15.4 Typical 'T' range: time/current characteristics (GEC Installation Equipment Ltd)

must be faster than the speed of operation of the motor starter or circuit-breaker to be protected; otherwise the apparatus under protection would be damaged or destroyed before the fuses had time to act.

Selection of fuses. When selecting suitable fuses for any particular situation the following factors should be considered:

1. *Short-circuit ratings of cartridge fuses.* Cartridge fuses for l.v. industrial applications are most commonly proved capable of interrupting fault currents up to 80 kA at 415 V, and some designs will safely interrupt such fault currents at 660 V. They therefore have more than adequate breaking capacities for the overwhelming majority of applications, but they must not be used at voltages above their assigned rating.
2. *Current rating of a fuse.* This is the maximum current which the fuse can carry continuously without deterioration. As already indicated the fuse current rating selected for any circuit should have a current rating not less than the full load current of the circuit.
3. *Overload protection.* A general purpose type fuse (gG) to BS 88: Part 2 will protect an associated PVC-insulated cable against overload if its current rating (I_N) is equal to, or less than, the current rating of the cable (I_Z). See Regulation 433-02-01 of BS 7671.
4. *Motor circuits.* In a motor circuit, the starter overload relays protect the associated cable against overload, and the fuses in circuit provide the required degree of short-circuit protection. Manufacturers usually make recommendations regarding the fuse ratings needed to cope with motor

starting surges, and also indicate minimum cable sizes needed to achieve short-circuit protection. Regulation 434-03-03 of the BS 7671 is applied to both fuse and cable selection in order to accomplish satisfactory protection.

In a well-engineered combination, the starter itself interrupts all overloads up to the stalled rotor condition, and the fuses should only operate in the event of an electrical fault. The starter manufacturers indicate the maximum fuse rating which may be used with a given starter to ensure that satisfactory protection is achieved.

The automatic circuit-breaker. Circuit-breaker equipment, while considerably more expensive than fuse gear protection, possesses important features which render it essential on all circuits where accurate and repetitive operation is required. It is usually employed on all main circuits, fuse gear being reserved for subcircuits.

Overload releases. The simplest automatic release used is the direct acting overcurrent coil, which is a solenoid energized by the current passing through the unit, and calibrated to operate when this reaches a predetermined value. The trip setting is adjustable, from normal full-load current up to 300% for instantaneous release and 200% for time-lag release. Therefore a setting as near as is desired to full-load current can be obtained. On a three-phase insulated system, coils in two phases will give full protection, since any fault must involve two phases: but with an earthed system, release coils must be provided in all three-phase, unless leakage protection is also provided, when two overcurrent coils will again suffice. In small current sizes it is usual to make the overcurrent coils direct series connected: but in larger sizes the coils are often operated from current transformers.

Time lags. If full advantage is to be taken of the facility to obtain close overcurrent settings that are offered by circuit-breakers, some form of restraining device is necessary to retard the action of the releases, as this is normally instantaneous. This retarding device, or time lag, usually takes the form of a piston and dashpot. Its purpose is to delay the action of the trips so as to allow for sudden fluctuations in the load, and also to give a measure of discrimination. In practice, a characteristic closely resembling that of a fuse is obtained, and a typical characteristic for an oil dashpot time lag – perhaps the most common type in general use – is given in *Figure 15.5*.

Use of time-limit fuses. When the overcurrent coils are transformer operated, it is possible to use a shunt fuse in place of the mechanical retarder. These fuses are designed to short circuit the trip coils, which, therefore, cannot operate until the fuse has blown. The fuse itself is usually of alloy-tin wire or some other non-deteriorating metal, and is accurately calibrated as regards rupturing current. When time-limit fuses are used, no advantage is to be gained by making the release coils themselves adjustable, and alternative trip settings are obtained by varying the rupturing current of the fuses.

Cartridge fuses and circuit-breakers. With a circuit-breaker, a certain time must elapse between the operation of the tripping mechanism and the actual breaking of the current, and with commercial types of breaker this time is usually of the order of 0.1 sec. On the other hand, a cartridge fuse is capable of clearing a very heavy fault in less than half a cycle; and, therefore, it will

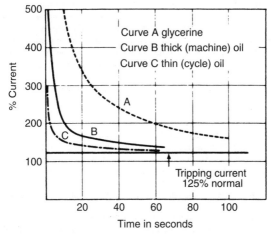

Figure 15.5 Characteristic curves for series trip coil with oil dashpot

operate in these conditions long before a breaker has had time even to start the opening operation.

Relays and Protective Gear

The protection of plant, distribution and transmission lines has reached a high state of perfection and faulty sections can now be automatically isolated before a fault or overload causes much damage to the section itself or the remainder of the system of which it is a part. Dependent on the cost and value of the switchgear and its role in the system of which it is a part, very sophisticated protection systems are available and there may be, in addition to individual circuit protection, systems of busbar protection for main and reserve busbars to which individual circuits may be selected.

In all the different methods the essential feature is that of isolating the faulty section, and in considering the principles of the various systems of protection it is usually understood that the actual isolation is carried out by circuit-breakers which, in turn, are operated by means of currents due to the action of the protective gear. Usually the required tripping current for the breaker is controlled by relays, which are in turn operated by the protective gear.

Similar principles are used for the protection of machines such as alternators, transformers, etc., as for overhead lines and cables. The essential difference is only a matter of adaptation, the most important difference being the fact that on overhead lines and feeders certain protective systems may require pilot wires connecting the protective apparatus at each end of the line, and these are sometimes undesirable. With machines this point is unimportant as the two sets of protection – one on each side – are not separated by any real distance.

With all systems one or both of the following two undesirable features are guarded against – namely, *overload* and *faulty insulation*. The overload conditions which make it necessary to disconnect the supply may be due to faulty apparatus or to an overload caused by connecting apparatus of too great a capacity for the line or machine. The faulty insulation or fault conditions may be either between the conductors or from one or all the conductors to earth.

In connection with all protective gear the following terms are generally used:

Stability ratio. This may be termed the measure of the discriminating power of the system. The stability is referred to as the maximum current which can flow without affecting the proper functioning of the protective gear. Stability ratio can also be defined as the ratio between stability and sensitivity.

Sensitivity is the current (in primary amperes), which will operate the protective gear. In the case of feeders and transmission lines this is a measure of the difference between the current entering the line and the current leaving it.

Overload relays. The simplest form of protection is that where the circuit-breaker is opened as soon as the current exceeds a predetermined value. For low voltage a.c. systems the overload trip may operate the circuit-breaker opening mechanism direct or through a protective relay usually mounted on the switchboard above the circuit breaker. For high voltage a.c. systems, oil, air, air-blast, SF_6 or vacuum circuit-breakers must be used and the trip coils of these breakers are operated by protective relays sometimes mounted on a separate panel.

The time-current characteristics of inverse definite-minimum-time-lag relays (i.d.m.t.l.) will be seen from the curves in *Figure 15.6*, which show how the time-lag between the overload occurring and the gear operating is inversely proportional to the magnitude of the overload. The setting or time-lag generally used to indicate how the relay has been adjusted is that of short-circuit conditions so that the settings of the relay for which the curves are drawn would be as marked.

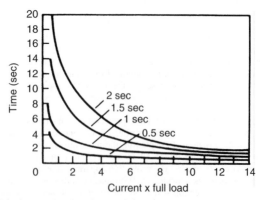

Figure 15.6 Inverse time-lag characteristics of an overcurrent relay

Figure 15.7 Grading of overcurrent relays for radial feeder

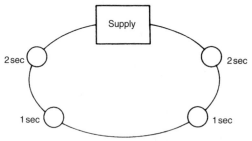

Figure 15.8 Grading for ring main

The principle of *grading* on a system is that of decreasing the setting as we proceed away from the source of supply. The settings of the relays for which the curves have been drawn would be as shown in the diagrams in *Figure 15.7* and *15.8*. *Figure 15.7* represents a distributor fed at one end, whereas *Figure 15.8* shows a ring main. A distributor fed at both ends is similar to *Figure 15.8*.

Electromechanical overcurrent relays are usually of the induction type. The action is similar to that of an induction wattmeter. A metal disc rotates against a spring, the angle of rotation being proportional to the current. As soon as the disc has rotated through a certain angle the trip circuit is closed. As the speed of rotation is proportional to the load the inverse time element is obtained and the time setting is adjustable by altering the angle through which the disc has to turn before making the trip circuit.

The latest generation of overcurrent and earth fault relays use microprocessor technology which has the following benefits:

(a) Selection of time-current characteristics built into the relays.
(b) Wider range of settings.
(c) Increased accuracy allowing shorter grading intervals.
(d) In situ relay testing by simply pressing a pushbutton.

Directional protection. In addition to overload protection it is often desirable for immediate interruption of supply in the case of a reverse current. In this case directional relays are used and these can be combined with overcurrent relays when required as shown in *Figure 15.9*. The contacts on the directional element are so connected that, under fault conditions and with the current flowing in the 'fault' direction, they complete the i.d.m.t.l. element operating coil circuit. The use of directional and non-directional relays is shown in the system reproduced in *Figure 15.10*.

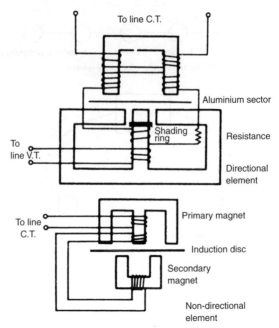

Figure 15.9 Directional overcurrent relay

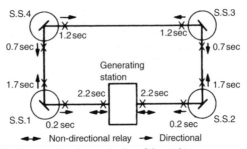

Figure 15.10 Discriminative time protection of ring main

Pilot wire systems. Protective systems requiring pilot wires for use on some overhead or feeder cable lines operate on the principle of unequal currents and are referred to as differential systems. The same methods are used for machine and transformer protection, the pilot wires being, of course, much shorter and thus not referred to as pilot wires in this case.

The essential principle is that if the current entering a conductor is the same as that leaving it there is no fault, but as soon as there is a difference more than is considered advisable the supply is disconnected.

Both *balanced voltage* and *balanced current* systems are in use. Current transformers are situated at each end of the conductor and either the voltage or current in the secondaries is balanced one against the other. For balanced voltage (*Figure 15.11*) these are placed back to back and no current flows for correct conditions. For balanced current, the secondaries are placed in series and current flows in the pilot wires. In this case fault conditions are determined by relays connected across the pilots as shown in *Figure 15.12*.

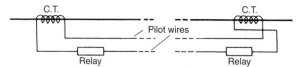

Figure 15.11 Balanced voltage system

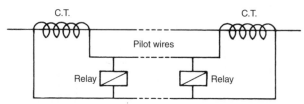

Figure 15.12 Balanced current system

The original balanced protection system was known as the Merz-Price system. There have been several modifications and most manufacturers have their own particular method of applying this principle and improving the actual performance compared with the original simple Merz-Price system.

A typical modification is the Translay S system. *Figure 15.13* shows the basic circuit arrangement for Translay S differential feeder protection from GEC Measurements.

A summation current transformer T1 at each line end produces a single-phase current proportional to the summated three-phase currents in the protected line. The neutral section of the summation winding is tapped to provide alternative sensitivities for earth faults.

The secondary winding supplies current to the relay and the pilot circuit in parallel with a non-linear resistor (RVD). The non-linear resistor can be considered non-conducting at load current levels, and under heavy fault conditions conducts an increasing current, thereby limiting the maximum secondary voltage.

At normal current levels the secondary current flows through the operation winding T_0 on transformer T2 and then divides into two separate paths, one through resistor R_0, and the other through the restraint winding T_r of T2, the pilot circuit and the resistor R_0 of the remote relay.

The resultant of the currents flowing in T_r and T_0 is delivered by the third winding on T2 to the phase comparator and is compared with the voltage across T_t of the transformer T1. The e.m.f. developed across T_t is in phase with that across the secondary winding T_s, which is in turn substantially the voltage across R_0.

356

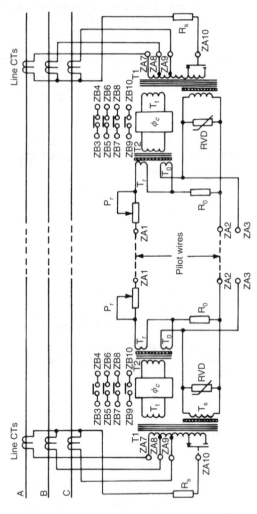

Figure 15.13 Circuit diagram for Translay S differential feeder protection (GEC Measurements)

Taking into account the relative values of the winding ratios and circuit resistance values, it can be shown that the quantities delivered for comparison in phase are:

$$(\bar{I}_A + 2\bar{I}_B) \quad \text{and} \quad (2\bar{I}_A + \bar{I}_B)$$

where I_A and I_B are the currents fed into the line at each end. For through faults $I_A = -I_B$. The expressions are of opposite sign for values of I_B which are negative relative to I_A and are between $I_A/2$ and $2I_A$ in value. The system is stable with this relative polarity and operates for all values of I_B outside the above limits.

If the pilot wires are open-circuited, current input will tend to operate the relay and conversely if they are short-circuited the relay will be restrained, holding the tripping contacts open.

In order to maintain the bias characteristic at the designed value it is necessary to pad the pilot loop resistance to 1000 ohms and a padding resistor P_r is provided in the relay for this purpose. However, when pilot isolation transformers are used the range of primary taps enables pilots of loop resistance up to 2500 ohms to be matched to the relay.

Reyrolle Solkor-R high-speed feeder protection. Solkor-R high-speed feeder protective systems belong to the circulating current class of differential protection in that the current transformer secondaries are connected so that a current circulates around the pilot loop under external fault conditions. The protective relay operating coils are connected in shunt with the pilots, across equipotential points when the current circulates around the pilots.

In this particular scheme equipotential relaying points exist at one end during one half cycle of fault current and at the other end during the next half cycle. During the half cycles when the relays at either end are not at the electrical mid-point of the pilots, the voltage appearing across them causes them to act in a restraining mode.

With the current transformers connected as shown in *Figure 15.14* a single-phase current is applied to the pilot circuit so that a comparison between currents at each end is effected. The tappings on the summation transformers have been selected to give an optimum balance between the demands of fault setting and stability.

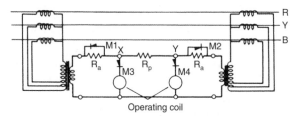

Figure 15.14 Connections of Solkor-R pilot wire feeder protection system

In the diagram the pilot wire resistance is shown as a single resistor R_p. The rest of the pilot loop is made up of two resistors R_a and two rectifiers M1 and M2. The operating coils are made unidirectional by the rectifiers M3 and M4 and are connected in shunt with the pilots at points X and Y.

During external fault conditions an alternating current circulates around the pilot loop. Thus on successive half cycles one or other of the resistors R_a at the end of the pilots is short-circuited by its associated rectifier. The total resistance in the pilot loop at any instant is therefore substantially constant and equal to $R_a + R_p$. The effective position of R_a alternates between both ends, being dependent upon the direction of the current. This change in the effective position of R_a makes the voltage distribution between the pilot cores different for successive half cycles. The resulting potential gradient between the pilot cores when R_a is equal to R_p is shown in *Figures 15.15(a)* and *15.15(b)*.

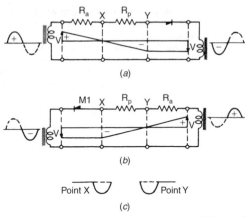

(a)

(b)

(c)

Figure 15.15 Behaviour of basic circuit under external fault conditions when $R_a = R_p$. Diagrams (a) and (b) show the effective circuit during successive half-cycles: diagram (c) indicates the voltage across relaying points X and Y during one cycle

From these two diagrams it will be seen that the voltage across the relays at points X and Y is either zero or in the reverse direction for conduction of current through the rectifiers M3 or M4. When R_a equals R_p a reverse voltage appears across a relay coil during one half cycle and zero voltage during the next half cycle. The voltage across each relaying point is shown in *Figure 15.15(c)*.

In practice, resistors R_a are of greater resistance than R_p, and this causes the zero potential to occur within resistors R_a as shown in *Figures 15.16(a)* and *15.16(b)*. The voltage across each relaying point X and Y throughout is shown in *Figure 15.16(c)*. From this it will be seen that instead of having zero voltage across each relay operating coil on alternate half cycles, there is throughout each cycle a reverse voltage applied. This voltage must be nullified and a positive voltage applied before relay operation can take place. This increases the stability of the system on through faults.

During internal fault conditions with the fault current fed equally from both ends, the effective circuit conditions during each half cycle are shown in *Figures 15.17(a)* and *15.17(b)*. Pulses of operating current pass through each relay on alternate half cycles, and the relay at each end operates.

With a fault fed from one end the relay at the end remote from the feed is energized in parallel with the relay at the near end. The relay at the feeding end

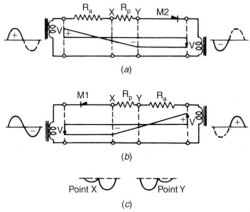

Figure 15.16 Behaviour of basic circuit under external fault conditions when $R_a > R_p$. Diagrams (a) and (b) show the effective circuits during successive half-cycles; diagram (c) indicates the voltage across relaying points X and Y during one cycle

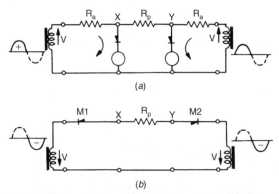

Figure 15.17 Behaviour of basic circuit under internal fault conditions, fault fed from both ends. Diagrams (a) and (b) show the effective circuits during successive half-cycles

operates at the setting current and the relay at the remote end at approximately $2\frac{1}{2}$ times the setting current. Thus the protection operates at both feeder ends with an internal fault fed from one end only provided the fault current is not less than $2\frac{1}{2}$ times the setting current.

In addition to the basic components mentioned above the complete protective system includes at each end two non-linear resistors, a tapped 'padding' resistor and a rectifier. The non-linear resistors are used to limit the voltage appearing across the pilots and the operating elements. The padding resistors

360

bring the total pilot loop resistance up to a standard value of 1000 ohms in all cases. The protection is therefore working under constant conditions and its performance is largely independent of the resistance of the pilot cable.

The rectifier connected across the operating coil is to smooth the current passing through the coil on internal faults, thereby providing maximum power for relay operation.

The operating element is of the attracted armature type with three pairs of contacts, each pair being brought out to separate terminals. They are suitable for direct connection to a circuit-breaker trip coil, thus no repeat relay is necessary.

The above description relates to the original design of Solkor R protection based on a half-wave rectification principle shown in *Figure 15.14*. The latest development is Solkor Rf which is based on a full-wave principle. This has the advantage of faster speed of operation, and the facility of including isolating transformers in the pilots to allow pilot voltages up to 15 kV. Without pilot isolating transformers Solkor Rf is suitable for use with 5 kV pilots as was the original Solkor R.

Current balance protection schemes can equally be applied to rotating machines and transformers. *Figure 15.18* shows a typical scheme for generators where an instantaneous unbiased relay can be used to give adequate protection (protection of large turbogenerators is covered in Chapter 7). The characteristics of transformers make them unsuitable for using an unbiased relay, because they cannot take into account tapchanging, magnetizing inrush currents and phase connection of the transformer. For these reasons the CTs are connected in phase opposition to the transformer connections and a biased relay is used as shown in *Figure 15.19*. In addition a harmonic bias feature is added to the relay to prevent operation under magnetizing inrush currents.

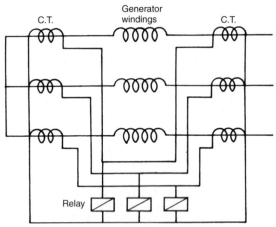

Figure 15.18 Generator protection

Distance protection. Pilot wire protective schemes are uneconomical for lines longer than 24 km to 32 km. For such lines it is necessary to use either

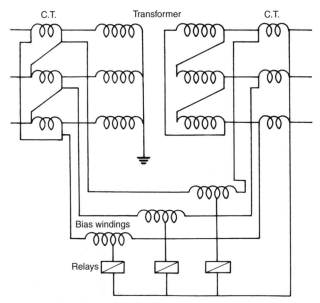

Figure 15.19 Transformer protection

distance protection or carrier current protection or a combination of both. These do not require pilot wires.

The principal feature of a distance scheme is the use of a relay that measures the fault impedance, which, if it is below a certain value, means that the fault is within the protected zone and the relay operates to trip out the circuit-breaker. The relay is energized by current and voltage and operates when the ratio of E/I gives an impedance below the relay setting.

Standard distance protection schemes protect the overhead line in three stages. The first stage has a zone of operation extending from the relaying point at the first substation to a point 80% of the way to the next substation. A fault occurring on this part of the line causes the protection at the first substation to trip immediately.

The second stage has a zone of operation extending from this 80% point past the second substation along the next feeder to the third substation. A fault in this second zone causes the circuit-breaker at the first substation to trip after Zone 2 time limit provided the protection on the second feeder has not operated.

The third stage has an operating zone extending a little way past the third substation and provides back-up protection with a longer delay.

The means by which these operating zones and times are obtained in the Reyrolle type THR range of distance protection schemes is shown typically on a single-phase basis in *Figure 15.20*. In this particular arrangement each fault type and each zone has an individual measuring element so that 18 elements are required to make up the scheme covering phase and earth faults in three zones. The first two zones are directional and are usually cross-polarized mho

362

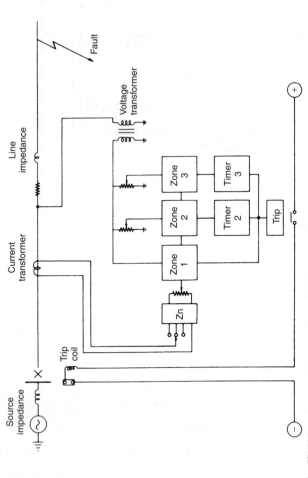

Figure 15.20 Simplified block diagram of 18 element type THR three-stage distance protection from NEI Reyrolle Protection

relays, but alternative characteristics for special requirements are obtained by directly replaceable modules. The third zone has an offset characteristic shaped against load ingression, or is plain offset mho and where switched schemes are employed acts as the starting relay. In a full scheme the Zone 3 relay is not required for starting but is included for power swing blocking and line check facilities.

The measuring elements of the three zones have definite settings relative to the line impedance which are referred to the primary side of the voltage and current transformers. Zone 1 is set to 80% of the line impedance by coarse and fine settings on the replica impedance ZN. Voltage dividers are used to set the Zone 2 to typically 120% and Zone 3 to between 200 and 300% of the line impedance.

If the line fault occurs within the first zone the polarized mho relay operates to trip the circuit-breaker at the first substation instantaneously. If its distance is in the zone of the second stage the Zone 2 times out, and the polarized relay trips the circuit-breaker. If the fault lies within the third zone of protection, the circuit-breaker at the first substation is tripped when the stage 3 timer has completed its operation.

16 Heating and refrigeration

Water Heating

Hot water systems for domestic, commercial and industrial premises are available in a variety of forms, the choice depending on the number and distribution of draw-off points, whether it is to be combined with space heating, and in some cases, the nature of the water supply. Electricity is particularly suitable as a sole energy source for water heating, especially so when advantage can be taken of off-peak tariffs. It is cheap and convenient, requires no fuel storage or flues, involves small capital cost, and operates at a high level of efficiency. When tanks or storage vessels are properly lagged, electric water heating is competitive with other forms of fuel, especially on off-peak tariffs. Various types of electric water heating appliances are available and the most suitable type for a given set of circumstances is best decided in consultation with the electricity supplier. The heating element is always completely immersed in the water to be heated so that heat transference is always 100% even if the element is encrusted with scale. For a given consumption of hot water the efficiency of a system depends only on the thermal losses from the storage vessel and the associated pipework.

Tariffs. In many cases it is possible to arrange for the electricity to be provided during the cheap off-peak period, usually during the night. This is best suited to those systems where water is stored in large quantities in suitably lagged vessels.

A popular tariff known as Economy 7 is one whereby the user has electricity at two rates, a day rate and a much cheaper night rate. Electricity used for 7–8 hours between around midnight and 7 or 8 a.m. is charged at the low rate. The precise times for the cheap night rate vary between suppliers. A special Economy 7 boiler has been designed for this tariff and is described on page 368.

Types of water heaters. Basically there are two types of water heaters: non-pressure and pressure. The non-pressure water heater is one in which the flow of water is controlled by a valve or tap in the inlet pipe, so arranged that the expanded water can overflow through the outlet pipe when the valve or tap is opened. This type of water heater can be connected direct to the main water supply.

The pressure type of water heater is supplied from a cistern in the normal way with provision for the expanded water to return to the feed cistern. A heater of this type must not be connected direct to a water main. Hot water tap connectors are made in the normal way to draw-off piping.

Non-pressure water heaters. Examples of non-pressure heaters are those that are installed over a sink and fitted with swivel outlets, *Figure 16.1*. Capacities range from 7 to 14 litres. They are available in cylindrical or rectangular shape, in various colours and in some cases are fitted with a mirror. A small under-the-sink storage water heater can be of the non-pressure type. It is plumbed with a choice of special taps to the wash basin or sink which has to be designed for use with it. Normal capacity is between 7 and 14 litres.

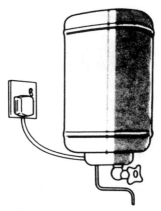

Figure 16.1 Oversink storage water heater

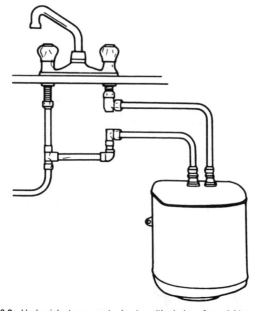

Figure 16.2 Undersink storage water heater with choice of special taps

This type is also available in either rectangular or cylindrical shape. Typical undersink or basin types are reproduced in *Figure 16.2*.

An instantaneous water heater of the flow type, fitted with a swivel outlet, is a further unit falling within the class of non-pressure. Typically a unit can

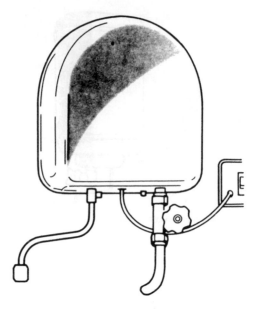

Figure 16.3 Instantaneous 'flow' type water heater

deliver approximately up to 0.23 litres per minute at a hand wash temperature of about 43°C, see *Figure 16.3*.

If a larger vessel with higher rated heating elements is employed than for the small instantaneous water heater it is possible to provide a self-contained instantaneous shower unit. This is fitted with an adjustable shower rose, water flow restrictor, pressure relief device and a thermal cutout. *Figure 16.4* shows

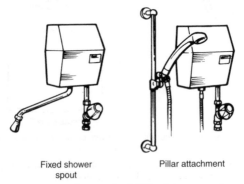

Fixed shower
spout

Pillar attachment

Figure 16.4 Instantaneous shower units: (left) fixed shower spout and (right) pillar attachment (Heatrae – Sadia Heating Ltd)

typical units which will deliver water at a temperature about 43°C at a rate of 0.47–0.57 litres per minute. Usually the precise outlet temperature is adjusted by a control knob which adjusts flow rate. Loading for a shower unit at 240 V would be about 6 kW and the weight of a unit about 3 kg.

Pressure cistern types. Cistern-type water heaters must be mounted above the highest tap otherwise the hot water will not flow. There must also be 200–300 mm clearance above the cistern to allow access to the ball valve for servicing. Because the hot water storage is of the pressure type, taps or valves can be fitted in an outlet pipe draw-off system. The ball-valved cistern is connected to the main water supply.

One of the most popular of the pressure types for use in homes is shown in *Figure 16.5*. The main characteristic of this unit is that it is fitted with two heating elements, one mounted horizontally near the base and the other horizontally but near the top of the vessel. The top heater keeps about 37 litres of water at the top of the tank hot for general daily use while the other is switched on when larger quantities of water are required. Total storage capacity is typically 108 litres. Both elements are controlled thermostatically. The heater can be stored under a draining board, alongside a working surface or in a corner. The associated plumbing must be vented back to a high level cistern, and a number of hot water taps can be served.

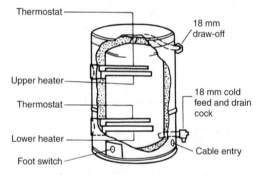

Outer casing 825 mm high 508 mm diameter

Figure 16.5 'Two-in-One' 180 litre heater

Larger capacity storage water heaters are available ranging from 182 litres to 455 litres as standard equipments, and even larger ones can be made to order fitted with two immersion heaters as in *Figure 16.5*; these are connected to off-peak supplies. Economy 7, for example, is described on page 368. A large water heater can be fitted with a number of elements, the associated thermostats being set, if necessary, to operate in a preselected manner to provide an ordered sequence of heating. When very large quantities of stored water are required it is preferable to install a number of smaller heaters than a few large ones. This permits the adoption of smaller pipework and component parts and also allows a heater to be taken out for servicing without affecting the overall performance too much.

Advantage can be taken of the benefits of the Economy 7 tariff (see page 364) by modifying as shown diagrammatically in *Figure 16.6*. The

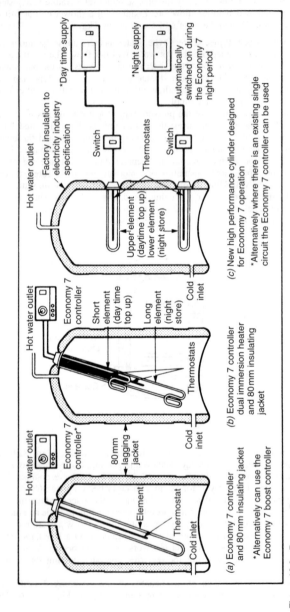

Figure 16.6 Economy 7 conversions

*Day time supply

*Night supply

Automatically switched on during the night period

Hot water outlet

Factory insulation to electricity industry specification

Switch

Thermostats

Switch

Upper element (daytime top up) lower element (night store)

*(c) New high performance cylinder designed for Economy 7 operation

*Alternatively where there is an existing single circuit the Economy 7 controller can be used

Hot water outlet

Economy 7 controller

Short element (day time top up)

Long element (night store)

Thermostats

Cold inlet

*(b) Economy 7 controller dual immersion heater and 80 mm insulating jacket

Hot water outlet

Economy 7 controller*

80 mm lagging jacket

Element

Thermostat

Cold inlet

*(a) Economy 7 controller and 80 mm insulating jacket

*Alternatively can use the Economy 7 boost controller

three main ways of converting to Economy 7 showing progressively improving running costs are as follows:

1. Installing an Economy 7 water heating controller or Economy 7 boost controller to control the existing 685 mm immersion heater and adding an 80 mm insulating jacket to the existing storage vessel (*Figure 16.6(a)*).
2. Installing an Economy 7 controller and a dual immersion heater and adding an 80 mm insulating jacket to the existing storage vessel (*Figure 16.6(b)*).
3. Installing a new high performance Economy 7 cylinder (*Figure 16.6(c)*).

Economy 7 water heating controllers are produced by a number of manufacturers but all operate in much the same way. The controller *Figure 16.7* incorporates a fixed cam timing arrangement and automatic changeover switch. In this way the controller automatically switches on the immersion heater during the night period while the customer can operate the immersion heater during the day period by means of the run-back or electronic boost timer, should the hot water become depleted.

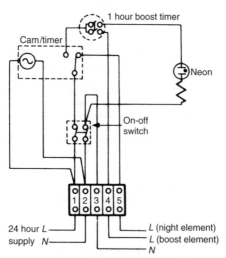

Figure 16.7 Schematic internal wiring diagram of Economy 7 water heating controller

With the dual immersion heater, the Economy 7 controller operates the long element overnight and the short element during the day using the boost timer. This allows a smaller quantity of water to be heated at the more expensive day rate than with the existing single immersion heater.

Two of the models of Economy 7 controller manufactured have a quartz timer with a rechargeable battery reserve of at least 100 hours. These controllers can be expected to keep reliable time without frequent inspection.

Where a specifier wants to ensure that the night-time electricity supply to the immersion heater is timed by the supply companies' timeswitch,

370

the Economy 7 boost control is an alternative. In this case the controller is normally sited at or near the electrical distribution board, e.g. in the kitchen or hall.

Although this control is wired with both an off-peak circuit (via the supply companies' timeswitch) and a 24 hour supply, a single immersion heater is controlled, normally fed from the former. If extra water is required during the day the customer turns a run-back timer and for the period required the immersion heater is boosted at day rates.

Where the hot water storage vessel is being replaced or a system is being installed in a new home a high performance hot water storage unit can be utilized as shown in *Figure 16.6(c)*. It may take the form of a cylindrical or rectangular combination unit or most commonly a replacement or new cylinder that will be plumbed into an existing cistern. Three sizes are available that will meet most needs, the BS type reference 7 (120 litre), 8 (144 litre) and 9E (210 litre), and for convenience they are called small, medium and standard. Type 9E is the preferred size for off-peak operation.

Typical arrangements of the pipework, cistern, heater and outlets for houses and flats are shown in *Figures 16.8, 16.9* and *16.10*. As can be seen the hot water draw-off pipe is continued upwards to turn over the side of the ball-valve cistern. The arrangement of pipework is very important if single-pipe circulation is to be avoided. Single-pipe circulation takes place in any pipe rising vertically from a hot water storage vessel due to the water in contact with the slightly cooler wall of the pipe descending being replaced by an internal column of hot water rising from the storage vessel. It is easily prevented by running the draw-off pipe horizontally immediately it leaves the top of the storage vessel for about 450 mm before any pipe, such as the vent pipe, rises vertically from it. *Figure 16.11* shows how this can be done.

Figure 16.10 is a self-contained cistern-type electric water heater which is made with an integral cold feed cistern controlled by a ball valve above the hot water tank. It is thus a complete water heating system in itself, and only needs connection to the cold feed, hot draw-off pipe, overflow and electrical supply.

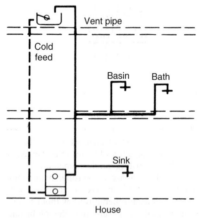

Figure 16.8 The use of dual element pressure-type electric water heaters (house)

371

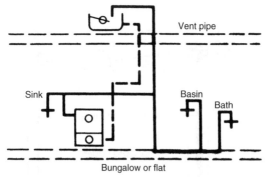

Figure 16.9 The use of dual element pressure-type electric water heaters (bungalow or flat)

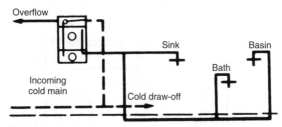

Figure 16.10 The use of self-contained cistern-type electric water heaters

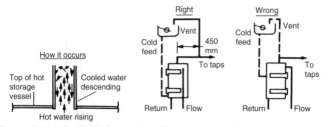

Figure 16.11 'Single pipe' circulation. The layout on the left will eliminate 'single pipe' circulation, whilst this type of circulation will result from the layout on the right

A packaged plumbing unit is made, suitable for providing both hot and cold water for bath, basin, kitchen sink and w.c. It comprises a cold water cistern, hot water container and pipework built into and within a rigid framework, *Figure 16.12*. The method of installation when serving two floors is shown in *Figure 16.13*. This also shows how it is used in conjunction with a back boiler. Many of the cistern-type systems discussed can be used in conjunction with fossil-fired boilers.

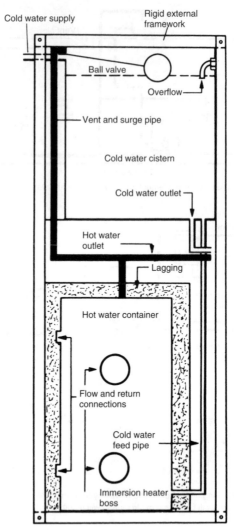

Figure 16.12 Packaged plumbing unit. Approx. dimensions: 1.98 m high × 0.7 m wide × 0.61 m front to back

Immersion heater. An immersion heater cannot heat water below itself. Water is heated by convection and not conduction unless there is fortuitous mixing. The installation of an immersion heater into an existing hot water tank is the cheapest method of converting a non-electric water heating system to

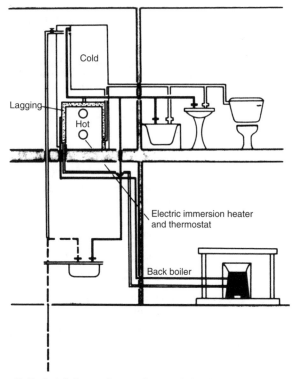

Figure 16.13 Installation serving two floors and showing alternative connection with back boiler

an electric method. It thus can augment an existing solid fuel system so that the latter can be shut down during the summer months, allowing the electrical element to provide the hot water required.

An immersion heater mounted horizontally at low level heats all the water above it, to a uniform temperature. A small amount of water cannot be produced quickly by this method, so when this is required a second heater is positioned near the top of the storage vessel.

An immersion heater mounted vertically cannot produce a tank full of hot water at uniform temperature. A temperature gradient is produced ranging from hot at the top to cold at the bottom. When a vertically mounted immersion heater is used it should be long enough to reach within 50–75 mm of the cold water inlet.

A compromise is the 'dual' or twin heat arrangement comprising a short and a full length element on one boss each with its own thermostat and interlocked to provide an alternative choice.

There are basically two types of immersion heater: a withdrawable and a non-withdrawable type. The withdrawable type is constructed so that the heating element can be withdrawn from the enclosing sheath without disturbing or breaking a water joint. This type of element is used mostly in industry. For the non-withdrawable type the heating element cannot be withdrawn without breaking the water joint.

Calculating energy required for water heating. It is a comparatively simple matter to calculate the amount of energy required to raise the temperature of water through a given number of degrees Centigrade.

$$\text{Energy (kWh)} = \frac{k \times \text{kg} \times \text{temperature rise in degrees C}}{\text{efficiency}}$$

It requires 4180 J to raise the temperature of 1 kg of water through $1°C$ and 4180 J equals 1 kcal. 1 kWh $= 3.6 \times 10^6$ J.

An example will illustrate how the energy is calculated:

An electric heater raises the temperature of 10 kg of water from $15°C$ to boiling point. Only 75% of the energy intake is used to heat the water, the rest is lost by radiation. Calculate the energy intake of the heater.

Heat energy of water content $= 10 \times (100 - 15) = 850$ kcal

Energy intake	$= 850/0.75$	$= 1133$ kcal
Energy in joules	$= 1133 \times 4180$	$= 4.75 \times 10^6$ J
Energy in kWh	$= \dfrac{4.75 \times 10^6}{3.6 \times 10^6}$	$= 1.32$ kWh

From the above formula it is easy to derive all other related quantities: it should be noted that k in the above formula is $4180/3.6 \times 10^6$.

$$\text{Efficiency} = \frac{4180 \times \text{kg} \times \text{temperature rise in degrees C}}{\text{kWh} \times 3.6 \times 10^6}$$

If it were required to know the time in hours it would take to raise a given quantity of water through a given temperature the formula would be:

$$\text{Time in hours} = \frac{4180 \times \text{kg} \times \text{temperature rise in degrees C}}{\text{loading in kW} \times 3.6 \times 10^6 \times \text{efficiency}}$$

Space Heating

Space heating can be classified into direct acting or storage systems. Storage systems are operated on lower cost off-peak supplies and can be designed to take full advantage of the thermal storage properties of the building and its contents. Direct acting appliances, which use electricity at daytime rates, have the advantage of immediate response to temperature changes and intermittent programming. The two systems can be used independently or may be complementary to each other. Storage heating would account for about 90% of the electricity consumed over the winter period in a combined installation.

Storage heater. Storage heaters operate on the principle of retaining the heat that is taken in during the off-peak period (usually overnight), and releasing it at a later time, i.e. during the day, when the supply is disconnected from them. Storage heaters are filled with a material that will absorb heat, and are so constructed that emission of heat is both gradual and continuous throughout the 24 hours. The basic types of off-peak systems are: storage fan heaters; Electricaire; Economy 7 boiler; hot water thermal storage vessel with radiators; and floor warming.

Storage radiators. The electric storage radiator consists of heating elements embedded in a storage core, surrounded by a layer of insulating material, the whole being in a metal casing. The heating element is automatically switched on during off-peak hours, and conserves its heat in the surrounding storage material and discharges it through the insulating layer for the full 24 hours. Control may be effected by a combination of charge controllers, external controllers and programme timers. Under ideal conditions the outer casing remains warm throughout the charging and discharging periods.

The rating of a storage radiator is the input in kilowatts taken by the heating element. It is of the order of 2–2.625 or 3.375 kW. The charge acceptance is the quantity of heat which the radiator can absorb and emit and is measured in kWh. Acceptance ratio is the ratio of actual to maximum charge acceptance. Most radiators on the market have an acceptance ratio of about 83%. The nearer this figure is to unity the more economical is the radiator to use.

Net storage capacity is the quantity of heat retained within the storage radiator when fully charged over an eight-hour period. Retention factor is the ratio of the net storage capacity to the eight-hour charge acceptance. For the average storage radiator this figure is about 66%.

There are two ways of obtaining a boosted output from a storage radiator. The first is by incorporating a damper controlled by a bi-metal strip which allows heat to be extracted direct from the radiator core by convection during the late afternoon. Radiators incorporating this are known as 'damper' models. The second way is to employ a fan to operate a closed air circuit between the core and the inside of the front panel, and a separate electricity supply is required for this.

Storage fan heaters. Storage fan heaters are basically similar to storage radiators except that they have more thermal insulation and incorporate a fan to discharge warm air into the room as required. A small amount of heat is given off from the heater all the time and this is sufficient to take the chill off a room. These heaters lend themselves to time and temperature control so that rooms can be quickly brought up to the desired heat and constant temperatures maintained.

Electricaire. Basically an Electricaire system consists of a central heater unit which supplies air to the rooms through a ducting installation. These systems are completely automatic, the electricity suppliers' time switch determining the period during which energy is supplied to the core and a core thermostat controlling the quantity. Air is discharged from the unit to the insulated ducts by a fan which gives the warm air a velocity of up to $7\,\mathrm{m^3/min}$. The fan is fed from an unrestricted supply source and is controlled by a wall-mounted thermostat. The location of this thermostat is important for successful operation of the system.

Storage capacities are available up to about 3.7×10^7 J per day and most models have a boost device to permit higher output over short periods of 20 minutes or so.

The core of an Electricaire unit is formed from a number of spirally wound elements of about 1 kW rating fitted to refractory formers. These are used to heat the storage blocks. These blocks are made of cast iron or refractory material and insulated to prevent loss of heat through the cabinet. A high temperature thermostat with manual adjustment controls the charge to the core. The outlet temperature of the unit is maintained automatically by a regulator which mixes the right amount of cool air with heated air from the store from which the distribution ducts originate. Fully automatic models are available requiring no manual attention from the user.

There are three basic layouts for the ductwork of an Electricaire system, stub-duct, radial-duct and extended plenum. With a stub-duct the Electricaire unit is usually installed in a central position and heat is conveyed to the rooms through short ducts. In some cases these ducts can be extended to upper rooms but quite often this is not necessary due to the natural convection of the air.

In a radial duct system warm air from the plenum chamber feeds through a number of radial ducts to the outlet registers. A riser for this type of system comes direct from the plenum chamber which allows air to be taken into a number of rooms in a larger or more complex building layout. A riser may be taken from any point on the plenum to serve upper floors.

Input loadings of Electricaire units range from 6 to 15 kW. The heater is considered to be exhausted when the temperature of the air leaving the heater falls to about 43°C. At this point there is still a useful quantity of heat left in the core and by arranging for a direct-acting element to supplement the output of the depleted core the active life is extended. The direct acting element is positioned across the air flow before it enters the plenum and can be switched on automatically by monitoring the air outlet temperature or the core temperature.

Economy 7 boiler. Hot water radiators are a familiar and long-established means of heating homes and one possible way to raise the temperature of the water is by an electric boiler specially designed for use with the Economy 7 cheap night rate tariff.

Most of the hot water needed to warm the radiators can be heated at the low cost night rate and then stored in the well-insulated electric boiler (see *Figure 16.6*, page 368, for the design). When the hot water is needed, a room thermostat controls the central heating pump which drives it around the radiators. Heaters in the top of the boiler can give an extra boost during the day if required or during spells of extreme or prolonged cold weather.

The boiler can be sited almost anywhere provided it is protected from the weather. A garage, outhouse, conservatory or a kitchen are all suitable. The boiler does not need a flue and requires little annual maintenance.

Size of boiler depends on the requirements of the home and its insulation level. A standard 680 litre boiler measuring $710 \times 813 \times 2134$ mm high is suitable for a 90 m^2 well-insulated detached house. A much larger house could use two standard units, or a special tailor-made unit to fit an available space.

Floor warming. Floor warming systems use elements directly embedded in the floor concrete. Most installations are of the embedded type. Capital outlay of a floor warming system is low compared with other forms of heating and it can be operated on the Economy 7 tariff. In principle the floor becomes a low-temperature radiating device. Concrete floors provide the heat storage medium and they may be of the following types of construction: hollow pot, hollow concrete beam, prestressed plank and in-situ reinforced concrete. Insulation of the edge and undersurface of the concrete floor is essential.

Floor warming elements have conductors made of chromium, iron, aluminium, copper, silicon or manganese alloys. Insulation can be asbestos, mineral, PVC, butyl, silicon rubber and nylon. Installations in the UK are designed for an average floor surface temperature of 24°C.

When installing, the elements should be arranged as far as possible to cover the whole of the floor area and in general the loading falls between $100\,W/m^2$ and $150\,W/m^2$. Elements should be chosen so that the spacing between adjacent elements does not exceed 100 mm or less than screed depth, so as to avoid temperature variations on the floor surface. The 'density' of the heating elements should vary according to location, with more heating elements being used near external walls, alternatively the floor area may be zoned, each zone being thermostatically controlled.

Some specialized outdoor areas may also be provided with this form of heating. They include car park ramps, airport runways and pedestrian ramps.

Direct-acting systems. A wide range of direct-acting heaters is available such as radiant fires, convectors, radiant/convectors, panel heaters, domestic fan heaters, industrial unit heaters and combined light and heat units. In addition ceiling heating, an extended panel system, deserves to be mentioned separately. All these systems are described in the following paragraphs.

Ceiling heating. Ceiling heating may be provided by prefabricated panels or installed as a complete entity. In the latter case the installation of the elements is carried out in a similar way to that of floor heating except that the element wiring is fixed in position on the underside of the ceiling before the ceiling board is erected.

The low thermal capacity of a ceiling heating installation enables it to respond more quickly to thermostatic and time controls than a thermal storage system. Where each room installation is individually controlled, response to changes in internal temperature, i.e. solar gains or heat gains from lighting or occupants, is rapid. Because of the economy possible in this way the system can be competitive though it uses on-peak electricity.

Ceiling surface temperatures are between 18°C and 21°C and heat emission approximately $3.6-4.3\,W/m^2$ deg C.

Radiant heaters. There is a wide variety of domestic radiant heaters ranging from 740 W bowl fires up to 3 kW with multiple heating bars which are switched separately. Often these fires incorporate a convector facility and are decoratively designed to provide an attractive fireplace.

They are portable and instantaneous in response.

Infrared heaters. The infrared heater consists of an iconel-sheathed element or a nickel-chrome spiral element in a silica glass tube mounted in front of a polished reflector. Loadings vary from 1 kW to 4 kW and comfort can be achieved with a mounting height of between 6 m and $7\,1/2$ m. These are intended for local heating in large areas.

A smaller rated version, 500 W or 750 W, is made suitable for bathrooms; such a unit may be combined with a lamp to form a combined heating and lighting unit.

Oil-filled radiators. Oil-filled radiators are very similar in appearance to conventional pressed-steel hot water radiators. They are equipped with heating elements to give loadings between 500 W and 2 kW. They can be free-standing or wall-mounted and are generally fitted with thermostatic control.

These radiators heat up more slowly than radiant or convector panels and retain their heat for longer periods after switch-off. Being on thermostatic control they avoid rapid temperature swings. Surface temperature is around 70°C, comparable to a hot water radiator.

Tubular heaters. These are comparatively low temperature heaters used to supplement the main heating in a building.

A heater is made from mild steel or aluminium solid drawn tubing with an outside diameter of around 50 mm and varying in length from 305 mm to 4575 mm. The tube is closed at one end and has a terminal chamber at the other. Elements are rated at 200 W to 260 W per metre run and the tube surface temperature is of the order of 88°C, slightly higher than the conventional low pressure hot water radiator.

Skirting heaters. The skirting heater is a refinement of the tubular heater but shaped like conventional skirting. Loadings vary widely from 80 W to 500 W per metre run.

Convectors and convector/radiant heaters. Convectors can be free-standing, wall-mounted or built in. They consist basically of a low temperature non-luminous wire heating element inside a metal cabinet. The element is insulated thermally and electrically from the case so that all the radiant heat is converted to warm air. Cool air enters at the bottom of the unit and warm air is expelled at the top at a temperature of between 82°C and 93°C and at a velocity of about 55 m/min. Integral thermostatic control is usually provided.

Combined convector and radiant heaters incorporate in addition to the convection element a radiant element usually of the silica tube type. Some models have three 1 kW silica tube elements together with a range of heats from the convector. Control allows a combination of convection and radiant heat to be obtained to suit a wide variety of requirements. Integral thermostatic control is also usually provided for these units.

For economical operation, particularly where convector or radiant convector panels comprise a control heating scheme, thermistor integral thermostats or wall-mounted remote air thermostats must be used to control them.

Fan heaters. Fan heaters operate on the same principle as convectors but employ a fan to circulate the warm air. Domestic types are available in loadings up to 3 kW. The fan is often mounted direct on the shaft of the small motor which may have two speeds. Fan blades are tangential in order to keep noise to an acceptable level. For domestic use these heaters are floor-standing models and can be as small 380 mm × 150 mm × 100 mm.

There is a design of fan heater which has both water/air and direct electric heating elements. It is suitable for use with all heating systems using water as the heating medium. The direct acting element can be used to provide supplementary heating to the central source or even independently if so desired.

Economics of electric space heating. In the 1980s trials conducted by the then Yorkshire Electricity Board showed that an electric boiler central heating system could be more economic than both oil and solid fuel. Although the relative costs of various prime fuels tend to vary rapidly and in cycles of a few years' duration, the situation in 2002 is that, even recognizing that there are no energy losses via exhaust flues and boiler ventilation, it is difficult to make an economic case for heating by electricity. The only exceptions to this might be in the case of fairly small flats where the provision of fuel service and gas outlet flue might present difficulties. Even in this case the economics, even in a small block of flats, would be in favour of a communal system using some other fuel.

In general, the advantages of all the types of electric space heating described above are in their convenience, from the point of view of ease of installation and modest capital cost, rather than in terms of their economics as a source of energy.

Thermostatic Temperature Control

Thermostatic switches are widely employed for automatic temperature control of water and space heating appliances. They generally operate on the bi-metal principle in which the unequal expansion between two metals causes distortion which actuates a switch.

Room thermostat. The TLM room thermostat from Satchwell Sunvic is designed for heavy current circuits suitable for the direct switching of heating loads up to 20 A. They are therefore suitable for the control of space temperature. Heaters, valves, drive motors and other heating and cooling equipment can be controlled with equal simplicity. The thermostat is an on-off control unit, the circuit closing with a fall in temperature.

Other models in the range can be fitted with changeover contacts for controlling motorized valves or for reverse acting applications. By installing the thermostat in a weatherproof case it is suitable for outside use.

The switch mechanism of the TLM thermostat ensures a high contact pressure up to the instant of opening so that even with loads up to 20 A any self-heat is reduced to a minimum. An accelerator heater is incorporated in the design so that rise in temperature is anticipated by the instrument when 'calling' for heat. Noticeable overshoot is thus eliminated ensuring precise maintenance of room temperature.

Immersion thermostat. The Satchwell Sunvic W immersion thermostat is a stem type model, *Figure 16.14*. It is designed for the control of the temperature of hot water installations, industrial processes and oil heaters. The temperature sensitive stem is made from aluminium brass and nickel-iron, and the resulting differential expansion is used to operate a micro-gap switch. Different stem lengths are provided according to the temperature range and sensitivity required. Large switch contacts are fitted to provide a long trouble-free life and the switch mechanism is enclosed in a dust-excluding moulded phenolic cover. Contacts 1 and 3 (*Figure 16.15*) break circuit with rise of temperature, contacts 2 and 3 make with rise of temperature.

Various modifications are available to the standard design to give a variety of operating techniques. A hand-resetting device can be fitted which ensures that once the control circuit is broken by the thermostat (contacts 1 and 3), it cannot be remade until the thermostat is manually reset by means of a pushbutton adjacent to the control knob. This facility allows the thermostat to

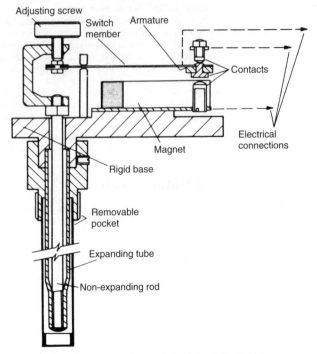

Figure 16.14 Type W immersion thermostat (Satchwell Sunvic Ltd)

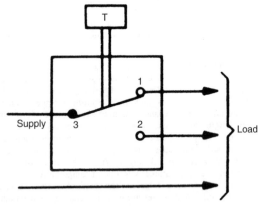

Figure 16.15 Connection diagram for thermostat shown in Figure 16.14

be used as a high limit control device with complete safety. Contacts 2 and 3 can be used to operate an alarm warning device if required.

Installation of thermostats. Room thermostats should be located in a position representative of the space to be controlled where it will readily be affected by changes in temperature. When used to control input to storage heating systems the accelerator heater should be left disconnected.

Knockouts are provided in the base of the TLM room thermostat for surface wiring allowing either top or bottom entry.

When ordering thermostats give the manufacturer as much information as possible about the installation in which it is to be used, particularly the control temperature and the differential required, i.e. the difference between the cut-in and cut-out points.

Electric Cookers

Modern electric cookers are usually fitted with an oven, a grill and usually four boiling rings. Apart from the conventional design, there are double-width and split-level cookers. Split-level cookers are those that have the grill mounted at eye level. The rings are generally of the radiant pattern and consist of a flat spiral tube with the heating element embedded in it. Dual rings are available in which the centre portion of the ring can be separately controlled from the outer section; alternatively the whole of the ring can be switched as one entity. A disc ring type or plate is available consisting of a flat circular metal plate, usually with a slightly recessed centre, enclosing the heating element. The whole unit is then sealed into the cooker hob.

A development of the split-level design is where the oven and grill units are separate from the hob and are built into the wall or into housing kitchen furniture at any position convenient to the user. Another development is the ceramic hob with four integral cooking areas.

Boiling rings are normally 180 mm in diameter but they are also available in 150 mm, 200 mm and 215 mm diameters. Electrical loadings from 1.2 kW to 2.5 kW are common. Rings are controllable from zero to full heat by means of rheostats operated from control knobs on the escutcheon plate of the cooker. Some models have thermostatic control of one or more of the rings, to keep the pan temperature at the desired setting.

Grills can either be fitted at waist level below the hob, at eye-level position above the rear panel or separate from the hob as mentioned above. Some eye-level grills are fitted with a motor-driven spit. Occasionally a second, separately controlled element is fitted below the spit for use when cooking casseroles and similar dishes. Heating elements for grills range from 2 to 3.5 kW. Control is in the same way as for boiling rings, i.e. with energy regulators. Some models have facilities for switching only one half of the grill on.

The latest designs in ovens include either stay-clean facilities or are of the self-clean type. In many cases the oven is arranged for complete removal to ease cleaning. Ovens are almost totally enclosed with only a small outlet for letting steam escape. The heating elements are usually fitted behind or on the side walls and sometimes beneath the bottom surface. Oven loadings are of the order of 2.5 kW. Oven controls allow temperature to be preset between

90°C and 260°C and maintained by thermostatic switching which is described later. An indicator light is often fitted to show when the oven is switched on. Modern cookers are generally fitted with a clock that can be used to automatically switch the cooker on and then off to provide delayed control.

Microwave cookers are now widely used. The electromagnetic radiation enables rapid cooking or reheating of food to be carried out. The waves, which range from 897 to 2450 MHz, penetrate the food to a depth of about 65 mm all round and cause the molecules to move around at high speed. The heat for cooking is produced by molecular friction inside the food.

Installation. It is important to install an electric cooker with due regard to electrical safety. Usual practice is to have a completely separate circuit for the cooker from all other circuits in the house. It is fused at either 45 A or 60 A although some older installations may be fitted with 30 A fuses. The supply from the consumer unit is terminated in a cooker control unit mounted just above or adjacent the cooker itself. A cable is generally internally connected to the cooker control unit and brought direct into the cooker itself with no intervening break.

Cooker control units are generally fitted with a neon indicator to show whether the main switch on the control unit is on, but if this neon is glowing it does not necessarily mean that the cooker is switched on. Many cooker control units incorporate a 13 A socket-outlet although current practice is to dispense with the socket-outlet and just have a main control switch. The view is held that the presence of a socket-outlet can encourage carelessness with trailing leads across the cooker hob with possible dangerous consequences. A cooker should never be installed near to a sink or tap; a distance of at least 1.8 m should be aimed at.

Earthing. For new installations the requirements of BS 7671 (Chapter 12) should be considered as mandatory. For an existing installation every effort should be made to ensure that the installation matches as closely as possible the current BS 7671 requirements with regard to earthing and bonding.

Cooker control. The most common method of heat control is the 'Simmerstat' controller. The 'Simmerstat' controller is shown in *Figure 16.16* and the associated wiring diagram in *Figure 16.17*. Here the heat control is obtained (given a certain amount of heat storage as in a boiling plate) by periodically interrupting the circuit. If a boiling plate is switched on for, say, 10 seconds and then switched off for 10 seconds repeatedly, over a period the amount of energy consumed is, of course, half that consumed if switched on full all the time giving half or 'medium' heat. If the ratio of on-and-off periods is altered to give 5 seconds on and 15 seconds off the result is 'low' or quarter heat. Two seconds on and 18 seconds off would be one tenth heat, and so on.

By turning the control knob up or down, any desired variation in energy fed to the boiling plate is achieved. The periodical operation of the switch contacts is effected by a bi-metallic strip which is made to move by a minute amount of heat applied to it by a very small heating element. When the strip has reached the end of its travel it opens a pair of contacts, breaking the circuit to the heating elements. The strip cools and resumes its former position when the contacts close again, and the cycle of movement is repeated indefinitely.

If a boiling plate is connected in parallel with the controlling heating element it, too, will be subject to the periodic application of energy and by varying the distance of travel of the bi-metallic strip, by turning the control

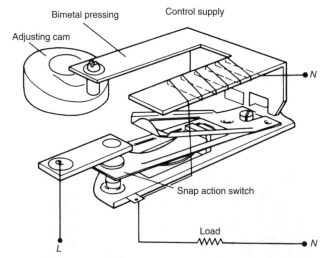

Figure 16.16 The Simmerstat (Satchwell Sunvic Ltd)

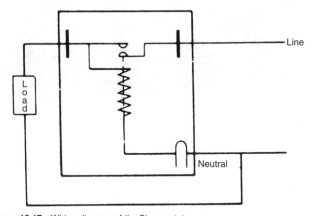

Figure 16.17 Wiring diagram of the Simmerstat

knob, any desired ratio of on-and-off periods can be obtained. The control knob is numbered from 0 to full with an 'off' position below the 0. Full heat is obtained when the control is turned to full on position, the switch contacts remaining together continuously. As soon as the control is turned to about 4 the contacts begin to interrupt the circuit periodically, and heat is reduced to about one half as variation is not required above this amount. An even and smooth reduction of heat is obtained by further turning the control towards zero, at which point the heat input is reduced to approximately 8%. This is

384

equivalent to about 150 W, with an 1800 W boiling plate, reducing heat input to a saucepan well below simmering point.

High Frequency Heating

Induction heating is used extensively today by industry in four main areas covering process heating applications: induction heat treatment, billet and bar heating, metal melting, and scientific equipment applications. A typical furnace is shown in *Figure 16.18*.

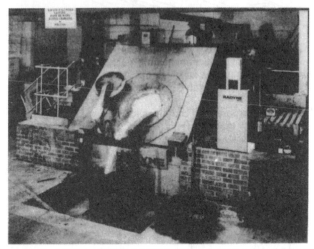

Figure 16.18 A Radyne 900 kW 1.5 tonne melting furnace powered by a 1 kHz solid-state generator in use at English China Clays, South West England

Where mass-produced parts require selective hardening to close tolerances, for example automobile engine and transmission components, induction heating is an ideal tool. The same process can be used for brazing, soldering, annealing, tube welding and many other applications.

The forging industry utilizes induction heating for heating billet and bar stock both continuously and selectively along the lengths of the bars concerned up to a temperature of 1250°C, prior to forging by mechanical presses or horizontal forging machinery.

Ferrous and non-ferrous materials are melted in induction furnaces with capacities from a few kilograms up to more than 70 tonnes. These furnaces can be small tilt, push-out, rollover, platform tilt and vacuum furnaces, depending on the particular requirement.

Scientific applications include epitaxial deposition on silicon for micro-circuits, crystal growing, zone refining, plasma analysis and etching, and carbon and sulphur determination.

Induction heating applications are normally categorized by frequency: r.f. (radio frequency) 250 kHz to 10 MHz; m.f. (medium frequency) 1–10 kHz; and l.f. (low frequency) 50–750 Hz. The frequency of the power source to be selected depends entirely upon the application; the higher the frequency the more shallow the heating effect achieved. For through-heating applications low frequencies are used.

The static inverter has greatly increased the scope for medium to low frequency induction heating applications because of its reliability, compactness and high efficiency, in some cases up to 95%. Efficiency of the valve generator for r.f. applications rarely exceeds 60%.

There are purpose built coating machines used in the paper and board industry that incorporate high velocity hot air dryers, sometimes with steam-heated drying cylinders. One of the most promising developments for non-contact drying is the use of radio frequency (r.f.) energy. The selective feature of r.f. energy, where most of the energy is absorbed in the web or coating, as well as the through-drying effect, gives a moisture profile levelling effect. This prevents moisture being trapped inside the material. Equipments having about 30 kW of r.f. output have been installed in the UK.

Both dielectric and microwave heating are being used in food processing. Both rely on the same phenomena, namely dipole agitation, caused by the alternating electric field. The main difference between the two techniques is that the former uses frequencies in the range 10–150 MHz while microwave heating employs frequencies above this range. The two bands currently used are 897–2450 MHz.

In dielectric heating, the material to be treated is placed between electrodes whereas in microwave heating waveguides are used for the same purpose. Greater power densities with less electrical stress on the food product can be achieved at microwave frequencies. This can be of particular advantage when used on relatively dry food products such as potato crisps.

One of the best established applications of dielectric heating is in the post-baking of biscuits and some bakery products. The process reduces the moisture content to a predetermined level and completes the inside cooking. There is usually no advantage in applying high frequency heating in the early stages of drying and baking where traditional methods are clearly effective. The best area of application is when the moisture level has dropped to around 15%, for at this level the conventional ovens cannot remove the moisture at anything like the rate required. Rapid drying down to very low moisture content levels is possible with dielectric or microwave heating.

Dielectric heating is widely used for setting synthetic resin adhesives, for 'welding' thermoplastics materials and for preheating rubber and moulding powders. Since the heat is developed evenly throughout the thickness of the material the temperature rise is uniform throughout so that the centre of the material is at the same temperature as the rest.

The welding of thermoplastics materials such as PVC is the largest active area for dielectric heating covering such applications as stationery items, medical products, inflatables, protective clothing, the shoe industry, leisure items, etc. A major user is the car industry for seating, door panels, sun visors and headlinings.

Electric Steam Boilers

There are many industrial plants where steam is required for process work and an electric boiler is worthy of consideration as an alternative to a fuel-fired boiler.

Comparison of boiler running costs should not be made on a fuel calorific value basis only. There are many other considerations to take into account, i.e. flue, fuel storage, is automatic boiler-plant operation wanted, summertime process loads, night-time and weekend loads. These are a few instances where electrically raised steam has advantages.

The principle of the electrode boiler is simple. An alternating current is passed from one electrode to another through the resistance of the water which surrounds the electrodes. A star point is formed in the water which is connected through the shell of the boiler to the supply system neutral, which, in accordance with the IEE recommendation, should be earthed. A circuit-breaker provided with the boiler gives overcurrent protection.

Generally electric boilers are connected to low voltage (415 V three-phase supply) but for very large loads high voltages may be used. *Figure 16.19* is a diagram of the GWB Jet Steam high voltage electrode steam boiler, but smaller steam process loads are provided from low voltage supplies. Loads up to 60 kW may be provided from immersion-heated steam boilers. Between 40 kW and 2300 kW electrode boilers are used, and *Figure 16.20* shows the GWB Autolec Junior range rated up to 170 kW.

For all electric steam boilers the specification includes the boiler complete with its mountings, a feed pump and circuit-breaker, which in the case of the immersion heated boiler is a contactor or thyristor directly controlled from water level and steam pressure.

All GWB electrode steam boilers are provided with a patented feed reservoir, which improves the performance of the boiler. All steam boilers require a feed tank. This is included in the Autolec Junior specification.

The principle of operation of low voltage electrode steam boilers is to control the steam output by varying the submerged depth of electrode in the water. This is achieved by releasing water from the boiler shell back to the feed tank through a solenoid valve. The valve is controlled as a function of the current flowing in the main electrode circuit or as a function of the boiler pressure. By these means a steady pressure is maintained under fluctuating steam demand conditions and the boiler is modulated to a virtual zero output when no steam at all is being drawn.

Water feed can form scale, or the electrical conductivity of the water can be increased by concentration of salts. These, in the case of the electrode boiler, may be removed from the shell through solenoid valves, and where the situation merits it a more sophisticated equipment automatic control may be fitted.

In addition to steam boilers GWB make electric steam superheaters ranging from 6 kW to 200 kW in a single shell. Operating pressures and temperature of superheaters are 20.4 bar and 350°C maximum. The control of the high voltage electrode steam boiler is by varying the number of water jets which issue from the jet column (item 4, *Figure 16.19*) on to the electrodes (item 11). The water not generated into steam falls back in the water in the bottom of the boiler shell. The control sleeve (item 7) is modulated as a function of current flowing in the electrode circuit and the boiler pressure.

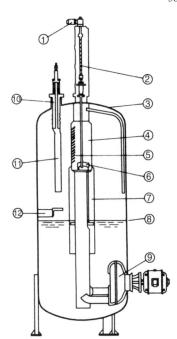

1. Control motor.
2. Control sleeve drive shaft.
3. Boiler shell.
4. Jet column.
5. Jets.
6. Control linkage.
7. Control sleeve.
8. Water level.
9. Circulating pump.
10. Insulator.
11. Electrode.
12. Counter electrode.

Figure 16.19 GWB Jet Steam h.v. electrode steam boiler

Electric Hot Water Boilers

Electric water boilers mainly provide hot water for central heating systems and the supply of domestic hot water. Where the hot water is to be used through an open outlet, i.e. taps or swimming pool heating, it is necessary to use a storage or non-storage heat exchanger (calorifier) (*Figure 16.23*).

Boilers may be connected to an 'on-peak' or 'off-peak' electrical system. In the latter case the energy is normally stored in thermal storage vessels in which the water is held at a high temperature and mixed to the required secondary system conditions. This principle is shown in *Figure 16.22*.

The actual heating load may vary from straightforward central heating and domestic hot water services to swimming pool heating, top-up heating for energy recovery systems and process solution heating.

The electric boiler operates at a very high thermal efficiency and provides a rapid response to varying demands which, in the case of a thermal storage plant, is by retaining the unused heat during the discharge cycle within the vessel, making the heat input less during the charging cycle.

Advantages of an electrical boiler plant are no need for space for fuel storage; no chimney requirement; no ash handling; no smoke or grit emission; minimum attention required for long plant life.

The principle of the electrode boiler is that from each phase of the three-phase supply connected to an electrode system a current flows to the neutral

388

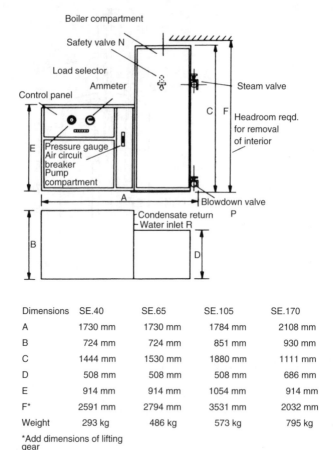

Dimensions	SE.40	SE.65	SE.105	SE.170
A	1730 mm	1730 mm	1784 mm	2108 mm
B	724 mm	724 mm	851 mm	930 mm
C	1444 mm	1530 mm	1880 mm	1111 mm
D	508 mm	508 mm	508 mm	686 mm
E	914 mm	914 mm	1054 mm	914 mm
F*	2591 mm	2794 mm	3531 mm	2032 mm
Weight	293 kg	486 kg	573 kg	795 kg

*Add dimensions of lifting gear

Figure 16.20 Autolec Junior packaged electrode steam boiler (GWB Ltd)

shield. This flow is controlled automatically by a porcelain control sleeve which is raised or lowered over the electrode to increase or decrease the current flow. The power input control is a function of the current flowing and of the temperature of the outlet water from the boiler. The electrical connection is star, and in accordance with the IEE recommendations the star point is connected to the system neutral and earthed at that point (Reg. 554-03-05). A circuit-breaker is provided with the boiler which gives overcurrent and no-voltage protection.

Figure 16.21 shows a cross-section of an electrode boiler. The gear motor unit (item 8) raises or lowers the load control shields (item 3) by rotating the load control shaft (item 1). When the control shields are in their minimum

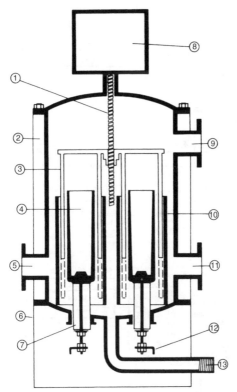

Figure 16.21 Electrode hot water boiler

1. Load control shaft.
2. Lagging.
3. Control shield.
4. Cast iron electrode.
5. Return or s/v correction.

6. Terminal guard.
7. Insulator.
8. Geared motor.
9. Flow connection point.

10. Shields.
11. Return or s/v connection.
12. Terminals.
13. Drain valve connection.

position, i.e. fully down over the electrodes, the boiler input is at a minimum and (dependent upon the feed-water temperature) will be in the order of 4% of the boiler rating.

Electric boilers are designed for three-phase low voltage (400 V) networks; but in certain circumstances, where the loading and tariff situations warrant, high voltage boilers can be made. *Figure 16.24* is a diagram of a high voltage electrode water boiler (6.6 to 18 kV). The kW rating of low voltage boilers ranges from 144 kW to 1600 kW and hot water boilers from 1 MW to 18 MW. For small heating requirements, ranging from 12 to 72 kW, resistance wire or immersion heated boilers are used. The elements are directly switched through a contactor.

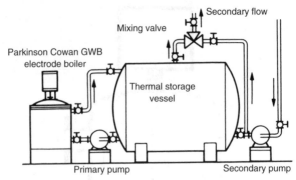

Figure 16.22 Thermal storage heating. The static head available is not sufficient for the required storage temperature. A false pressure head is created by pressurization plant. The water drawn from the vessel is mixed and circulated through the secondary system

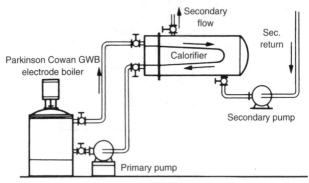

Figure 16.23 Swimming pool heating and hot water services. The calorifier is connected to the primary side of the electrode boiler. The water to be heated flows through the secondary side of the calorifier

As in all cases of boiler plant, the water used should be as free of scale-forming properties as possible and the conductivity of the water within the design range for the boiler and system. For low voltage boilers this will probably be below 1500 Dionic units and for high voltage boilers 200 Dionic units.

Lamp Ovens for Industry

Many items of furniture and domestic appliances are finished with cellulose lacquer. This finishing process usually consists of several distinct operations including coating and recoating with lacquer with drying periods between

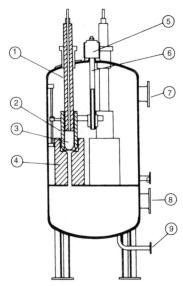

Figure 16.24 Diagrammatic cross-section of h.v. electrode hot water boiler

1. Bushing insulator.
2. Load control shield.
3. Electrode.
4. Jet tube.
5. Load control mechanism.

6. Load control shaft.
7. Flow connection.
8. Return connection.
9. Lower drain.

each coat. Rapid drying of the lacquer can be achieved by using radiant heating ovens.

The main source of heat in an oven of this nature is tungsten lamps which emit a considerable proportion of the energy consumed in the form of short-wave infrared radiation. Short-wave radiant heat passes through air and most gases in a manner similar to light, without any appreciable loss of energy. It is only when the radiant heat falls upon an object in its path that energy is absorbed and the temperature of the object rises.

The efficiency of a radiant heat oven is very high and with quartz glass heat lamps, temperatures in excess of 1300°C can be achieved.

Advantages associated with this form of heating are speed, compactness, rapid response, ease of control, and cleanliness.

Heat lamps divide into two basic types. The first is substantially a parabolic-shaped bulb having an integral reflector inside the envelope. In a typical range from Philips Electrical there are three ratings suitable for industrial applications: 150 W, 300 W and 375 W. These are used widely in industry for paint baking and low temperature process heating up to about 300°C.

The second type of heat lamp, suitable for temperatures up to about 1000°C, is a tubular quartz glass type. Elements range typically from 500 W to

3 kW and are used with external reflectors. Lamps with reflectorized coatings are also available. Other lamps with ratings up to 20 kW having a halogen additive are suitable for temperatures above 1000°C.

In general, plane or flat reflectors are the best when treating flat or sheet materials. When solid objects are to be processed or where radiant energy is to be concentrated on to a small area parabolic or elliptical reflectors are better.

For drying and paint baking applications it is necessary to have an extraction plant to remove the saturated air. Since there is no contamination of the air by combustion products it is possible to recover any expensive solvents being carried away in the air stream.

Radiant heat ovens can be built into existing production plants with the minimum of modifications.

Refrigeration and Air Conditioning

While the electrical engineer's interest in the field of refrigeration may generally remain confined to the design, installation, maintenance or repair of electrical equipment already familiar to him from other applications, knowledge of the 'production of cold' will be of assistance to him.

The basis of all refrigeration is that heat is extracted either directly or indirectly from the item to be cooled (usually food) and used to evaporate a liquid known as the refrigerant. All liquids require heat to cause evaporation, for example water evaporates at 100°C and ammonia at $-33°C$ at atmospheric pressure. This heat is known as the latent heat of the substance and is made use of in refrigeration. A low boiling refrigerant is contained in a component called an evaporator and heat is taken from the item to be refrigerated to boil the refrigerant.

In many cases an intervening medium is used to act as the refrigerating agent. It may be blown over the coils of an evaporator in an air cooler, or it may be employed in a domestic refrigerator or display cabinet. In other cases the medium may be water in an ice flake machine or fish in a contact freezer. This extraction of heat from the ambience is called the 'production of cold'.

Refrigerating agents have received much attention in the 1990s and early twenty-first century. This is because of the damage done to the environment – depletion of the ozone layer – by the materials, known as halon, chlorofluorocarbons (CFCs) and hydrochlorofluorocarbons (HCFCs) and which until that time were the most popular, widely known and used refrigerants.

These were formerly used in both domestic and commercial refrigerators and in air conditioning installations. However, since the Montreal Protocol of 1987 there has been international agreement to phase out their use, originally by the year 2000. By 1990 their use in most new equipment had been banned but some use in certain types of existing equipment will be permitted until about 2010 when their use will be banned entirely.

The speed with which this international agreement was reached and the restricting legislation introduced have tended to take the industry by surprise so that ideal replacements were not really available. There has been some increase in the use of ammonia, which has always had an application in industrial installations. Some manufacturers are producing hydrocarbon-based fluids, which have the disadvantage of being flammable. Hydrofluorocarbons

(HFCs) are also being developed. These contain no chlorine and have zero ODP (ozone depletion potential). There has also been some return to the use of carbon dioxide, which was one of the first refrigerants to be used. This has the disadvantage of requiring a high operating pressure. Other manufacturers are aiming to develop fluorocarbons without an ozone depleting effect. There are, in the early 2000s, no clear front-runners in the race to find an acceptable alternative to CFCs and HCFCs.

To perform its duty economically in a closed cycle, the vapour formed by evaporation of the refrigerant and loaded with the heat picked up at a low temperature must be allowed to discharge the heat and thereby re-liquefy or condense. For this purpose, the refrigerant must be brought into heat exchanging contact with a heatsink such as atmospheric air or cooling water at ambient temperature. This means that the refrigerant leaving the evaporator must be elevated to a temperature higher than that of the heatsink.

In the vapour compression cycle this duty is assigned to a compressor, which for small and medium installations may be a reciprocating piston type or a rotary type. For large installations a centrifugal compressor is generally employed. The useful part of the energy spent results in a pressure and temperature rise of the vapour. During subsequent dissipation of heat the refrigerant is liquefied but still remains at high pressure. For example, for ammonia condensing at $25°C$ the saturation pressure is 10 atmospheres. To return the liquid to the evaporator and thus complete the cycle, the high pressure liquid must be expanded to low pressure whence cooling takes place. A throttling device is used and it may take the form of a simple capillary tube or a throttle valve, the action of which may be electrically controlled in response to either pressure or temperature difference. A further method may be a metering device which opens or closes a small bore orifice in response to a liquid level.

Compressors may be of the open or closed type. For the former the shaft, sealed against loss of refrigerant and lubricating oil by a gland, protrudes through the casing. The driving motor is either directly coupled to the compressor shaft or fitted with a belt drive.

In a closed-type compressor, the rotor is directly coupled to, or mounted on, the compressor shaft and thus the motor windings are exposed to the refrigerant which is mixed with oil vapours. The assembly forms either a hermetically sealed or a semi-hermetic unit. The latter is a unit which can be opened in a factory by undoing a flange connection while the former can only be opened by cutting.

Though based on the same principle of evaporation of a refrigerant at low temperature, a different form of refrigerating cycle is obtained; when replacing the mechanical energy (of the compressor) by heat energy and removing the vapour in the evaporator by absorbing it in an appropriate liquid to which it has chemical affinity. That liquid is contained in the absorber. Ammonia may be used with water as the absorbent, or water vapour may be used with lithium bromide as the absorbent. As the solution becomes stronger and saturated, it is pumped to the higher pressure of a boiler from which the refrigerant (ammonia or water) is boiled off. It is subsequently condensed, drained and expanded through a throttling device into the evaporator.

Absorption systems are extensively found in large size air conditioning installations where the steam generated in winter for heating purposes can be used in summer for heating the strong solution and expelling the vapour of the refrigerating system. Likewise this system is found in the chemical industry, where the steam from back-pressure turbines is used, and in freeze-drying installations for food processing or preparation of instant coffee or tea. It does, however, require a large area for installation of the plant as compared with the space occupied by a vapour compression cycle. For this reason absorption plants are not in common use.

While the electrical equipment, such as motor starters, controls, relays and circuitry, may vary according to the manufacturer or country of origin, the principles of construction are mainly those described in the appropriate sections of this book.

Microprocessors are playing an increasing part in the control of refrigerating plants, particularly in the starting and stopping of compressors in a multi-compressor plant. Such control is designed to reduce the amount of manual attention required.

A fairly recent newcomer to the refrigeration field is the screw compressor. This consists of two large screws or helical gears meshing together which pump the refrigerant and oil into a separator and thence into a condenser. Having no reciprocating parts it is smooth running and can be operated at speeds of 2500–3000 rev/min. It is compact for its output and has a wide range of duties from $-29°C$ to $-1°C$. This type of compressor is particularly useful for dealing with large duties in air conditioning and with low temperature, and its output can be controlled down to as low as 10% of full load.

Another type of compressor quite recently introduced into this field uses only one screw and two star rotors. It has the advantage of eliminating the need for an oil pump by generating a pressure difference between the suction and delivery lines to circulate the oil and the refrigerant.

Suffice it to say that an important role is assigned to low voltage electrical heating matting in the floor of low temperature freezing chambers and storage rooms built at ground floor level. Even the best type of thermal insulating material applied under the floor of such an installation will, within the lifetime of the cold store, allow its low temperature to spread gradually into the soil and, with its high moisture content, particularly in the presence of a high water table, will gradually form a solid block of ice. The expanding mass of ice will exert great pressure on the subfloor with its thermal insulation and on the concrete and granolithic flooring.

This is known as 'frost heave' and must be prevented otherwise it would cause irreparable damage to the floor. One method consists in electrically heating the subfloor to a controlled temperature above freezing point, say $+1$ to $+3°C$. Depending on the temperature of the chamber and on the thermal conductivity and thickness of the insulating material, the specific heating energy required is of the order of 12.5 W/m^2 of floor area.

Air Conditioning and Ventilation

Air conditioning in the UK is not applied to buildings on the same scale as in the USA, but its use is becoming more common in the large buildings now being erected. The amount of fresh air required in a given building will depend on the activity therein, and the IHVE guide sets out the recommended amounts for given occupations.

Air conditioning takes various forms depending upon the requirements, but essentially it is a system of delivering cooled air to a given space and maintaining the area at a set temperature, and sometimes humidity. This can be achieved by a large plant cooling water that is circulated to cooling coils within the rooms of the building. Air is circulated over the coils, and the temperature is controlled by a thermostat in the room which cuts off the water supply as required. Humidity is more difficult to control within fine limits, but this can be done by injecting water spray into the air stream.

Another method is to have a large plant cooling air that is conveyed by trunking to the various spaces. Temperature is regulated by controlling the amount of air delivered to the space. This system is known as variable air volume. Such plants are particularly suitable for incorporating centrifugal or screw-type compressors.

Where a comparatively small room is to be air conditioned – or three or four offices of average size – individual air conditioning units may be used in each room, each independently controlled by a room temperature thermostat. These incorporate their own refrigerating plant and are designed to look like standard electric convector heaters, although they are slightly bulkier.

17 Building automation systems

Realizing the Potential of Building Management Systems

For the past 25 years, the focus of the building control and automation industry has been almost exclusively on technology. First, direct digital control was introduced. Then minicomputers were replaced by personal computers and recently the focus has been on connectivity and standard protocols. These changes have allowed more data to be collected by a Building Management System (BMS) and reduced the cost of systems (minicomputer to PC). The BMS features have remained virtually unchanged over this time.

Over the next ten years, computer and communications technology will have a major impact on BMS. Costs will continue to drop, though not as dramatically as in the past. The amount of data available to a BMS will continue to increase as more and more equipment manufacturers add intelligence and standard communications capabilities. The biggest impact, however, will be in the area of new software applications. The Internet, ease of third party software integration and the increasing demands of a more sophisticated breed of facilities manager will force the BMS industry to develop tailored solutions that will dramatically increase the usefulness of a BMS.

Hardware and networks are only a platform, they do not provide benefits to the facilities manager. It is the software applications, the BMS features, which deliver benefits. The benefit of a BMS is maximized building performance, which has little to do with processors or protocols – it is about energy, operations and maintenance. The dimensions of energy, operations and maintenance can be further split into control issues and monitoring issues (*Figure 17.1*).

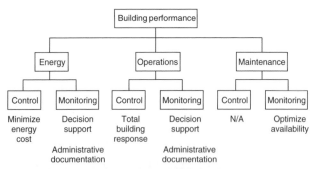

Figure 17.1 Roles of a BMS in maximizing building performance

Energy control issues. Time scheduling is the most commonly used energy control feature of most BMS, but some BMS also implement chiller sequencing. Unfortunately, the list stops here for almost all facilities. In almost all cases, the BMS installed in a facility already has the hardware required to implement more advanced energy control features, such as:

- Duty cycling
- Demand limiting
- Chilled water reset
- VAV static pressure reset

These strategies can result in significant savings with almost no hardware investment by the facilities manager. If this is the case, why have they not been implemented in buildings? The answer is that these features are best set up when the facility is operational, not when it is under construction. To help set up these features and realize the energy savings, the facilities manager can leverage the expertise of the BMS company or bring in a consultant.

Take, for example, projects that use thermal storage systems. Thermal storage is the principle of cooling a mass (water to ice) during times of low cooling demand (night) and then using the stored energy during times of high demand (afternoon). By levelling the demand for external power over time, money is saved by minimizing purchases of high cost energy during high demand times and by reducing the chiller peak capacity needs of the building. The problem of controlling a thermal storage system is quite complex as the savings from thermal storage will depend on three variables – how much ice is made, when the ice is used, and how fast it is used. With enough information, time, and computing power, a near-optimal solution can be determined.

Energy monitoring issues. The Building Management System can collect the data necessary to support the energy decisions faced by many facilities managers.

There is a management adage that says, 'If you can't measure it, you can't manage it' and this is true for energy management. There are three main ways in which energy is measured as part of the energy management process:

- *Utility profile* – this is a standard report from the BMS that organizes consumption and demand according to the tariff structure used by the utility. Information from this report can be mapped into data points and multiplied by the appropriate rates to calculate the total utility bill cost. The utility profile is a useful way of checking utility bills and providing a real-time answer to the question, 'What would be my utility bill if it were to be issued today?'
- *Energy auditor* – this may be a spreadsheet computer program that uses detailed data from the BMS to provide a quick and simple estimate of energy savings. The multiple regression tools of the computer program are used to create an energy model for the facility and provide a real-time answer to the question, 'How much did the energy conservation measure that I implemented last week save me?'
- *Load profiling* – the BMS can sample energy consumption throughout the day and present the results in a graphical format. By studying the daily graph, the facilities manager can identify times to turn off equipment to reduce facility demand charges. By comparing graphs from different days, he or she can identify abnormal energy consumption.

A BMS can also provide the administrative documentation support for energy cost allocation by location or tenant billing for use of HVAC outside of normal operating hours. In other words, the BMS can identify the hours of overtime air conditioning requested by a tenant.

Operations control issues. The role of a Building Management System in supporting operations control issues is to provide a 'total building response' to the state of a zone. A zone can be in a number of states such as:

- Occupied
- Unoccupied
- Standby
- Security alarm
- Fire alarm
- Power failure

As shown in *Figure 17.2*, the Building Management System can co-ordinate the activities of all building systems to optimize the operation of the building, while ensuring that it is simple to override the current state of a zone (i.e. by using a phone, browser, etc.). If possible, the status of a zone should be detected automatically or passed from another system.

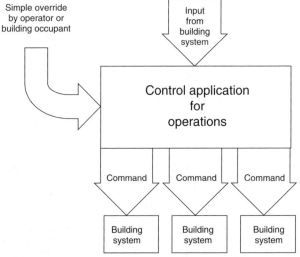

Figure 17.2 Implementing a total building response using a BMS

Examples of how a Building Management System may implement a Total Building Response to an event include:

- In the event of a security alarm, turn on lights in the area to scare away the intruder, ensure a good image on CCTV camera, and protect the investigating security guard.
- In the event of a fire, turn on lights on the floor of incidence to aid evacuation, and inhibit unwanted HVAC and security alarms to allow the operator to focus on the fire.

- In the event of a power failure, certain 'essential' equipment may be turned off based on the current occupancy pattern.

Ways of doing business are changing and the successful buildings will be those that can support these changes. One significant change for buildings is the shift in occupancy patterns. In the past, buildings were either fully occupied or fully unoccupied. In the future, buildings will be partly occupied for longer periods of the day (*Figure 17.3*).

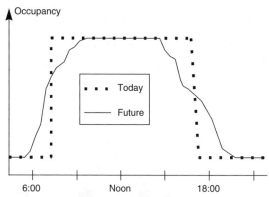

Figure 17.3 Changing occupancy pattern in offices

Some of the technologies that can be used to detect the occupancy state of a zone include:

- Time schedule
- Occupancy sensor
- Input from telephone
- Card access
- Door lock indicator
- Input from browser

When a zone is occupied, the building systems can be programmed to operate as follows:

- Sensitivity of smoke detectors set to LOW to avoid unwanted alarms due to dust and smoke generated in the space.
- Alarm reporting inhibited for selected security devices in the zone.
- Temperature set points changed to the occupied level.
- Ventilation requirements (fresh air per person) set to occupied level.
- Lights in the zone turned on.

In the past decade or so, indoor air quality (IAQ) has become an important issue with facilities managers. Though most of the attention has been focused on litigation associated with sick building syndrome (SBS), it is also understood that there is a direct correlation between the amount of fresh air and the productivity of the building occupants. (Increasing the amount of fresh air also increases building equipment energy costs.)

After an extensive study of the problem in the US, ASHRAE has defined a standard, 62-1989, which requires occupancy-based ventilation control. Demand-based ventilation control (i.e. controlling fresh air dampers based on return air CO_2 level) is described in the ASHRAE standard as an optional addition to occupancy-based ventilation control, but is not a substitute for occupancy-based ventilation control.

The challenge of occupancy-based ventilation control is finding a way of accurately and cost effectively measuring the amount of fresh air being introduced into a real-world AHU. In almost all cases, it is not practical to insert an airflow monitoring station into the fresh air supply. A practical solution to this problem has recently been developed (see *Figure 17.4*) which uses a single CO_2 sensor to sample air streams from three locations in the AHU. The percentage of fresh air in the supply air can then be calculated using Equation (17.1) below.

$$\%FA = \frac{X_{RA} - X_{SA}}{X_{RA} - X_{FA}} \quad (17.1)$$

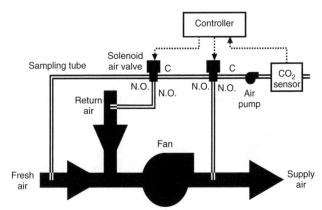

Figure 17.4 Multipoint CO_2 sensing

where X_{RA} = measured CO_2 level for return airstream
X_{SA} = measured CO_2 level for supply airstream
X_{FA} = measured CO_2 level for fresh airstream.

Though CO_2 sensors are subject to drift over time, this drift does not affect the accuracy of the calculation of percentage of fresh air. When multi-point CO_2 sensing is used, the measured CO_2 level for the return air stream can be used to implement optional demand-based ventilation control to complement the occupancy-based ventilation control.

Operations monitoring issues. With the trend toward performance-based agreements between occupants and facilities managers and between facilities managers and third party partners, it is important to be able to monitor service levels of key performance indicators (KPI). This can be done at multiple levels such as:

1. *Raw data* – current value of KPI (i.e. current space temperature).
2. *Current period* – service level of performance indicator in current measurement period (i.e. mean, variance, percentage of time within comfort zone).
3. *Historical periods* – service level of performance indicator in previous measurement periods (text and bar graph formats).
4. *Modelling* – an analysis of the historical periods using pattern matching (i.e. diagnostic, with confidence level).
5. *Forecasting* – extends the model to future periods.

For more advanced applications, the role of the Building Management System may be to simply store the data in a file for analysis by a third party application.

A BMS monitors and collects a huge amount of data. The operators must make sense of all of this data and make decisions from it. Data visualization is a technique that helps the system operator see how the building is performing by transforming data into information. A single display can show the user the status of the entire facility, helping the operator sort out the building systems that need the most attention, and direct resources to those systems.

Data visualization puts data into context rather than simply displaying it in columns of text. For example, an operator could create a custom report showing all of the temperatures and humidities in the building, but displaying the information on a psychometric chart as shown in *Figure 17.5*. This provides an instant analysis of the state of the various spaces in the facility.

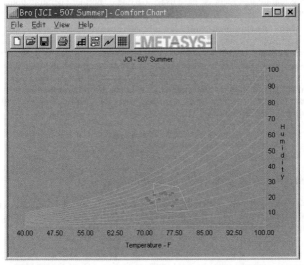

Figure 17.5 Comfort chart

A recent survey of consultants, contractors and end-users revealed that the biggest problem they had with their current BMS was 'ease of use'.

At first sight, this answer is surprising. Almost all BMS in the industry today use the industry standard Microsoft Windows graphical user interface. It takes less than an hour to train an operator on how to access the system, display information, issue commands and respond to alarms. If basic user functions can be mastered in less than an hour, why should 'ease of use' be the single biggest problem with today's BMS?

On closer examination, it becomes clear that the problem is not in getting the data on to the BMS screen but rather in understanding how to use the data that the BMS collects. This is where *data visualization* comes into play.

Data visualization tools give a much greater insight into the comfort, energy and control conditions in the building. This insight is key in the facilities manager's quest to manage 'total building energy'.

There will be some diversity in documentation requirements such as:

- Ability to provide data mining of environmental measurements. For example, in an office, records may be maintained of IAQ, equipment start/stop time and temperature/humidity conditions. In a factory, there may be ISO requirements to maintain records of environmental conditions that affect the manufacturing process.
- Ability to provide an integrated audit trail of conditions, events and activities covering all building systems.

Business practices will reflect the increasingly variable occupancy pattern leading to 'pay as you go' arrangements between building occupants and facilities management. To support this model, it is critical to provide a tenant-billing feature that records space usage outside of normal operating hours and calculates the associated cost.

Maintenance monitoring issues. There is growing recognition that taking a process-oriented approach to maintenance offers significant benefits. Preventive and predictive maintenance focus on optimizing the mean time between failure (MTBF) of equipment. A process-oriented approach to operations and maintenance focuses on optimizing equipment availability. This approach to operations and maintenance looks at both mean time between failure (MTBF) and mean time to repair (MTTR). The relationship between availability, MTBF and MTTR, is as follows:

$$\text{Availability} = \frac{\text{MTBF}}{\text{MTBF} + \text{MTTR}}$$

Many facilities expect 98% availability

The current paradigm among maintenance professionals is that 'repair is a process, not an event'. The repair process can be subdivided into three processes that are executed sequentially:

- *Response process* – becoming aware of a problem and getting the right person, equipped with the right tools, to the location of the problem.
- *Stabilization process* – overriding or bypassing a system to provide partial availability until a permanent solution can be implemented.
- *Rectification process* – implementing the permanent solution.

The root cause analysis process involves an engineering study after a failure to fine-tune and improve the maintenance strategies (*Figure 17.6*).

A Building Management System can support the maintenance process by alerting the operator when maintenance is required, based on calendar time or equipment run time, or when a monitored condition exceeds a threshold level, such as the monitoring of a dirty filter.

The Building Management System can be programmed as part of the response process. Alarm events can be triggered by a variety of conditions such as:

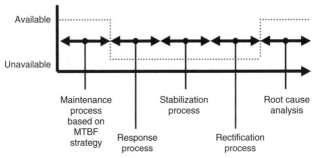

Figure 17.6 Optimizing equipment availability

- Fault condition detected in equipment.
- Temperature value outside of normal operating range.

Once an alarm event has been triggered, the Building Management System can implement an automated response process as shown in *Figure 17.7*.

The ways in which a Building Management System can support a stabilization process are:

- Providing rapid access to building information.
- Acting as a platform to allow in-house training of facilities management staff on building systems.
- Allowing access to remote expertise using the Internet (both accessing information on the Internet and allowing remote experts to access the BMS using the Internet).

Elements of a rectification strategy include handling of spare part inventory and standard pricing with suppliers for key components. There is little scope for the BMS in this area.

Root cause analysis is the feedback component in the maintenance process. Root cause analysis relies on historical data collected before the failure. For example, if a pump is maintaining the water level in a tank and a tank level alarm is received, a root cause analysis application might analyse the recent pattern of operation of the pump to suggest the possible source of the problem:

- *Low tank level, no recent pump operation* – problem may be with pump cut-in controls, power supply to pump or hand/off/auto switch left in off position.

404

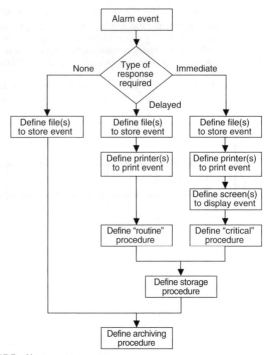

Figure 17.7 Alarm response process

- *Low tank level, recent pump operation* – problem may be upstream to pump (i.e. water supply to pump), leak in piping, or leak in tank.
- *High tank level, no recent pump operation* – problem may be with level sensor.
- *High tank level, recent pump operation* – problem may be with pump cut-out controls or hand/off/auto switch left in hand position.

Typically in the past, a BMS did not have sufficient inputs to do a complete root cause analysis. High level data interfaces now provide a broad range of data ideal for root cause analysis applications.

Conclusion. Maximizing building performance is a multifaceted issue. A Building Management System can be an invaluable tool to address both the monitoring and control issues involved.

The building automation section was prepared by Johnson Control Systems Ltd.

18 Instruments and meters

Most instruments and meters are based on either the magnetic principle or that of electromagnetic induction. Some of these meters like phase angle, wattmeter, frequency meter and VAr meters use an electronic converter and a moving coil indicator rather than a direct induction effect. This should be remembered when reading the appropriate sections. In addition there are a number of other instruments which make use of electrostatic, heating and chemical effects each of these having a special application or applications.

Ammeters and Voltmeters

The normal instruments for measuring current and voltage are essentially the same in principle, as in most cases the deflection is proportional to the current passing through the instrument. These meters are therefore all ammeters, but in the case of a voltmeter the addition of a series resistance can make the reading proportional to the voltage across the terminals. Types are:

1. Moving iron (suitable for a.c. and some suitable for d.c.).
2. Permanent magnet moving coil (suitable for d.c. only).
3a. Air cored dynamometers – moving coil (suitable for a.c. and d.c.) particularly used for high precision measurements.
3b. Iron cored dynamometers – moving coil (generally only suitable for a.c.) particularly used for high precision measurements.
4. Electrostatic (for a.c. and d.c. voltmeters).
5. Induction (suitable for a.c. only).
6. Hot wire (suitable for a.c. and d.c.).

All these instruments measure r.m.s. values except the moving coil which measures average values. Therefore when measuring rectified and/or thyristor-controlled supplies it is important to know whether the load current, for example, as registered is the r.m.s. or average value. Heating or power loads generally require r.m.s. indication while battery charging and electrochemical processes depend on average values.

Although some of the above square-law instruments will read d.c. it should be appreciated that it is still the r.m.s. value that is indicated. This will only be the same as the average value indicated by a moving coil instrument if the supply is d.c. without any ripple. The difference between the moving iron and moving coil instrument connected in the same unsmoothed single-phase rectified supply could be as much as 11%.

Note that the permanent-magnet moving-coil instrument can be used for a.c., provided a rectifier is incorporated, and this is described in another section.

Accuracy. The accuracy of instruments used for measuring current and voltage will naturally vary with the type and quality of manufacture, and various grades of meters have been scheduled by the BSI. These are set out in BS 89.

There are nine accuracy classes in BS 89 designated in 'Class Indices' which enable an engineer to purchase instruments with an accuracy sufficient for the purpose for which the instrument is intended. For normal use instruments of Accuracy class 1.5 and for switchboard and panel instruments Accuracy class 2.5 are usually sufficiently accurate.

When specifying an instrument for a particular purpose it is important to realize that the accuracy relating to the Class Index may only be obtained under closely specified conditions. Variations of temperature, frequency, magnetic field, etc. may cause the error to exceed the Class Index. The additional errors permitted are related to the Class Index and should be considered in the light of a particular application.

Many manufacturers are now producing indicating instruments, such as ammeters, voltmeters and wattmeters, with a platform scale. With this type of instrument the pointer traverses an arc in the same plane as the actual scale markings, thus eliminating side shadow and parallax error.

Moving-iron instruments. This type of instrument is in general use in industry owing to its low first cost and its reliability. There are three types of moving-iron instruments – the attraction type, the repulsion type and some using a combination of both principles. Short scales, i.e. 90° deflection, are generally associated with repulsion meters, and long scales, i.e. 180–240° with combined repulsion and attraction types (*Figure 18.1*).

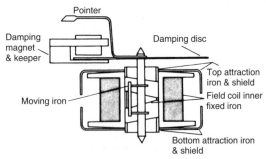

Figure 18.1 Diagrammatic sketch of moving-iron movement. This operates on the combined repulsion-attraction principle. The field coil is enclosed in nickel-iron screens, the inner edges of which project into the bore of the moulded plastics field coil bobbin forming attraction elements. The magnetic field provided by the coil polarizes these elements and also the fixed and moving irons

The principle of operation is that a coil of wire carrying the current to be measured attracts or repels an armature of 'soft' iron, which operates the indicating needle or pointer. With the attraction type the iron is drawn into the coil by means of the current; and in the repulsion type there are two pieces of iron inside the coil, one of these being fixed and the other movable. Both are magnetized by the current and the repulsion between the two causes the movable unit to operate the pointer.

It will be seen, therefore, that the direction of current in the coil does not matter, making the instrument suitable for measuring any form of current, either d.c. or a.c., including rectified a.c. Accuracy is not affected by the distorted waveforms generally encountered on commercial supply systems, but peaky waveforms can cause saturation of the iron in the magnetic circuit resulting in incorrect readings.

Some causes of error

(a) *Stray magnetic fields.* Due to the fact that the deflection is proportional to the magnetic field inside the operating coil, magnetic fields due to any outside source will affect the deflection. Errors due to this are reduced by suitable magnetic screening of the mechanism.

(b) *Hysteresis.* The use of nickel iron in the magnetic circuit of modern instruments has reduced the error due to hysteresis such that an instrument of Class Index 0.5 can be used on both a.c. and d.c. within that class. Due to other factors the use of nickel iron is not usually possible on long-scale instruments. These are not therefore suitable for use on d.c. in view of the large hysteresis error.

(c) *Frequency.* Ammeters are not affected by quite large variations in frequency, an error less than 1% between 50 Hz and 400 Hz being quite usual. Errors do occur on voltmeters with change of frequency due to alteration in the reactance of the coil which forms an appreciable part of the total impedance. Where necessary the error can be substantially reduced by connecting a suitably sized capacitor across the series resistance as compensation.

Permanent-magnet moving-coil instrument. This is essentially a d.c. instrument and it will not operate on a.c. circuits unless a rectifier is incorporated to permit only unidirectional current to pass through the moving coil.

The operation of a moving-coil instrument will be seen from *Figure 18.2*. By means of a suitably designed permanent-magnet and a soft iron cylinder between the poles, a circular air-gap is formed through which a pivoted coil can move. This pivoted coil carries the current to be measured (or some proportion of it), and as the field is uniform in the air-gap the torque will be proportional to the current. The deflection is controlled by means of a torsion spring with a suitable adjustment device, and the current is taken to the moving coil by means of these or other coil springs.

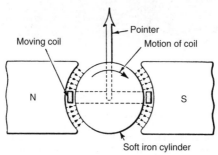

Figure 18.2 Permanent-magnet moving-coil instrument (90 degree scale)

Permanent-magnet moving-coil instruments are only suitable for small currents for actuating purposes, and thus both shunts and series resistances are used to a fair extent. Stock instruments usually have a full-scale deflection of 1 mA for voltmeters (1000 ohm/V) and 75 mV for ammeters operated from shunts. These shunts may be contained inside the instruments for ranges up to around 60 A and external to the instrument for higher current ranges.

The actual calculations for values of shunts and series resistances are dealt with subsequently, and an example is included of an instrument of this type.

Damping is obtained by the eddy currents induced in the metal former which is used for the moving coil. With the introduction of new magnetic materials flux densities have increased and stability is greatly improved. Silicone fluid damping provides a smoother effect free from overshoot, and permits reduction in the weight of the moving element. It is now possible with taut-band suspension (see below) to produce instruments having full-scale deflection with current values as low as $10\,\mu\text{A}$.

Although used extensively for monitoring a.c. and d.c. power supplies, many taut-band instruments are being employed for the readout on electronic equipment, as transducers for electrical and physical quantities, electrical tachometers, and remote position indicators, for example. *Figure 18.3* shows a typical scale of a taut-band silicone fluid damped instrument. The scale is substantially uniform and can be made to extend over arcs up to $250°$.

Figure 18.3 A microamp panel meter incorporating taut-band suspension (Crompton Instruments)

Taut-band suspension. This form of suspension used in instruments was introduced by Crompton Instruments in 1962 and it eliminates the need for pivots, bearings, and control springs. In this system the moving element is suspended between two metal ribbons, one at each end of the movement. These spring-tensioned ribbons, by their twisting during deflection, provide the controlling and restoring forces. Bearing friction is completely eliminated and there is nothing to wear out or go wrong. The upper portion of a taut-band movement is shown in *Figure 18.4*.

The spigots on the end of the hub at each end of a taut-band moving element work in clearance holes in a bush-forming part of the fixed frame

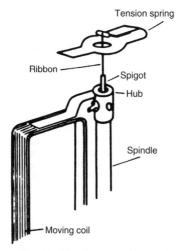

Figure 18.4 Upper portion of taut-band movement (Crompton Instruments)

assembly. By introducing oil of suitable viscosity into the gap between the spigots and the clearance holes it is possible to control resonance in the moving assembly. The method adopted by Crompton Instruments to contain the oil in these two areas is shown in *Figure 18.5*. The spigot is increased in length and an annular pad or washer of expanded polythene is fixed to the outer hub. A similar washer is fixed to the facing end of the mounting bush. The damping

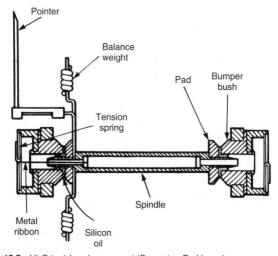

Figure 18.5 Hi-Q taut-band movement (Crompton Parkinson)

oil is inserted between the pads and the spigot. The oil serves a number of other functions. It provides all the damping action needed to bring the moving system quickly to rest in its deflected position. No other form of damping is required resulting in a very light assembly. The torque/weight ratio is therefore improved and so instruments of a higher sensitivity can be produced. Response time is also considerably reduced compared with pivoted instruments.

An important factor is that the oil-retaining pads act as built-in shock absorbers. In the latest form of Hi-Q movement, *Figure 18.5*, the joints between the ribbons at each end and the spindle have been eliminated. A single ribbon runs effectively from one tension spring to the other forming, with a metal locating and stiffening plate, the central core of an injection moulded nylon spindle. Aluminium end caps fitted to the spindle carry the shock absorbing damping pads and, at one end, the pointer and balance weight arms. In moving-coil instruments they also carry the coil mounting brackets, while in the moving-iron system, the injection moulding spindle is radially extended to provide a fixing for the moving iron.

Dynamometer moving-coil instruments. Instead of the permanent magnet of the previous type of moving-coil instrument, the necessary magnetic field may be set up by passing the current through fixed coils as shown in *Figure 18.6*. There are thus two sets of coils – one fixed and one pivoted – the torque being provided proportional to the square of the current, making the instrument suitable for either d.c. or a.c.

In the case of an ammeter, the two systems are usually connected in parallel with suitable resistances, and in the case of a voltmeter they are in series. The leads to the moving coil are in the form of coil springs, which also act as the torque control, and in this case damping has to be provided by some type of damping device such as an aluminium disc with a braking magnet.

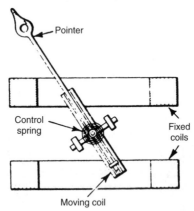

Figure 18.6 Dynamometer moving-coil instrument

Owing to the necessity of keeping the current in the moving coil low, the torque obtainable may be small, and these instruments are not used to any extent for general industrial purposes. They are particularly useful in the laboratory for d.c./a.c. transfer instruments. The calibration can be checked on d.c.

using a potentiometer and then, if required, in conjunction with known accuracy transformers can be used as a standard for checking other a.c. instruments, i.e. moving iron where the d.c./a.c. error may be large or unknown. When used care must be taken to see that they are not affected by stray magnetic fields.

The dynamometer instrument is, however, much used for wattmeters, and its use for this purpose is described later.

Electrostatic voltmeters. Electrostatic instruments are essentially voltmeters, as they operate by means of the attraction or repulsion of two charged bodies. They have certain applications in the laboratory as they are unaffected by conditions and variations which give rise to errors in many other instruments. These include hysteresis and eddy current errors, errors due to variations in frequency and waveform which do not cause incorrect readings with electrostatic instruments.

For ordinary voltages the torque is small and multi-disc instruments have to be used. On this account, however, they are ideal for measuring high voltages and form practically the only method of indication where the voltage exceeds 100 kV.

For lower voltages from 400 V to a few thousand volts the quadrant type is used, the principle being as in *Figure 18.7*. The moving vane is pivoted and is either repelled or attracted or both by the charges on the vane and the quadrants – these charges being proportional to the potential due to their connection to the supply.

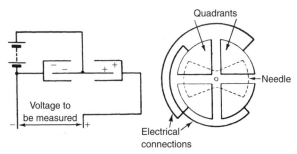

Figure 18.7 Quadrant voltmeter

In commercial models several sets of vanes are used in parallel to obtain sufficient torque. Voltages of 100 kV and over are measured by means of two shaped discs with an air-space between them, one disc being fixed and the other movable axially. By means of a balance the force between the two is measured and the scale of the balance marked in kilovolts. These voltmeters are not, of course, accurate to a degree that makes them suitable for indicating line voltages or for switchboard purposes. They are, however, ideal for testing purposes where cables and other apparatus are subject to high voltage and tests to destruction. In this case they form a visual indication of the applied voltage and are a check on the value indicated by the instruments on the voltage *stepping-up* apparatus.

Electrostatic instruments for laboratory use are termed *electrometers* and are usually of the suspension type with mirror-operated scales.

As these instruments consume negligible power on a.c. they have the advantage that they do not affect the state of any circuit to which they are connected. The exception is of course in the case of radio-frequency measurements.

Hot-wire instruments. The current to be measured, or a known fraction of it, is passed through a fine wire and, due to the heating effect of the current, the wire expands. If the resistance and coefficient of expansion of the wire is constant, then the heating and consequent expansion of the wire are both directly proportional to the square of the current. If the expansion is sufficient, by suitable connections as shown in *Figure 18.8* it can be made to move the pointer. This movement is also proportional to the square of the current. By suitable scaling the current can thus be determined.

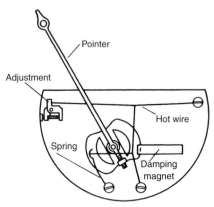

Figure 18.8 Hot-wire instrument

Since such instruments obey the square law they are suitable for both d.c. and a.c. systems. Furthermore, since they only operate on the heating effect, the r.m.s. value of an alternating current is measured irrespective of frequency or waveform. They are also unaffected by stray magnetic fields.

Induction instruments. These instruments, which will function only on a.c., may be used for ammeters and voltmeters, but their use for the measurement of these two quantities is much less than for wattmeters and energy meters – these applications being described fully below.

In all induction instruments the torque of the moving system is due to the reaction of a flux produced by the current to be measured on the flux produced by eddy currents flowing in a metal disc or cylinder. This eddy current flux also being due to the current is arranged to be out of phase with the flux of the current being measured.

There are two methods by which these fluxes are obtained. With a cylindrical rotor two sets of coils can be used at right angles, or with a disc rotor the shaded pole principle is used (*Figure 18.9*) an alternative being the use of two magnetic fields acting on the disc.

Although extremely simple, with no connections to the rotor, these instruments have many disadvantages which prevent their use for general purposes

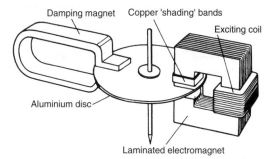

Figure 18.9   Principle of shaded-pole induction meter

as ammeters and voltmeters. The points in their favour are a long scale, good damping and freedom from stray field effects. Their disadvantages include fairly serious errors due to variation in frequency and temperature, high power consumption, and high cost. The former may be reduced by suitable compensation, but in ordinary commercial instruments the variation is still important, although Class 1 instruments can be supplied if required. See the reference to electronic converters at the start of the chapter.

Wattmeters

Dynamometer wattmeters. If of the air-cored type the wattmeter will give correct readings both on d.c. and a.c. It consists of a stationary circuit carrying the current and a moving circuit representing the voltage of the circuit. For laboratory instruments the meter may be either of the suspended coil or pivoted coil type. The former is used as a standard wattmeter, but the pivoted coil has a much wider scope as it is suitable for direct indication (*Figure 18.10*). Long-scale wattmeters are of the iron-cored type and are not generally suitable for use on d.c. due to the large hysteresis errors.

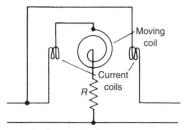

Figure 18.10   Dynamometer-type wattmeter

For accurate measurements on low range wattmeters it is necessary to compensate for the flow of current in the other circuit by adding a compensating winding to the current circuit.

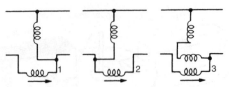

Figure 18.11 Uncompensated (1, 2) and compensated (3) wattmeters

This will be seen from *Figure 18.11*, where the connections (1) and (2) will introduce errors due to the extra current in the current coil in (1) and the volt drop in the current coil in (2). It will be seen that the arrangement in (3) overcomes these errors.

Dynamometer wattmeters are affected by stray fields. This is true for both d.c. and a.c. measurements, but on a.c. only alternating current stray fields affect readings. This effect is avoided by using a static construction for laboratory instruments and by shielding for portable types. The use of a nickel-iron of high permeability has enabled modern wattmeters to be practically unaffected by stray fields in general use.

When purchasing wattmeters of this type it is desirable to see that a high factor of safety is used as regards the rating of the coils of the instrument. This is desirable from the point of possible overload on any of the ranges of a multi-range instrument, but it also enables the indication to be made towards full scale even at low power factors.

For three-phase wattmeters, adequate shielding between the two sections is necessary, but even then they do not give the same accuracy as single-phase models.

Induction wattmeters. These can only be used on a.c. and are similar to energy meters of the induction type, the rotating disc in this case operating against a torsion spring. The two circuits are similar to the induction energy meter and the connections are as shown in *Figure 18.12*.

Although affected by variations in temperature and frequency, the former is compensated for by the variation in resistance of the rotating disc (which is opposite in effect to the effect on the windings), and the latter does not vary considerably with the variations in frequency which usually obtain. Induction

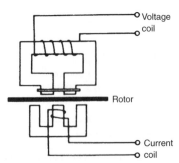

Figure 18.12 Induction-type wattmeter. The construction and principle of operation are similar to the energy meter

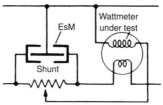

Figure 18.13 Electrostatic wattmeter used for calibration purposes

wattmeters must not, of course, be used on any other frequency than that for which they are designed unless specially arranged with suitable tappings.

The general construction of induction wattmeters renders them reliable and robust. They have a definite advantage for switchboard use in that the scale may be over an arc of 300° or so. As with dynamometer types, the three-phase two-element wattmeter is not as accurate as the single-phase type.

Three-phase wattmeters. These operate on the two wattmeter principle, the two rotors being mechanically coupled to give the sum of the torques of the two elements. As already stated, it is important to avoid any interference between the two sections, and there are several methods of preventing this. One is to use a compensating resistance in the connections to the voltage coils, and another, due to Drysdale, is to mount the two moving coils (of the dynamometer type) at right angles.

Measurement of three-phase power. The connections of temporarily connected instruments for measuring three-phase power are given on page 19. Most portable wattmeters are designed for a maximum current of 5 A, and are used with current transformers. If two such wattmeters with transformers are connected as in *Figure 18.14*, the algebraic sum of the wattmeter readings multiplied by the transformer ratio will give the total power in a three-phase circuit.

In most tests of the power consumption of three-phase motors, one wattmeter will give sufficient accuracy if it is properly connected, because motor loads are very approximately balanced. In this case, the total three-phase power is three times the wattmeter reading multiplied by the transformer

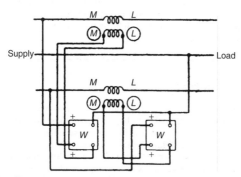

Figure 18.14 Two wattmeter method of measuring three-phase power

ratio. If the supply is 3-wire without a neutral and unbalanced, two current transformers are required. Three-phase supplies usually include a neutral for lighting circuits. If the power installation is supplied from a local substation in the works, the earthed point of the neutral will be, electrically, very near the measuring position. In this condition three-phase power may be measured in balanced circuits by connecting the wattmeter voltage circuit between the line supplying the current coil of the instrument and earth. The earth connection is electrically equivalent to a connection to the supply neutral. When the power installation is some distance from the substation, there may be considerable differences of potential between the neutral of the supply and earth. In this case, the following method is very convenient. Connect one voltage terminal of the wattmeter to the line in the usual way; to the other terminal join a single lead connected to one of the terminals of a lampholder adapter. By inserting the adapter into a switched-off lampholder, the voltage circuit can be joined to the supply neutral in the holder.

Power transducer/indicator systems. The advent of the electrical power converting transducer used with a local or remote moving-coil instrument can indicate such parameters as frequency, watts, VArs, phase angle, a.c. current and voltage, position, weight, temperature, etc.

Load-independent transducers provide a remote signal facility for indicators, data loggers, computers, control systems, etc.

Neon plasma analogue instruments. The latest in CMOS circuitry has been coupled to a neon plasma display by Crompton Instruments to give an analogue indication of the quantity being measured. It has been applied to a wide range of meters including ammeters, voltmeters, kilowatt meters and angular displays.

Readability is excellent, even in poor ambient lighting conditions and is achieved by arranging a series of neon plasma bars in a strip and illuminating them consecutively so that the resultant column is proportional to the input signal. The column appears as a continuous bar and is clearly visible over a wide viewing angle.

An indicator-controller version is available with high and low adjustable set points. When the input signal moves outside the chosen limits this is clearly shown in the display, and alarm and control equipment can be operated by the output relays.

These instruments are well suited to most metering applications where accuracy and readability are important, and are ideal where vibration or dust can be a problem or where ambient lighting may be low. Accuracy is Class 1 to BS 89, DIN 43780 and IEC 51.

Valve Voltmeters

These essentially consist of a thermionic valve which has a milliammeter connected in its anode circuit. The voltage which is to be measured is normally applied to its control grid circuit, which imposes very little load on the circuit, even at a high frequency. Although the basic arrangement of a valve voltmeter has a limited range, this can be extended by the use of a potential divider.

As already noted above, the valve voltmeter takes practically no power at all from the source under test, and this factor is an important one in the measurement of voltages in a radio circuit. An ordinary moving-coil meter, no matter how sensitive it may be, always draws some power from the circuit under test. In circuits where there is plenty of power available this is not serious, but when dealing with a circuit in which even a load of a few microamperes would seriously affect the accuracy of the reading, a valve voltmeter should be used.

When comparing the load imposed upon an a.c. circuit by a moving-coil voltmeter, it is also necessary to consider the frequency, since a moving-coil meter is very frequency sensitive. By this is meant that the instrument in the first place is calibrated at a particular frequency and will only measure alternating current accurately at this frequency. An important advantage of the valve voltmeter is that it may be designed to cover practically any frequency. Commercially manufactured instruments are suitable for frequencies up to 50 MHz and above.

Second, the valve voltmeter will often be used on circuits where the voltage is not very high so that its input impedance can only be compared with a moving-coil meter which has been adjusted to the appropriate range. This means that it is usually impossible to determine accurately a low voltage on the high voltage range of a moving-coil meter. On the other hand, the valve voltmeter maintains a high impedance over all its ranges and is, in fact, the only type of instrument that can be used for low voltage r.f. measurements.

A further application of the thermionic valve to voltage measurement is to be found in the diode peak voltmeter. When an instrument of this type is connected to an a.c. source, rectification takes place in the diode each half-wave surge charging a capacitor in the output circuit to the peak value of the wave. Provided the voltmeter in the circuit is of sufficiently high impedance and is large enough to prevent a loss of charge through the meter when the rectifier is not passing current, the reading on the meter will indicate the peak voltage, irrespective of the waveform.

It is an important point to note that either a.c. only, d.c. only, or a.c. and d.c. can be measured on the various types of valve voltmeters in use today. An a.c. valve voltmeter which has a blocking capacitor incorporated in its input circuit cannot be used for measuring d.c. However, it is not a difficult matter to short circuit the capacitor when it is necessary for a d.c. measurement to be taken.

Because of the steady potential on the grid in the case of d.c. the calibration will be affected. In this case a resistance is sometimes inserted between the d.c. input terminal and the range potential divider to compensate for the rise in sensitivity so that the calibration of the meter is similar for both a.c. and d.c. measurements.

Shunts and Series Resistances

The control or reduction of the actual currents flowing in the various circuits of an instrument or meter may be obtained either by means of resistances (for both d.c. and a.c.) or by the use of instrument transformers (a.c. only).

Shunts. Non-inductive resistances for increasing the range of ammeters are termed shunts and are connected as shown in *Figure 18.15*. The relative values to give any required result can be calculated as follows:

418

Let R = resistance of ammeter

r = resistance of shunt

I = total current in circuit

i_a = current in ammeter

i_s = current in shunt

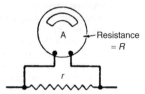

Figure 18.15 Ammeter shunt

We have $I = i_a + i_s$, and as the volt drop across the meter and the shunt are the same we get

$$i_a R = i_s r$$

$$\therefore \quad i_a = i_s \frac{r}{R} = I \frac{r}{R + r}$$

or $\quad I = i_a \frac{R + r}{r} = i_a \frac{R}{r} + 1$

The expression $\dfrac{R}{r} + 1$ is termed the *multiplying power* of the shunt.

As an example, take the case of a meter reading up to 5 A having a resistance of 0.02 ohm. Find a suitable shunt for use on circuits up to 100 A.

As the total current has to be $100/5 = 20$ times that through the meter, the shunt must carry 19 times. Thus its resistance must be 1/19 that of the meter, giving a shunt whose resistance is 0.02/19 or 0.00105 ohm.

Series resistances. In the case of voltmeters it is often impracticable to allow the whole voltage to be taken to the coil or coils of the instrument. For instance, in the case of a moving-coil voltmeter the current in the moving coil will be in the nature of 0.01 A and its resistance will probably be only 100 ohms. If this instrument is to be used for, say, 100 V, a series resistance will be essential – connected as shown in *Figure 18.16*.

The relation between total voltage and that of the meter is simple as the two voltages are in series. Thus in the above case the full-scale current will be 0.01 A, so that the total resistance for 100 V will have to be 100/0.01 or 10 000 ohms. If the resistance of the meter circuit and its connecting leads is 100 ohms, then the added or series resistance must be $10\,000 - 100 = 9900$ ohms.

The multiplying power is the ratio between the total voltage across the instrument and that across the coil only. If

$$R = \text{resistance of meter}$$

$$r = \text{value of series resistance}$$

then multiplying power $= \dfrac{R + r}{R}$

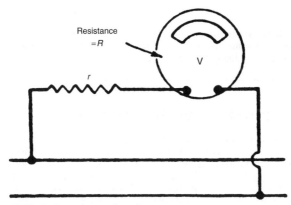

Figure 18.16 Series resistance for voltmeter

Construction of shunts and series resistances. Shunts usually consist of manganin strips soldered to two terminal blocks at each end and arranged so that air will circulate between the strips for cooling.

As with an ammeter shunt, the impedance of the series resistance used in a voltmeter must remain constant for differing frequencies, i.e. the inductance must not vary. For this reason the resistance coils (of manganin) are often wound upon flat mica strips to reduce the area enclosed by the wire, and hence reduce the enclosed flux for a given current.

Voltage dividers. Voltage dividers or volt-boxes can only be used with accuracy with testing equipment or measuring instruments which do not take any current, or with electrostatic instruments. The former state of affairs refers to tests using the 'null' or zero deflection for balancing or taking a reading while the electrostatic meter actually does not take any current.

The principle of the voltage divider is seen in *Figure 18.17*. The voltage to be measured is connected across a resistance R and the connections

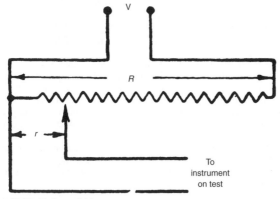

Figure 18.17 Voltage divider

are taken off some fraction of this resistance as shown. The ratio is given by R/r.

A similar arrangement is possible with capacitors for use with electrostatic instruments, but the method is somewhat complicated owing to the variation in capacitance of the instrument as the vane or vanes move.

Current and Voltage Transformers

The above principles of measurement are just as applicable to the high voltage, high power, systems used in electricity supply and in industry as they are for domestic users taking a supply of up to 50 A at 230 V. For these high voltage, high power, systems voltages and currents must be reduced to levels which can be input to the measuring instruments. (Similar means must also, of course, be used to reduce voltage and current signals to suitable levels to be input to protective relays.)

Almost invariably this is done using voltage transformers (VTs) or current transformers (CTs) as appropriate. These are basically similar to other transformers with the exception that they are specifically designed to have very accurate ratios. In the case of voltage transformers the designs are optimized to provide accurate voltage ratio, while for current transformers accurate current ratio is the priority.

Current transformers are specified in accordance with BS EN 60044-1 and voltage transformers to BS 3941. In the case of current transformers ratio accuracies from 0.1 to 3% are specified according to intended use, with phase displacement between primary and secondary quantities between 10 and 120 minutes. For voltage transformers similar ratio errors are permitted but the permitted phase displacement errors are slightly less. In both cases the highest accuracy devices are used for energy metering purposes while lesser accuracy is generally required for protection purposes.

Regardless of the primary voltage, voltage transformer secondary voltage is invariably standardized on 110 V for a VT used on a single-phase system. For a three-phase VT the ratio is invariably primary line voltage to $110/\sqrt{3}$. The primary voltage is also generally made divisible by 11 so that the ratio will be a whole integer, for example, even for a 23 kV generator voltage system, voltage transformers will have the ratio $22\,000/\sqrt{3}$ to $110/\sqrt{3}$, i.e. 200 : 1.

Similarly, modern core materials have enabled current transformer designs that are standardized as having primary currents in multiples of 100 A with either 1 A or 5 A secondaries so that these might be 500 : 5 or 2000 : 1 with instruments and relays designed to have either a 1 A or 5 A input.

Where accurate measurement is required below 100 A a primary wound CT is required, where the conductor is a physical part of the CT design with accurate winding and spacing around the section of busbar.

Retrofit and upgrade installations can now benefit from a split core CT design, where the core, normally a continuous magnetic ring, may be split and then remade surrounding the conductor, without necessitating the disconnection/dismantling of the installation.

Energy Meters

Use of d.c. meters in the UK is very limited and demands in overseas markets are diminishing. They are therefore not described.

Induction meters. The popular single-phase induction meter consists fundamentally of two electromagnets, an aluminium disc, a revolution counter, and a brake magnet, connected as shown in *Figure 18.18*.

In the Ferranti type F2Q-100 meter the voltage element is positioned above the rotor disc and comprises a three-limb core with the coil mounted on the centre limb which also supports the friction compensating device. The power factor compensator consisting of a copper loop encircles the centre limb and is attached to a vertically moving rack carried on the main frame.

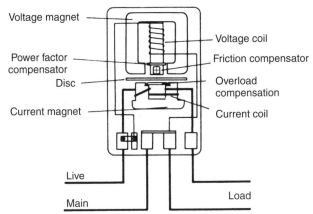

Figure 18.18 Electrical circuits of a typical induction meter connection to MMLL to BS 5685 (Ferranti Ltd)

The current element located beneath the disc has a U-shaped core with a coil supported on an insulated former. A magnetic shunt between the core and limbs compensates for the effect of current flux braking at high loads.

The braking system consists of two anisotropic magnets located above and below the disc respectively, and enclosed in a die-cast housing. Fine adjustment is provided by a steel screw which shunts part of the magnetic flux. Coarse adjustment is achieved by sliding the whole magnet system sideways. The rotor disc is supported by a magnetic suspension system.

Voltage and current magnets create two alternating magnetic fluxes which are proportional to the supply voltage and currents being measured respectively. Each of these fluxes sets up eddy currents in the disc; due to the interaction of the voltage flux with the eddy currents created by the current flux, and to the interaction of the current flux with the eddy currents created by the voltage flux, a torque is exerted on the rotor disc causing it to rotate. The permanent-magnet system acts as an eddy current brake on the motion of the rotor disc and makes the speed proportional to the power being measured. When installed in a consumer's premises the voltage coil of the meter is always on circuit. To prevent the disc from rotating when the consumer's load is switched off, the disc is provided with an 'anti-creep' device. In the meter described, protection against creep is provided by cutting two small radial slots in the periphery of the disc diametrically opposite to each other. This is only one of a number of ways in which compensating devices can be arranged to provide similar results. Although the construction of a single-phase induction meter is not complex it can be calibrated to a high level of accuracy which it will sustain for many years.

In metering two- or three-phase systems, two or three elements are arranged to drive a single rotor system.

Testing of Meters

It is not practicable to test meters under working conditions and it is therefore customary to use 'phantom loading' for test purposes.

Single-phase meters are tested on transformer circuits such as that shown in *Figure 18.19*. The current and voltage are supplied from separate sources, enabling the current to be used at a low voltage with consequent saving in energy. It also allows the test equipment design to be such as to permit variation in the phase angle between current and voltage so as to create conditions of inductive loading required in the calibration of meters. Polyphase test set circuits are more complex as they must provide facilities for testing of two or three elements both individually and collectively.

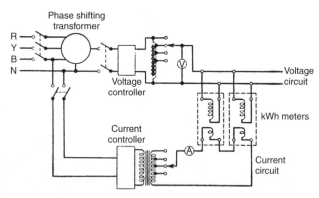

Figure 18.19 Simplified diagram for single-phase meter testing

Testing requirements for meters are given in detail in BS 5685, but in the UK the provisions of the Electricity Supply (Meters) Act 1936 govern the testing of all types of meters installed in all domestic premises and the majority of industrial premises. Appendix B of this Act sets out the methods to be used in testing meters and the actual tests to be employed by the meter examiner prior to certification of a meter.

A summary of the three methods is given below.

Method A

Long-period dial tests using substandard rotating meters. The load on any substandard rotating meter to be not less than one quarter of or more than one and one quarter times its full load. The duration of any such test to correspond to not less than five complete revolutions of the pointer to the last dial of the meter under test.

Method B

(a) *Tests (other than long-period dial tests) using substandard rotating meters.* The load on any substandard rotating meter to be not less than one quarter of or more than one and one quarter times its full load. The duration of any such test to correspond to the number of complete revolutions of the disc of the meter under test ascertained as follows, namely, by multiplying 40 by the percentage of the marked current of the meter at which the test is being made and dividing by 100; provided that the duration of any test shall correspond to not less than five complete revolutions and need not be greater than a period corresponding to 25 complete revolutions.

(b) *One long-period dial test.* To be made in accordance with and of duration not less than that prescribed under Method A.

Method C

(a) *Tests by substandard indicating instruments and stopwatch.* The load on any substandard indicating instrument to be such as to give a reading of not less than 40% of its full-scale reading. The duration of any such test to correspond to not less than three complete revolutions of the disc of the meter under test, or not less than 100 seconds, whichever be the longer period.

(b) *One long-period dial test.* To be made in accordance with and of a duration not less than that prescribed under Method A. (Method C alone to be used for testing d.c. motor meters.)

Actual tests. 1. Every motor meter other than a single-phase two-wire alternating whole current pattern to be tested:

(a) at 5% of its marked current, or in the case of a d.c. meter having a marked current of less than 100 A at 10% of its marked current;
(b) at one intermediate load; and
(c) at 100% of its marked current, or in the case of an a.c. meter, at 100 or 125% of its marked current.

Provided that in cases where the tests are carried out in accordance with Method B or Method C, the prescribed long-period dial test shall be at one of the loads specified for the three foregoing tests and shall be additional to such test.

Every single-phase two-wire alternating whole current pattern meter to be tested at high, intermediate and low load at unity power factor; each of the loads to be within the ranges specified in *Table 18.1*.

Provided such tests are carried out in accordance with Method B or Method C the prescribed long period dial test shall be at one of the loads specified in *Table 18.1* and shall be in addition to such tests.

Table 18.1 Tests on single-phase two-wire whole current meters

Testing load	Load ranges for the four meter categories in terms of marked current			
	Basic maximum	Maximum continuous rating	Long range	Short range
High	50% max to 100% max	50% to 100% (1/2–1/1)	100%–200% (1/1–2/1)	100%–125% (1/1–1 1/4)
Intermediate	One intermediate load to be taken at any part of the curve between the high and low loads actually taken			
Low	5% basic to 10% basic	1.67%–3.33% (1/60th–1/30th)	5%–10% (1/20th–1/10th)	5%–10% (1/20th–1/10th)

Provided such tests are carried out in accordance with Method B or Method C the prescribed long period dial test shall be at one of the loads specified in *Table 18.1* and shall be in addition to such tests.

2. Every alternating current meter also to be tested at its marked current and marked voltage at 0.5 power factor (lagging), subject to a tolerance of plus or minus 10% in the power factor.

3. Every watt-hour meter also to be tested for 'creep' with its main circuit open and with a voltage of 10% in excess of its marked voltage applied to its voltage circuit.

4. (a) A three-wire d.c. or single-phase a.c. meter may be tested as a two-wire meter with the current circuits of the two elements in series and the voltage circuits in parallel.
   (b) Every three-wire d.c. meter to be tested for balance of the elements (in the case of watt-hour meters with the voltage circuits energized at the marked voltage):
      (i) with the marked current flowing in either of the two current circuits of the meter and no current flowing in the other current circuit; and
      (ii) with half the marked current flowing in both current circuits.

Tests on polyphase meters

(a) Every polyphase meter to be tested as such on a circuit of the type for which the meter was designed.
(b) Every polyphase meter to be tested for balance of the elements with the voltage circuits of all the elements energized at the marked voltage with the marked current flowing in the current circuit of one element and no current flowing in the other circuit or circuits; such test to be carried out on each of the elements at unity power factor and also at 0.5 power factor (lagging) subject in the latter case to a tolerance of plus or minus 10% in power factor.

Meters operated in conjunction with transformer or shunts.
Every meter intended for use with a transformer or shunt to be tested with such transformer or shunt and with the connecting leads between the meter and such transformer or shunt.

5. Multi-rate meters. All multi-rate meters to be tested in accordance with the foregoing paragraphs, on the rate indicated when the relay is energized. In addition with the register operating in the deenergized position, and where applicable on the high rate, a further test to be taken at low load with a Method A test at high load. For polyphase meters these additional tests shall apply only to polyphase balanced load conditions.

Transducer Systems

The use of transducers to provide signals proportional to the quantities to be measured is becoming increasingly widespread.

The use of the transducer dramatically reduces the need for a.c. instruments, typically moving-iron, air- and iron-cored dynamometer and hot-wire instruments. The transducer produces a d.c. mA signal proportional to the measured quantity and therefore this signal can be measured by means of a moving-coil milliammeter. The indicator need not be located at the source of the measurement but at distances up to 1 kilometre away. Not only is the measurement more accurate but the device is linear and can measure bi-directional quantities such as electrical power. Short-range transducers often used for voltage measurements can be achieved using electronic offset techniques.

In terms of manufacture, the number of types of instrument is greatly reduced. The signals can also be conveniently measured using digital indicators and also made available to readily interface with computing and SCADA systems. The advantage of digital indicators is the high input impedance which means that a voltage measurement does not affect the circuit being measured, hence the measurement is accurate.

Another advantage of digital techniques is that transducers need not be solely a single function device but multifunction, often with up to six measurements available. The d.c. mA output from a single function transducer normally trends the measurement with a small delay of about 50–100 milliseconds. Where multifunction devices are used the measurement tends to be updated every 1–10 seconds dependent on the operational requirements. The use of multifunction transducers reduces the installation costs.

Typical electrical measurements which can be made using transducers are active and reactive power, voltage, current, power factor and frequency. Transducers can also be used to measure other quantities, typically weight, temperature and pressure.

Recent years have seen dramatic developments in electrical instrumentation, the instruments becoming extremely accurate and stable.

Improvements in electronic measurement techniques have resulted in the production of electronic energy meters, again not single function devices with high levels of accuracy but multifunctional devices capable of measuring active, reactive power/energy as well as voltage, current and power factor. Typically meters are commercially available with an accuracy of 0.2% for use on high accuracy electrical tariff metering systems associated with generation and distribution systems. Although the cost of a 0.2% meter is high, the benefits from an accurate measurement can save companies several million pounds. The meters produce far more information than the energy measurements, they can provide instantaneous measurements of circuit parameters, the values outputted either digitally or in an analogue form acting therefore as a transducer. The meter has the ability to store energy demand data (half-hourly) using large capacity electronic storage facilities for long periods. The data from these meters can be remotely accessed using the standard tele-phone or mobile phone networks (integral modems), mobile phone or radio communication systems.

The cost of electronic energy meters has greatly reduced and these are now replacing the induction-type energy meters in factories and homes with similar accuracy levels. The electronic meters are generally more stable and therefore need minimal checks on their calibration. They usually fail completely and it is often more economic to replace rather than repair.

This does not mean that the induction meter is being replaced completely. Although the numbers are reduced there is still a large demand, based on the fact that the cost of the induction meter is low and the failure rate is very small.

Multifunction Instruments

As an alternative to the multifunction transducers described above, basic transducer signals can be provided to multifunction instruments such as the Crompton Instruments Integra Family of devices. These can provide an indica-tion of the measured parameters on either an LED or LCD digital display, and communicate the parameter status using digital interfaces for as many as 100 plus circuit parameters. In addition to basic quantities such as current, voltage and watts, frequency, power factor, more recently total harmonic distortion can be indicated either as digital displays or input to SCADA systems.

Total harmonic distortion. The harmonic content in an electrical network is becoming an increasing problem to system designers. Growing usage of modern technologies in load control systems and efficient switched power supply systems has resulted in the generation of distortion by introducing harmonic frequencies to the fundamental. In particular those frequencies at factor 3 harmonics, i.e. 3rd, 5th, 9th, 15th, etc., will propagate freely around the network, and often outside the installation which can cause problems to neighbouring power users.

A too high harmonic distortion level causes significant stressing of equip-ment, with neutral conductors carrying more current than possibly the rating, and assets with wound components being at risk of overheating, such as trans-formers and generator-set alternators.

There are moves within the regulatory bodies to standardize the permis-sible levels of distortion allowed back into the network. G5/4 (UK) is now making recommendations and it is likely that mandatory maximum levels will be set within the EU EMC Directive.

THD (Total Harmonic Distortion) is defined as the r.m.s. value of the waveform remaining when the fundamental is removed, where a perfect sine wave = 100%.

$$\text{THD}\% = \frac{(\text{r.m.s. of total} - \text{r.m.s. of fundamental})}{\text{r.m.s. of total waveform}}$$

A measurement of THD% as indicated by the Crompton Integra 1540 Digital Metering Instrument enables observation of a problem before it exceeds acceptable limits.

19 Electric welding

There are three main types of welding, namely arc welding, resistance welding and radiation welding. These can again be divided into groups as follows.

(a) Flux-shielded arc welding: This method comprises four forms:
    (i) manual metal arc;
    (ii) automatic submerged arc;
    (iii) electro-slag;
    (iv) electro gas.
(b) Gas-shielded arc welding. There are three principal types:
    (i) inert gas tungsten arc (TIG);
    (ii) plasma arc;
    (iii) gas metal arc.
(c) Unshielded and short-time arc processes: Two main types are:
    (i) percussion;
    (ii) arc stud welding.
(d) Resistance welding. The four main types are:
    (i)   spot;
    (ii) seam;
    (iii) projection;
    (iv) flash.
(e) Radiation welding. Three methods fall into this category:
    (i) arc image;
    (ii) laser beam;
    (iii) electron beam.

Flux-shielded Arc Welding

In this important series of welding processes the electric arc supplies the heat for fusion while a flux is responsible for the shielding and cleaning functions, and often also for metallurgical control. The most widely used form is a manual process known as manual metal arc welding. In this method electrodes, in the form of short lengths of flux-covered rods, are held by hand. The flux on the electrode is of considerable thickness and is applied by extrusion.

The covering has other functions to perform besides that of a flux. It stabilizes the arc, provides a gas and a flux layer to protect the arc and metal from atmospheric contamination, controls weld-metal reactions and permits alloying elements to be added to the weld metal. Finally, the slag left behind on the surface of the weld should assist the formation of a weld bead of the proper shape.

Metal arc welding electrodes are made with core wire diameter from 1.2 mm up to 9.5 mm. For all but exceptional circumstances the useful range is 2.5 mm up to 6 mm. The length of an electrode depends on the diameter; for smaller diameters, where the manipulation of the electrode calls for the greatest control, the length may be only about 300 mm. They are usually about

428

450 mm long and are consumed at a burn-off rate of 200–250 mm/min. The core wire is made of mild steel.

The power source which supplies the current for the arc may be either a.c. or d.c. Regardless of the source means must be provided for controlling the current to the arc. This may be by means of a regulator across the field of a d.c. motor generator or a choke in the output side of a mains-fed transformer.

Automatic submerged arc. In manual metal arc welding with covered electrodes, the current must pass through the core wire, and there is a limit to the length of electrode which can be used because of the resistance heating effect of the current itself. In the submerged arc process a bare wire is employed and the flux added in the form of powder which covers completely the weld pool and the end of the electrode wire. Extremely high welding currents can be used, giving deep penetration.

Submerged arc welding can either use a single electrode or multiple electrodes. In parallel electrode welding two electrodes are usually spaced between 6 mm and 12 mm apart and are connected to the same power source. If the electrodes are used in tandem, one following the other, an increase in welding speed of up to 50% is possible. When used side by side wider grooves can be filled. For multiple electrode welding an a.c. power source is preferred because multiple d.c. arcs tend to pull together.

Another solution to the problem of introducing current to a continuous fluxed wire is to put the flux into tubular steel electrodes drawn from strip or enclosed in a folded metal strip. A method which overcomes the disadvantage of the bulk of the cored electrode wire and the lack of visibility of the submerged arc process uses a magnetizable flux. This is in a stream of carbon dioxide which emerges through a nozzle surroundings the bare electrode wire. Because of the magnetic field surrounding the electrode wire when current is passing the finely powdered flux is attracted to and adheres to the wire, forming a coating. *Figure 19.1* shows the various arrangements described.

Electro-slag welding. As the thickness of the metal welded increases, multi-run techniques become uneconomical. The use of automatic welding, however, with high current large passes in the flat position can give a weld pool so large that it runs ahead of the electrode out of control resulting in inadequate

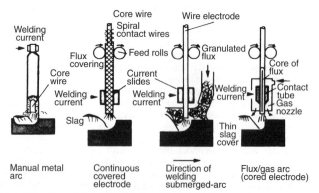

Figure 19.1 Manual and automatic flux-shielded welding

fusion. This is overcome by turning the plates into the vertical position and arranging a gap between them so that the welding process becomes rather like continuous casting. Weld pools of almost any size can be handled in this way provided that there is sufficient energy input and some form of water-cooled dam to close the gap between the plates and prevent the weld pool and the slag running away. This is the essential feature of electro-slag welding. *Figure 19.2* shows the basic arrangement for the most common, wire electrode type.

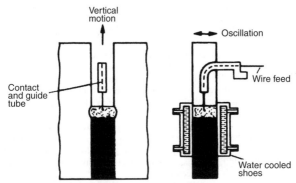

Figure 19.2   Electro-slag welding with wire electrode

For thick plates, i.e. over 100 mm, it is necessary to traverse the electrode so that uniform fusion of the joint faces through the thickness of the point is obtained. For thick materials, more than one electrode wire may be employed.

There are a number of variations of electro-slag welding mainly to do with the shape of the electrode. One method, which uses a consumable guide, is much simpler than the wire electrode in that the welding head and wire feed mechanism do not need to be moved up the joint as the weld is made. Another type is where plate or bar electrodes are suspended in the joint gap and lowered slowly as they are melted away by the slag bath. A three-phase power supply can be used with three plate electrodes connected in star fashion.

Electro-gas welding. In this form of welding heat is generated by an electric arc which is struck from a flux-cored electrode to the molten weld pool. The flux from a flux-cored electrode forms a thin protective layer but does not give a deep slag bath as in electro-slag welding.

Additional shielding of the weld pool is provided by carbon dioxide or argon-rich gas which is fed over the area through the top of each copper shoe. Mechanically the equipment is similar to that for the wire electrode type of electro-slag welding.

Gas-shielded Arc Welding

Essentially gas-shielded arc welding employs a gaseous shielding medium to protect both the electric arc and the weld metal from contamination by the atmosphere.

430

Inert gas tungsten arc welding. Tungsten electrodes are used for this form of welding as being the least eroded in service. Other refractory metals have been tried without success. The rate of erosion of tungsten electrodes is so small as to enable them to be considered non-combustible. The gas which surrounds the arc and weld pool also protects the electrode. Argon or helium gas is used in preference to hydrogen because it raises the arc voltage and requires a high open-circuit voltage. It can also be absorbed by some metals affecting their characteristics.

It is usual to make the electrode negative thus raising the permissible current to eight times that with a positive electrode system. The chief advantage of having the d.c. electrode positive is the cleaning action exerted by the arc on the work.

Where this cleaning action is required and the current is over 100 A, an a.c. power supply must be used. With a.c., however, the reversals of voltage and current introduce the problem of reignition twice in every cycle. When the electrode is negative there is no problem but when the electrode becomes positive the restrike is not automatic. Three ways are available to assist reignition of the arc under these conditions.

The first is with a well-designed transformer with low electrical inertia that will provide the high voltage required to cause the arc to restrike. The second is by means of a high frequency spark unit. This may be switched off by a relay once welding begins, or alternatively it may be operated from the open-circuit voltage so that it ceases to operate when the voltage drops to that of the arc. Both methods suffer from the disadvantage that the open-circuit voltage tends to be high and the power factor has to be low because this high voltage must be available at current zero.

In the third method of arc maintenance a voltage surge is injected into the power circuit to supply the reignition peak. This is done by discharging a capacitor through a switch which is tripped automatically by the power circuit, see *Figure 19.3*. When the arc is extinguished at the end of the negative half

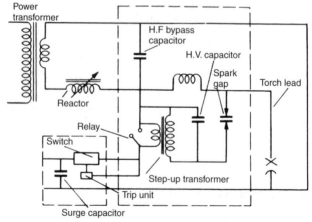

Figure 19.3 Arc maintenance circuit for inert gas tungsten arc welding. A voltage surge is injected into the power circuit to supply the reignition peak

cycle the reignition peak begins to develop and itself fires a gas discharge valve which discharges the capacitor. Welding can be carried out at less than 50 V r.m.s. as opposed to 100 V by the first two methods.

There are several variations of the tungsten arc method for special purposes. The pulsed tungsten arc is suitable for welding thin sheet material using a d.c. power supply and a modulated current wave. The twin tungsten arc provides a wide weld bead and is used, for example, in the manufacture of cable sheathing from strip. Tungsten arc spot welding employs the same equipment as for conventional manual tungsten arc welding except that the control system includes a timing device and the torch is modified. The process is used to join overlapping sheets of equal or unequal thicknesses.

Plasma arc welding. A plasma arc heat source may be considered as a development of the inert tungsten arc. There are two types, the non-transferred and the transferred or constricted arc. If the inert gas tungsten arc torch is provided with a separately insulated water-cooled nozzle that forms a chamber round the electrode and the arc is struck from the electrode to this chamber, the arc plasma is expelled from the nozzle in the form of a flame. This is the non-transferred arc which with a powder feed into the plasma is used for metal spraying.

When the arc is struck from the electrode to the workpiece the arc is contracted as it passes through the orifice in the nozzle. This is the transferred plasma arc and it is used for cutting because of the high energy density and velocity of the plasma. To start an arc of this kind requires a pilot arc from the electrode to the nozzle. The transferred arc can also be used for welding: at low currents for welding sheet metal less than 1.5 mm thick and at currents up to 400 A for welding thick metal. When using the plasma arc for welding additional inert-gas shielding is required.

Gas metal arc welding. The electrode used in this form of welding is of the consumable type and a d.c. supply is employed. The electrode end melts and molten particles are detached and transported across the arc to the work by magneto-dynamic forces and gaseous streams. Size and frequency of the droplet transfer is related to the wire size and current; voltage has only a limited effect, except in one case. The best performance is obtained with small diameter wires and high current densities.

Unshielded and Short-time Processes

Use of a flux or gas shield in welding is to exclude air from contact with the arc and weld pool and so prevent contamination of the weld. In unshielded welding methods, correction of the consequences of atmospheric contamination is made by adding deoxidizers such as silicon, manganese, aluminium or cerium. The welding method is the same as for gas metal arc except that no gas is employed.

Two short-time and discontinuous unshielded arc processes are percussion welding and stud welding. Shielding is not required because either the whole process is carried out so rapidly that contamination is negligible or the molten metal is squeezed out of the joint. A common feature of the two processes is that they permit the joining in a single operation of parts with small cross-sectional areas to other similar parts or more massive pieces.

Percussion welding appears in several forms but in all a short-time high intensity arc is formed by the sudden release of energy stored generally, but not always, in capacitors. Subsequent very rapid or percussive impacting of the workpieces forms the weld. Three ways in which the arc is initiated are:

1. Low voltage with drawn arc.
2. High voltage breakdown.
3. Ionization by a fusing tip.

With each method the energy source may be a bank of capacitors, which is charged by a variable voltage transformer/rectifier unit.

Stud welding. One of the most useful developments in electric welding is stud welding. This method of fixing studs into metal components offers several advantages over the older method which entailed drilling and tapping operations to permit the studs to be screwed into position. By obviating drilling and tapping, stud welding simplifies the design of products, structures and fittings, giving a better appearance, improved construction and economy of materials.

There are two methods of stud welding, drawn-arc and capacitor discharge – both may be described as modified forms of arc welding. In both cases a special welding gun is required and also an automatic head or controller for timing the duration of the welding current.

The principle employed in the drawn-arc process is that the lower end of the stud is first placed in contact with the base metal on which it is to be welded. A small current is then passed through the stud to burn away any scale and establish an ionized path before the full welding current is allowed to flow through the gun; a solenoid in the gun raises the stud slightly to form

Figure 19.4 A Cromp-Arc TC12 electronic welding system being used to install a new anti-skid surface at a ferry terminal (Crompton Parkinson)

an arc. After a small interval the current is automatically cut off and a spring in the gun drives the stud into the pool of molten metal which has been formed by the arc.

With the capacitor-discharge method the stud has a small pip or spigot at its lower end. In the welding operation this spigot is placed in contact with the base. When the current is switched on the spigot is fused to create the arc and the stud is pressed into the molten metal.

To provide the necessary welding current for the drawn-arc process a d.c. supply from an arc-welding generator or from a static rectifier is required. Equipments of the transformer type using alternating current in the weld have also been developed. An automatic controller is included in stud welding equipment for timing the weld, the operation cycle being preset by the operator. *Figure 19.4* shows a Cromp-Arc stud welding equipment being used to install a new anti-skid surface at a ferry terminal.

Resistance Welding

All welding methods described in the preceding pages employ an arc to produce heat in the metal to be welded. Another way of producing heat in the metal, also using current, is to pass the current through the workpiece. This method has the advantage that heat is produced throughout the entire section of the metal. The current may be introduced into the material either by electrodes making contact with the metal or by induction using a fluctuating magnetic field surrounding the metal. All resistance welding methods require physical contact between the current-carrying electrodes and the parts to be joined. Pressure is also required to place the parts in contact and consolidate the joint.

Resistance spot welding. In this process, overlapping sheets are jointed by local fusion caused by the concentration of current between cylindrical electrodes. *Figure 19.5* shows the method diagrammatically. The work is clamped between the electrodes by pressure applied through levers or by pneumatically operated pistons. On small welding machines springs may be used. Current is generally applied through a step-down transformer. The electrodes and arms

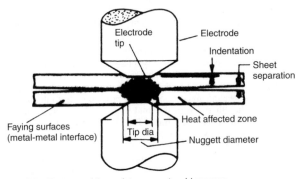

Figure 19.5 Features of the resistance spot weld process

of the machine usually form part of the secondary circuit of the transformer. A spot-welded joint consists of one or more discreet fused areas or spots between the workpieces.

Welding current can be controlled either by changing the turns ratio of the transformer or by phase shift means. For the former method with a single-phase supply the welding current value can be altered by changing the tapping ratio of the primary side of the transformer.

In phase shift control the transformer primary is supplied through thyristors which act simply as high-speed switches. At the beginning of each half cycle of primary current the thyristors is made non-conducting, but later in the half cycle it is fired. Current flows until the next current zero. Thus only a portion of the available power in the half cycle is passed through the workpiece. The thyristors are connected in inverse-parallel or back-to-back configuration, and can be used for both single-phase and three-phase working.

The time the current is flowing is important and can be controlled either by mechanical or electronic means. Electronic timers work without regard to the phase of the welding current and because an exact number of cycles cannot be guaranteed they are not used for times less than ten cycles.

Resistance seam welding. Two methods of continuous seam welding are available. Using spot welding equipment a series of overlapping spots can be made, a process known as stitch welding. Alternatively the electrodes may be replaced by wheels or rollers so that work may be moved through the welder continuously without the necessity for raising and lowering the head between welds. Rollers are power driven and may or may not be stopped while individual welds are made. Current is generally passed intermittently while the electrodes are stationary, although continuous current is also used to a limited extent. A common technique is known as step-by-step seam welding, because while each weld is being made rotation is stopped and the current is then switched off while the rollers move to the next position. The amount of overlap between spots depends on peripheral roller speed and the ratio of current on-and-off time. A normal overlap would be 25–50%.

Adjustment of timing can be made to produce not a continuous seam but a series of individual welds. When this is done the process is called roller-spot welding.

Generally seam welds are made on overlapped sheets to give a lap joint but sometimes a butt joint is required. There are several methods of obtaining this and space does not allow a full description of them. Methods include mash-seam welding, foil butt-seam welding, resistance butt-seam welding, high frequency resistance welding and high frequency induction welding.

Projection welding. In this process current concentration is achieved by shaping the workpiece so that when the two halves are bought together in the welding machine, current flows through limited points of contact. With lap joints in sheet a projection is raised in one sheet through which the current flows to cause local heating and collapse of the projection. Both the projection and the metal on the other side of the joint with which it makes contact are fused so that a localized weld is formed. The shaped electrodes used in spot welding can be replaced by flat-surfaced platens which give support so that there is no deflection except at the projection.

The method is not limited to sheet-sheet joints and any two mild steel surfaces that can be brought together to give line or point contact can be projection welded. This process is used extensively for making attachments to sheet and pressings and for joining small solid components to forging or machined parts.

Flash welding. Flash welding is a development from resistance butt welding and it uses similar equipment. This comprises one fixed and one movable clamp, so that workpieces may be gripped and forced together; a heavy duty single-phase transformer having a single turn secondary; and equipment for controlling the current. The parts to be joined are gripped by the clamps and brought into contact to complete the secondary circuit, *Figure 19.6*. When the welding voltage of up to about 10 V is applied at the clamps current flows through the initial points of contact causing them to melt. These molten bridges are then ruptured and small short-lived arcs are formed. The platen on which the movable clamp is mounted moves forward while this takes place and fresh contacts are then made elsewhere so that the cycle of events is then repeated. This intermittent process continues until the surfaces to be joined are uniformly heated or molten.

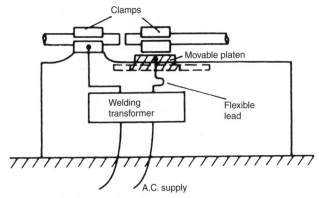

Figure 19.6   Basic arrangement for flash welding

By this time the moving platen will have advanced, at an increasing rate, to close the gap as metal is expelled. The total distance up to the point of upset is known as the 'flashing allowance'. At this point the rate of movement of the platen is rapidly increased and a high force applied to forge the parts together and expel the molten metal on the surfaces. The expelled metal forms a ragged fin or flash round the joint. Ideally all the contaminated metal produced during flashing should be removed in this way to produce a high quality joint having many features of a solid-phase weld.

Radiation Welding

A small group of processes employ energy for welding in the form of radiation. Two processes use energy as electromagnetic radiation while another uses an electron beam. Radiation welding is unique in that the energy for welding can be focused on the workpiece, heat energy being generated only where the beam strikes.

Unlike arc or flame sources the work is not brought into contact with any heated media, gas or metal vapour, and the process may be carried out in a vacuum or low pressure system for cleanliness. In contrast to arc welding the melted pool is subjected to only negligible pressure.

Arc image welding. Fusion is accomplished by focusing the image of a high temperature source on the workpiece. Mirrors are used for this purpose. High pressure plasma arc sources have been developed as the heat source and outputs above 10 kW have been used for welding and brazing. Optical systems with top surface mirrors of high accuracy are necessary for focusing, but even so losses occur by dispersion because the source is finite and emits in all directions. Although the method has no future for terrestrial welding it may eventually be employed in space using the sun as the heat source.

Laser welding. A laser is a device which, when irradiated by light from an intense source, is capable of amplifying radiation in certain wavebands and emitting this as a coherent parallel beam in which all waves are in phase. This beam can be focused through lenses to produce spots in which the energy density is equalled only by the electron beam. The word laser stands for light amplification by stimulated emission of radiation.

The laser has been used for micro-spot welding and this appears to be its most promising application, particularly where minimum spread of heat is required.

Electron beam welding. Electron beam welding is a process involving melting in which the energy is supplied by the impact of a focused beam of electrons. The essential components of a welding equipment are the electron gun, the focusing and beam control system and the working chamber which operates at low pressure.

There is a variety of types of electron gun but they all work on the same principle. Electrons are produced by a heated filament or cathode and are given direction and acceleration by a high potential between cathode and an anode placed some distance away. In a limited number of electron beam welding plants the anode is the workpiece. In other cases a perforated plate acts as the anode and the beam passes through this to the workpiece behind. The former system is known as 'work-accelerated' and the latter 'self-accelerated'. The self-accelerated type is the one used for welding.

Magnetic focusing is usually employed, this taking place after the beam has passed through the anode, *Figure 19.7*. In this type the beam current is controlled by a negative bias on an electrode placed round the cathode which is called the control electrode or bias cup. This electrode behaves as a grid in that it is able to control the flow of electrons from the cathode.

Tungsten inert gas (TIG) welding. The Argonarc welding process is a method of joining non-ferrous metals and stainless and high alloy steels without using flux. An electric arc struck between an electrode and the workpiece produces the heat to fuse the material, and the electrode and the weld area are shielded from the atmosphere by means of an inert gas (argon). Standard d.c. or a.c. welding equipment may be used provided the open-circuit voltage is around 100 V for a.c. and 70 V for d.c.

For welding aluminium, magnesium and their alloys, stainless and high alloy steels, nickel alloys and copper alloys, up to 3 mm thick, a.c. is suitable.

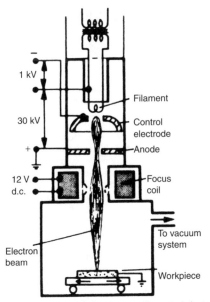

Figure 19.7 Outline of triode electron beam gun with control electrode or bias cup

D.C. may be used for all other common metals, and it is essential for the welding of copper and stainless steels and alloys over 3 mm thick.

Method of welding. The argon is supplied to the torch through PVC tubing, where it is directed round the tungsten electrode and over the weld area by either a ceramic or water-cooled copper shield. The torch is held (or clamped) so that the electrode is at approximately 80° to the weld and travels in a leftward direction. Filler wire may be added to reinforce the weld, the angle being approximately 20° to the seam. It is important that the end of the filler wire should be pressed on to the workpiece and be fed into the front edge of the weld pool.

When comparatively long straight seams are to be welded, e.g. joining sheets, strip, or fabricating cylindrical vessels, Argonarc machine welding may be used. Much of this type of work is done without using filler wire, i.e. the butted edges are simply fused together. Where maximum strength or a flush surface is required or when material over 3 mm thick is being welded, filler wire may be added.

Argonarc spot welding. This is a supplementary process to resistance spot welding, and is used to make spot welds in positions inaccessible to resistance spot welders.

In principle, a local fusion takes place between the sheets to be joined. The torch, which is hand controlled, contains a tungsten electrode recessed within a water-cooled shield. Argon gas is used to shield the electrode and weld area.

20 Battery electric vehicles

The classification 'battery electric vehicles' can now be subdivided into three categories: commercial and industrial vehicles, which includes milk floats and fork-lift trucks; battery-driven light cars, falling mainly into the 'development' class; and, most recently, hybrid electric cars using an internal combustion engine and battery/electric motor drive train in combination.

Use of the first category started soon after the Second World War, initially with the development of fork-lift and platform trucks. Diversification followed fast and today there is a wide variety of special purpose electric vehicles available. The list includes milk floats, delivery vans, overhead maintenance vehicles, tractors and trucks both driverless and rider controlled, fork-lift trucks, mini-buses, mobile shops, articulated goods trucks, ambulances and interworks transporters. These are the types of vehicles which can most benefit from the advantages of electrical motive power, and which are least inconvenienced by the disadvantages.

The advantages of battery electric vehicles are clear: they are non-polluting, they use no energy at all between tasks when idle, they are virtually silent, the motors have a high starting torque so that initial acceleration is high and they are quickly up to operating speed. Maintenance requirement for an electric motor is also very much less than for an internal combustion engine (ICE), although modern ICE equipped vehicles require far less maintenance than those of a decade ago. The use of off-peak electricity for overnight recharging is almost always feasible, hence in most instances running costs can also be kept quite low.

The disadvantages of battery-driven vehicles include the fact that the batteries themselves are heavy and bulky, thus greatly reducing the payload of the vehicles. The range between recharges is low. After the brisk initial acceleration cruising speed is modest and when cruising, further acceleration such as might be required for safe overtaking is low.

Battery-driven Light Cars

At the present time, one major advantage of battery driven vehicles is that they do not produce any of the carbon dioxide, carbon monoxide, oxides of nitrogen or particulates produced by ICEs. At the point where the electricity is produced, the necessary environmental measures can be put in place to limit the production of pollutants. This has provided the main inducement to continue with the development of battery-driven light cars and similar vehicles despite the fact that for this use, the disadvantages identified above are at present quite significant.

Initial attempts to produce a commercially viable battery-driven light car date back to the mid-1970s. At that time the incentive was the rapid escalation in the cost of oil which gave added attraction to the low running costs that could be obtained from electricity. In the UK, in 1973, the then Electricity Council bought 60 Enfield 8000 electric commuter cars from Enfield

Automotive for trials to be carried out by the area electricity boards, in an attempt to demonstrate their commercial viability. Maximum speed of these vehicles was 64 km/h (40 miles per hour) and their range was 96 km (60 miles), more than many designs of such vehicles could achieve at that time. A year or two later Bedford Commercial Vehicles and Freight Rover supplied a large numbers of 1 tonne vans for fleet use to the UK electricity supply industry. These had drive trains developed by Lucas Chloride Electric Vehicle Systems.

Despite these various attempts to give impetus to the development of battery-driven vehicles there has been little, if any, increase in their use over the last 30 years. This is due to the inability to find anything better than the lead-acid battery as a means of power storage, and the fact that only modest improvements have been made in the performance of this type of battery in terms of power stored per kg of mass.

Improvements in control. One area in which there have been significant strides since the mid-1970s has been in the area of the peripherals associated with the drive systems, that is, solid-state thyristor inverter/rectifiers for the drives/chargers and computers for control of the regenerative braking and charging systems. These have enabled the highest possible efficiencies to be obtained from the drives, and optimization of the energy gained from any deceleration cycle.

Advances in vehicle design. Although not unique to electric vehicles, international pressures to reduce pollution and for the reduction of greenhouse gases have led to significant advances in the design of the vehicles themselves. As well as aiming to achieve ICEs with forever greater and greater fuel economy, efforts have been directed to reduction in 'drag', or resistance of the vehicle as it passes through the air, and also reduction in the weight of the vehicle. This low aerodynamic loss leads to increased range for an electric vehicle, and the reduced weight means better ability to carry the weight of the battery as well as improved acceleration.

Advances in battery technology. As indicated above, many authorities consider that the lead-acid battery will not be surpassed as a means of power storage for electric vehicles. However, in recent years there have been detailed improvements in lead-acid batteries (see Chapter 21). Sealed valve-regulated cells now provide improved storage density combined with rugged handling and operating capabilities very much superior to those of lead-acid batteries of the 1970s and early 1980s. Nickel-metal hydride batteries are also being developed with improved storage capability combined with longer life when compared with sealed lead-acid cells.

As a result of all of the factors identified above it is now possible to produce a battery-driven 'commuter vehicle' which is becoming much closer to being economic than those attempted in the last decade. In the USA General Motors are now announcing their 'second generation' electric vehicle, EV_1, which they aim to sell in a competitive market place. According to General Motors' literature it has a range of between 55 and 95 miles, depending on terrain, driving style and temperature. It is powered by a 100 kW (137 h.p.) three-phase a.c. induction motor and has a single ratio reduction gear set of 10.946 : 1. It uses a combination of regenerative and conventional braking.

The car is available with an optional alternative battery pack using nickel-metal hydride batteries which increase the range to 75 to 130 miles. Charging is at 220 V and a full charge can be delivered in 5.5 to 6 hours for lead-acid and 6 to 8 hours for nickel-metal hydride. The vehicle owes much of its performance to an advanced body design with a very low drag aerodynamic shape and composite construction having exceedingly light weight.

Hybrid Vehicles

Even with vehicles having such advanced characteristics as the General Motors EV_1 design described above, it is not possible to obtain a performance which allows the car to be used for all purposes. That is, the owner of such a car would still need an alternative vehicle for long journeys. Even though many people nowadays own two cars, this is seen as a serious disadvantage.

Hybrid vehicles represent an attempt to overcome this problem. They have drive trains which use a combination of ICE and battery/electric motor propulsion. The ICE is generally fairly modest in size, 1000 or 1100 cc, and usually petrol driven. This will operate in conjunction with the electric drive when maximum performance is demanded. At times of light loading the car is driven solely by the electric motor. When the control system detects that the battery requires charging, the ICE will run to perform this duty regardless of the traction power requirements. Hybrid vehicles do not rely on any external battery charging facility. As in the case for pure battery-driven vehicles, hybrids invariably conserve power by use, at least in part, of regenerative braking.

As will be apparent from the above, the hybrid system owes its feasibility to the availability of computers to perform the complex drive/charge management functions.

Toyota with their 'Prius' and Honda with the 'Insight' were the first to get hybrid vehicles into commercial operation at the end of the 1990s. However, most of the major motor manufacturers are developing hybrid vehicles, including Daimler Chrysler, Ford, General Motors, Kia, Mercedes, Mitsubishi, Nissan, Peugeot, Renault, Volkswagen and Volvo.

As in the case of the pure battery-driven models, those hybrid vehicles which are in production owe many aspects of their excellent performance not simply to the basic technology of the hybrid drive system and its control computers but also to the advanced features of the vehicles by way of low drag and light weight. In addition the ICE will probably be of advance design with multiple valves per cylinder and variable valve timing.

Although the hybrid vehicle is to be taken very seriously – Toyota claim that by 2002 they had built 100 000 of these vehicles (a relatively small quantity by Toyota standards) – like the pure battery-driven vehicle it has a serious disadvantage. This is the high cost and strange logic of producing a vehicle which effectively has two drive systems. The protagonists would argue that these are, in effect, two half-drive systems, and the manufacturers aim to use components for dual and sometimes triple functions in order to optimize the economics. For example, the drive motor will also be the starter motor, as well as the generator.

Fuel Cell Drives

Much informed opinion considers that in view of the shortcomings of both pure battery and hybrid-type drives described above, the ultimate electric vehicles will utilize fuel cell drives.

Fuel cells. The principle of the device which we know as a fuel cell has been around since the earliest beginnings of experimentation with electricity. It converts chemical energy directly into electricity and in this respect it is simply a type of battery. The essential difference compared to the batteries which are in every day use is that the hydrogen 'fuel' can be replenished as it is used up, so that the cell does not have a limited short life like a disposable battery. The other greatly attractive feature of the fuel cell as a source of electrical energy is that the by-product is simply pure water, so that it is entirely non-polluting.

In the 1830s, in the UK, Sir William Robert Grove had experimented with the electrolysis of water. He argued that if the passage of an electric current through an electrolyte could produce hydrogen and oxygen, it must be possible to reverse the process, that is create the conditions under which hydrogen and oxygen can be combined to generate an electric current. Grove did, in fact, succeed in doing this by immersing two separate containers, one filled with hydrogen and one with oxygen in an electrolyte of dilute sulphuric acid and causing a current to flow between two electrodes. In 1890, also in the UK, two chemists, Ludwig Mond and Charles Langer, attempting to build a practical device using coal gas and air, coined the name fuel cell.

At around this time the development of early internal combustion engines killed any incentive for further development of fuel cells and they remained little more than a scientific curiosity for more than half a century until the American space programme created a revival of interest. In the 1960s, NASA, looking for a source of power aboard their spacecraft, and ruling out conventional batteries because they were too heavy, solar energy because it was impractical and expensive, and nuclear power because of the associated risks, saw fuel cells as an ideal solution. They therefore funded a number of research projects into developing practical fuel cells. In the mid-1970s the oil crisis and escalation of petrol prices provided an incentive for further research directed to their use in motor vehicles, and several of the large manufacturers began to invest in experimental fuel cell powered electric vehicles.

Despite the extensive research outlined above, economic practical fuel cells still appear to be little nearer. The problems which remain to be solved on an economic commercial basis are the development of suitable porous/permeable membrane material which will allow the gases to be brought into close contact to enable the reaction to take place, and the production of electrodes with acceptable corrosion resistance. Nevertheless many major car manufacturers who, no doubt, feel that they cannot afford to be left behind should a competitor achieve a major breakthrough, are continuing with the development of fuel cell powered vehicles. These include Daimler Chrysler, Nissan, Peugeot, and Toyota.

Industrial Vehicles

Delivery vehicles. The most popular vehicle in this class is used for the delivery of milk. Three- or four-wheeled vehicles are available with capacities up to 2 tonnes. Battery capacity varies from 235 Ah to 441 Ah at the 5 hour rate. Three-wheeler vehicles are for light duty and are extremely manoeuvrable. Three types of controller are available: resistance, parallel-series system of battery connections and solid-state pulse type. The Pulsomatic thyristor controller provides smooth control of speed either forward or reverse. Power is drawn from the battery in short bursts, depending on the firing of the thyristors, and the absence of resistors eliminates the losses associated with resistance control methods.

Figure 20.1 CAV motor in wheel

The Commando dairy vehicle available from Electricars is typical of the modern type of this transport. A heavy duty, enclosed, ventilated and protected series wound traction motor is employed. Control is either by a simple resistance which increases the voltage applied to the motors in four steps, forward or reverse, or by thyristor. Control contactors have magnetic blowout facilities and they are operated by a footswitch incorporating electronically controlled time delay. The 300 A plug and docking sockets completely isolate the traction and auxiliary circuits during charging. A Layrub propeller shaft with universal joints provides transmission. A two-stage drive axle consists of a first stage helical spur and the second a spiral bevel mounted on taper roller bearings.

The Electricars F85/48 TS electric two-speed dairy vehicle has been developed for situations where greater range and speed are required. For travel to and from an urban area the high-speed mode can be used, while for urban delivery the speed is halved.

Platform trucks. Fixed or fork-lift-type platform trucks are available for pedestrian control or for carrying a driver either standing or seated. Pallet and stillage trucks are designed for the horizontal movement of unit loads on pallets, stillages and skids. They are self-loading and Lansing Linde has models capable of handling loads exceeding 3000 kg. They are particularly useful as auxiliaries to fork-lift or reach trucks.

Two basic types of pallet transporters made by Lansing Linde are:

1. The pedestrian controlled truck employed where duties are intermittent or distances are not excessive.
2. Stand-on transporters allowing the operator to ride on the truck and thereby work over much longer distances.

Stillage and skid handling trucks are also available in both forms.

The Electricars company has a range of driven-type platform trucks with payloads up to 2000 kg and fixed or elevating platforms. On an elevating model a fully loaded platform can be raised in 20 secs. Controller can be resistance or thyristor system depending on the model.

Fork-lift trucks. These are available in three-wheel or four-wheel rider-operated counterbalanced reach trucks. Montgomerie Reid's Series 3 range is capable of handling payloads from 1000 to 1500 kg, while the Series 4, four-wheel model will lift loads from 1700 to 2750 kg (*Figure 20.2*). The largest is the MR4.27, which has a maximum lifting height of 7.77 m. Forward and backward tilts of the forks of 3° and 10° respectively are provided to assist picking up and dropping loads. A 72 V battery has a capacity of 431 Ah at the 5 h rate.

Figure 20.2 Montgomerie Reid Series 4 forklift truck

The Montgomerie Reid 25 P counterbalanced three-wheel pedestrian-controlled fork-lift truck has a capacity of 1200 kg at 3.3 m lifting height. High transmission efficiency is obtained by using a motor-in-wheel unit, and a diode contactor control system makes the best use of the battery power. Capacity of the 24 V battery at the 5 h rate is 324 Ah. Gradeability on a continuous rating is 2% laden and 5% unladen, rising to 8% and 12% respectively on a 5-minute rating.

Pedestrian-controlled, stand-on operated or seated-rider types of fork-lift trucks are designed for working in narrow aisles and are capable of high lifts. The FAER 5.1 turret truck made by Lansing Linde is a narrow aisle model fitted with a rotating turret and reach mechanism which can turn through 180°, enabling it to pick up and deposit loads on either side of the operating aisle. It is available in three capacities, 1000 kg, 1500 kg and 2000 kg, with lifting heights up to 9 metres. A solid-state electronic control system enables the truck to be accelerated from standstill to full speed in a smooth manner. The twin accelerator pedals, one for each direction, also control the truck's electrical braking system.

To speed operation times and increase operating safety a rail-guided steering system within the racking is recommended. The truck is steered by side rollers working in rails fixed to either side of the racks, enabling the operator to release the steering wheel and concentrate on the lifting and stacking operations. This type of truck is ideal for large warehouses and stores where space utilization and fast access to goods are of major importance.

Lansing Linde was among the first to supply a fork-lift truck with an integral motor/wheel arrangement. Another, provided by CAV, is available in two sizes and is designed for mounting directly onto the vehicle frame, eliminating the need for a transmission system. The d.c. motor is built inside the hub of the wheel unit and for the 330 mm diameter wheel is driven by a 24 V battery, and for the 381 mm diameter wheel a 36 V battery. The wheel can also be used in reach trucks, tow tractors and other battery electric vehicles, see *Figure 20.1.*

Tow tractors. Tractors or tugs do not themselves carry a load unless working as an articulated unit. Their purpose is to pull trailers or trolleys, either as single units or coupled together to form trains. The tractive resistance of the trailers can differ considerably depending on the type of trailer, its wheels and the road surface. Most manufacturers provide information on the speed attainable for given drawbar pulls. For the Electricars Model B2000, the speed with a trailed load of 2000 kg is 8.8 km/h.

Aisle width depends on the number of trailers making up a train. Typically for one trailer the aisle width should be about 2.2 m and for four trailers 3.1 m.

Some three-wheel models are obtainable having a single front wheel drive to give, it is claimed, increased manoeuvrability. Generally four-wheel tractors are favoured. Ranges vary according to battery capacity.

Motor operating voltages range from 24 V to 72 V and ratings extend up to 12 kW. Series motors are generally employed and control may be by one of the three types mentioned for other battery electric vehicles. Drive may be a direct coupled propeller shaft and universal joint to a double reduction heavy duty automotive-type rear axle, or by a duplex primary chain with a final gear reduction to the drive wheel.

21 Battery systems

Applications

There are many differing applications for batteries in today's industrial and commercial world. The type of battery system selected is dependent on the duties of the application. These fall into four basic categories as set out below:

(i) Stationary – batteries in a fixed location, not habitually moved from place to place, and which are permanently connected to the load and to the d.c. supply. One of the most important categories within this group is batteries used in the provision of UPS systems but it also includes batteries used in the operation of electrical switchgear and batteries for telecommunications equipment.
(ii) Traction – batteries used as power sources for electric vehicles or materials handling equipment.
(iii) Starter – batteries for starting, lighting, and auxiliary equipment for internal combustion engine vehicles, e.g. cars, commercial or industrial vehicles.
(iv) Batteries of the valve regulated lead-acid or sealed nickel-cadmium type for cyclic application in portable equipment, tools, toys, etc.

Two types of cells can normally be considered for most of the above applications: lead-acid and nickel-cadmium. The latter has one clear advantage, this is its ability to be left idle for long periods in any state of charge coupled with its rapid recovery after neglect by topping up and recharging.

This advantage is usually strongly outweighed by the disadvantages, namely:

(i) It is not possible to check the state of charge without carrying out a discharge test.
(ii) The electrolyte requires replacement several times within the life of a battery, otherwise the high rate discharge capability is greatly impaired.
(iii) The internal resistance is higher than for a lead-acid cell. This is a significant disadvantage for motor starting applications. For telecommunications applications it can lead to greater levels of background noise.
(iv) The cost of nickel-cadmium is greater than lead-acid of equivalent performance.
(v) Maintenance costs for nickel-cadmium tend to be higher than for lead-acid because of the greater number of cells required to achieve a given voltage and the greater need for cleanliness to prevent tracking between poles due to close spacing.

Lead-Acid Batteries

Operating principles. The active components of lead-acid cells are the lead dioxide, PbO_2, of the positive plates (chocolate brown), the spongy lead, Pb, of the negative plates (grey) and the sulphuric acid, H_2SO_4, of the electrolyte. During discharge there is a partial reduction of the positive-plate material with oxidation of the negative-plate material and combination of the product in each case with sulphuric acid. The result is a transformation of part of the material of both plates into lead sulphate, $PbSO_4$, accompanied by a lowering in concentration of the electrolyte. The reaction may be summarized as:

$$PbO_2 + Pb + 2H_2SO_4 \rightarrow 2PbSO_4 + 2H_2O$$

This means that the lead dioxide of the positive plate and the sponge lead of the negative plate react with the acid in the electrolyte to form lead sulphate and water. The discharge reaction is electrochemically 100% efficient. This applies to both discharge through an external load or self-discharge, i.e. internally within the cell.

The recharge reaction is the reverse of discharge, and may be summarized thus:

$$2PbSO_4 + 2H_2O \rightarrow PbO_2 + Pb + 2H_2SO_4$$

In this case lead sulphate on both the positive and negative plates reacts with water to form lead dioxide and sponge lead on the positive and negative plates respectively, with acid in the electrolyte.

The efficiency of the recharge action can be 100% but this is related to the state of charge of the battery or cell. When the recharge reaction becomes less than 100% efficient, the gassing reaction (electrolysis of water) takes place. This may be summarized as:

$$2H_2O \rightarrow 2H_2 + O_2$$

In this case electrolysis of water takes place with an evolution of hydrogen from the negative plate and oxygen from the positive plate. The variation of recharge efficiency with state of charge is generally as follows:

(i) Above typically 80% state of charge (SOC), the required ratio of recharge in ampere-hours to discharge in ampere-hours is of the order of $9:1$.
(ii) This ratio is capable of fully recharging a lead-acid battery provided the only discharge reaction occurring between the 80% SOC and the fully charged state is that arising from the self-discharge process.
(iii) If the reservation in (ii) does not apply, then it is highly probable that the battery will not be able to recharge and will never reach its fully charged state.

Cell construction. Lead-acid batteries appear in three generic types of construction. Each type reflects the type of positive plate used in the cell, and it is the type of plate which in turn determines the essential electrical characteristics of the product. The three types of construction, illustrated in *Figure 21.1*, are:

(i) The Planté cell.
(ii) The pasted plate cell.
(iii) The tubular cell.

The Planté cell. The Planté positive plate is a thick sheet of lamellated pure lead on to which a very fine layer of active material has been deposited.

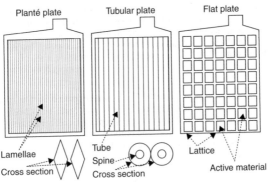

Figure 21.1 Types of lead-acid positive plates

The purpose of the lamellated surface is to create a large surface area within the plates, and this in turn relates to the capacity of the plate.

The plates are 6–8 mm thick overall, and have a long life, but the energy density of the cell is low, except at very high rates of discharge. The construction of the Planté cell allows the cell to have a generous volume of electrolyte, which requires a low electrolyte specific gravity (1.220 kg/l).

Pasted or flat plate cell. The pasted or flat positive plate comprises a lead alloy frame with an internal lattice or grid, which is filled with active material. The word 'flat plate' describes what it looks like and 'pasted' describes how it is manufactured.

Some flat plate cells are described as lead-antimony cells. This is because of the alloy used in the plate frame and lattice. The antimony hardens the lead and its content for the frame and lattice varies between 0.5 and 6.0% antimony, though for long life and low maintenance periods the antimony level is kept as low as possible. Other flat plate cells use an alloy containing up to 1% calcium for the frame and lattice.

In comparison to the Planté cell the surface area of the pasted plate is much higher and this is reflected in the higher energy density, particularly at the longer rates of discharge. This higher energy density requires electrolyte specific gravities of the order of 1.240 kg/l.

Tubular positive cells. The tubular positive plate is a fabric gauntlet of tubes containing active material and conducting lead-antimony spines. Antimony content of the spine alloy varies between 0.5 and 8.0%. Tubular positive plates achieve similar surface areas to pasted plates and therefore similar energy densities are obtained, again particularly at the longer rates of discharge. This also requires that the electrolyte specific gravity be of the order of 1.240 kg/l.

A battery is built from a series of alternating negative and positive plates assembled within a plastic container made from a material such as styrene acrylonitrile. *Separators* are placed between alternate plates and these may simply serve to keep the plates apart or, in some types of construction, they may help to retain the active material within the plate structure. The most commonly used separator material is microporous PVC. This has a high degree of diffusability, low electrical resistance in acid and high durability.

Cell venting arrangements – vented cells. As indicated above, towards the end of the charging cycle of a battery oxygen and hydrogen are evolved and these gases are generally released to the atmosphere. This represents a loss of water from the electrolyte making it necessary to periodically 'top up' the cells with distilled water. The simplest venting arrangement consists of a removable vent plug which allows gas to escape and can be removed to enable topping up to be carried out.

Valve regulated lead-acid cells. These utilize pasted or flat plates with frame and lattice in either lead-calcium alloy or pure lead. Their main feature is that the vent plug of the conventional cell is replaced with a Bunsen-type valve typically as shown in *Figure 21.2*. The purpose of the valve is to encourage the recombination within the cell of the electrolysis products and thus reduce the loss of water while at the same time releasing to atmosphere small quantities of excess gas produced during normal operation. The design of the valve also prevents ambient atmosphere from entering the cell. Valve regulated cells use a micro glass-fibre separator, which absorbs the electrolyte, and by careful adjustment of the degree of saturation, can assist the recombination process.

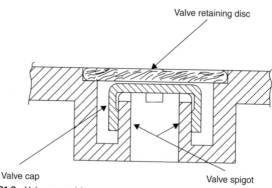

Valve retaining disc

Valve cap

Valve spigot

Figure 21.2 Valve assembly

It should be noted that very often valve regulated lead-acid cells are referred to as sealed lead-acid cells. The use of the term valve regulated, however, is an attempt to draw attention to the fact that this type of cell is not hermetically sealed and the term sealed-cell should therefore be avoided.

Characteristics

Charging and discharging. Stationary batteries, unlike the other three principal types listed above, generally operate permanently connected to their charger. In this type of installation the charger operates in such a manner as to maintain the battery constantly in a fully charged condition. This is termed operation on 'float charge'. There are two subcategories for this type of battery and charger arrangement:

(i) Standby applications.
(ii) Working applications.

In the former, the main purpose of the battery is to maintain supplies to vital equipment in the event of loss of the main supply of power. This might be for an uninterruptible power supply system (dealt with in more detail in Chapter 7) or it might simply be to maintain supplies for communications equipment or emergency lighting for which a short interruption might be acceptable. With this type of application there is sufficient time to return the battery to a fully charged condition following the restoration of mains supplies.

In the latter case, the duty of the battery involves primarily 'load stabilization'. That is, the battery must supply sudden short demands such as the operation of switchgear solenoids or the starting of d.c. drives. This could also include meeting the demands of a system supplied from solar energy for which long-term availability is fairly predictable but for which short-term predictability is uncertain. For this type of application the main design criterion is to ensure that the battery can be maintained in a fully charged state for sufficient proportion of the time to maintain it in good condition. In the case of a battery for switchgear operation the solution may be to give it a regular freshening or 'boost-charge'.

Selection of battery type. For larger stationary applications of the standby type identified above, Planté type cells have a clear advantage in that their performance under high discharge conditions for relatively short periods, say up to one hour, is greatly superior to that of pasted plate or tubular cells. Typical rates of discharge are from 5 min up to 30 min, with the most popular requirement for a standby period of 15 min.

Valve regulated cells can be very attractive for many stationary batteries, particularly for telecommunications and similar applications where their virtual leak-proof characteristics enable them to be installed complete with their chargers within the panels housing the electronic equipment for which they provide a standby supply. None of the special provisions normally necessary for battery rooms are required in this case.

For the other, non stationary, applications pasted plate or tubular cells may be used with special features incorporated to meet specific demands of the duty.

For example, engine starting batteries usually employ pasted plates. The active material is held in lead-calcium-tin alloy grids. The grids are pasted with lead oxide-sulphuric acid pastes which are afterwards converted electrolytically to the required active materials. Plate thicknesses depend upon the work for which they are designed and vary from about 1.5 mm for aircraft applications to about 2 mm for car starting and about 5 mm for starting diesel locomotive engines.

Because of the high currents required for starting engines the plates are arranged as close as possible to each other to keep the internal resistance to a minimum. This is achieved by using very thin diaphragm separators made from microporous PVC or rubber, occasionally with glass wool mats. Such batteries are assembled in polypropylene containers, for car batteries, or hard rubber-based materials where greater durability is required.

For motive power duties such as fork-lift trucks, milk floats and coal mine locomotives, heavier plates combined with devices to assist in retaining the paste on the positive grid are employed. Cells containing heavy flat plate positives, up to approximately 6 mm thick, usually employ glass wool mats together with microporous PVC or rubber separators of substantial thickness and pasted negatives of up to approximately 5 mm thick in a compact assembly. However, the vast majority of these batteries in the UK now use cells containing tubular positive plates up to 9 mm thick separated from flat pasted negative plates by their microporous envelopes of either a multi-tubular sleeve of Terylene or individual tubes of glass fibre with an outer layer of perforated PVC. The space between the tube (retainer) and lead alloy spine (conductor) is packed tight with positive material. This type of plate construction gives high mechanical strength and a high degree of porosity for acid penetration.

Costs and service life. In very general terms, comparison of cost between the three generic types of lead-acid cell, used in stationary batteries operated in a float charge situation, is given in *Table 21.1*. The principal point to note here is that if costs are amortized over the expected life of the cell, then the preferred order would be first the Planté cell, then the tubular, and finally the pasted plate cell. In a real situation, however, the pasted plate cell can be the preferred option for many applications. The reason for this is that service life requirements in most UPS and telecommunications installations do not exceed 10 years and in many instances may be less. Under these circumstances amortizing the cost of the cell over its expected service life is unimportant, and selection is determined by the product with the lowest cost.

Table 21.1 Cost comparison of lead-acid cells

Type	Plante	Tubular	Pasted plate
Capacity (Ah)	400	400	400
Internal cost*	100	80	75
Life (years)[†]	20	14	12
Annual cost	5	5.7	6.3

*The cost figures are arbitary units with the Plante cell attributed with a value of 100.

[†]The life values assume operation in a true floating battery situation. The significance of this observation becomes clear when, later in the chapter, UPS systems are considered which do not operate the battery in a true float situation.

Installation. When stands are supplied with a lead-acid battery they consist of two or more epoxy-coated steel end frames, together with longitudinal runners and tie bars. Before commencing erection, smear all the screw heads with petroleum jelly. Then follow the manufacturer's instructions regarding the method of erecting the stand. Usually rubber channels are provided upon which the individual cells stand and these should be placed in position and the stand height adjusted so that it is level. Ensure all the boxes and lids of the cells are thoroughly clean and dry. Also clean the flat contact-making surfaces of the terminal pillars. If there is evidence that acid has been spilled the whole length of the pillars should be wiped with a rag which has been dipped in a non-caustic alkali solution, preferably dilute ammonia. This will neutralize any acid on these parts. Do not allow any of this solution to get into the cells.

Wipe the pillars clean and dry and cover the whole length of each pillar down to the lid with petroleum jelly, e.g. Vaseline. Do not scrape the surface of any lead-plated copper pillar or connector because the plating may be damaged.

For cells in self-contained cabinets installation is simply a matter of placing the cells on the plastic sheets on the shelves according to the maker's instructions, starting with the lower shelves to ensure stability of the cabinet. When all the cells are in position the connecting straps may be installed.

First charge. Manufacturer's instructions must be adhered to when charging batteries, and it is only possible to consider the important points to be observed. Very pure acid must be used in this connection.

Dry and uncharged batteries have to be given a first charge in which the strength of acid for filling, time of soaking before commencing charge, charging rate and temperature on charge are always carefully specified. A faulty cell may show up in the first charge by its tardiness in reaching the gassing stage and by overheating. The fully charged readings and strength of electrolyte for final adjustment vary with different types, but first charge is never complete until the voltage (on charge) is in the region of 2.6–2.7 V per cell and all the plates are gassing freely.

Once the strength of the electrolyte has been adjusted at this stage, no further adjustment should be necessary throughout the life of the cell. Many automotive batteries are dry charged and only require filling with electrolyte of the appropriate relative density before being put into service. Vent plugs should be kept in place during first charging and also recharging and should only be removed for topping up and other maintenance operations.

Maintenance. Batteries and battery equipment should be kept scrupulously clean, as impurities inside and moisture outside the battery cause rapid self-discharge and deterioration. Vent plugs must be free from obstruction to avoid internal gas pressure and naked lights or sparking near batteries will ignite the evolved gases, causing violent explosions. Never allow the electrolyte to get so low that either the separators or plate tops are uncovered. Mechanical damage follows the chemical changes that occur under these conditions. Topping up is best done when cells are on recharge or on float charge and only distilled or deionized water should be used. Lower than normal relative densities indicate that the cell or battery is not in a fully charged state. A refreshing charge should be given in these circumstances to ensure a fully charged condition. Low densities in odd cells are an indication of internal short-circuit, or the cells could have been left open-circuited for a long period of time. Many modern designs are of the sealed variety and no maintenance of the electrolyte is necessary.

Maintenance testing. Where the importance of the battery warrants, those responsible for its maintenance should periodically test its condition. In order to do this properly it should be tested on the equipment for which it is intended. A typical example of this would be where a battery is required to supply power for emergency lighting for a specific period or is required for starting up an engine. If a battery is supplying power for emergency lighting, then it should supply the necessary power for the appropriate standby time, and where it is required to start an engine, test starting should be carried out at regular intervals.

Cell voltages. 1. There is little value attached to cell voltage readings when no current is passing through the cells.

2. When a lead-acid cell voltage on charge does not rise above 2.3 or 2.4 V after all the others are up at about 2.7 V per cell, that cell may have an internal short-circuit. These short-circuits are often hard to find, but the

voltage shows it up at once. Rapid cell-voltage drop will occur if there is a short-circuit, particularly on discharge or in service, and this will invariably be accompanied by a fall in the relative density of the electrolyte. Such short-circuits may also be detected on batteries that are on float charge by a cell that is short-circuited having a lower floating voltage than the others. It will also be accompanied by a lower relative density.

3. If the voltage of any lead cell rises to 3 V when first put on charge, with recommended charge rate flowing, it could well be sulphated and must be charged at a very low rate until its voltage has fallen and then commences to rise again.

4. Do not attempt to add acid to any lead-acid cell that appears to contain only water. Put it on charge at about half normal rate for a time. If then the voltage is seen to rise slowly, the rate may be increased slowly, for the cell is only overdischarged. Only adjust the relative density of the acid after charging has ended; this will be indicated by the relative density and cell voltage remaining constant for three consecutive hourly readings.

Additional information. For standby applications where many cells may be used for a battery, it is best to have a battery room or cabinet in which the cells may be placed. Checks should be carried out at regular intervals to ensure that electrolyte levels are maintained, the battery is kept fully charged and the charging equipment is in good order. Each battery should have its own hydrometer and instruction booklet, together with a quantity of Vaseline or petroleum jelly to keep the cell connections well protected. Where the battery has to perform an important function, a thermometer, record book and cell-testing voltmeter are useful additions. Electrolyte examination and testing is not required or possible on sealed designs, sometimes called 'maintenance free'. However, cell connections need checking as outlined above.

For motive power uses, the operator should be conversant with the maintenance operations required, i.e. the need for regular topping up and adequate recharging following the battery discharge. Again a hydrometer should be available and a cell-testing voltmeter, thermometer and record book are valuable aids.

For automotive applications, it is necessary to keep the battery electrolyte level correct and occasionally check the relative density with a hydrometer.

For all applications, a supply of distilled water or deionized water should be available. Tap water should never be used for topping up cells as many domestic water supplies contain additives and impurities that may be harmful.

Where both lead-acid and alkaline cells are employed, it is best to have separate battery rooms and under no account should an alkali be added to a lead-acid cell or vice versa. No equipment such as hydrometer or thermometer should be used on both lead-acid and alkaline batteries to avoid contamination.

Nickel-Cadmium Alkaline Cells

Chemical principle. The active material of the positive plate consists of nickel hydrate with a conducting admixture of pure graphite. The active material of the negative plate is cadmium oxide with an admixture of a special oxide of iron. The electrolyte is generally a 21% solution of potassium hydroxide of a high degree of purity.

For the purposes of practical battery operation the reaction may be written as follows:

$$2NiOOH + 2H_2O + Cd \; discharge \rightarrow 2Ni(OH)_2$$

$$+ Cd(OH)_2 \leftarrow charge$$

In a fully charged battery the nickel hydrate is at a high degree of oxidation and the negative material is reduced to pure cadmium; on discharge the nickel hydrate is reduced to a lower degree of oxidation and the cadmium in the negative plate is oxidized.

The reaction thus consists of the transfer of oxygen from one plate to the other and the electrolyte acts as an ionized conductor only and does not react with either plate in any way. The relative density does not change throughout charge or discharge. At normal temperature the cell is inert on open-circuit.

Construction. There are a number of different forms of construction for nickel-cadmium cells. In the nickel-cadmium pocket plate cell the positive material consists of a mixture of nickel hydroxide and graphite. The mixture is compressed into flat pellets and supported in channels made from perforated nickel-plated steel strips. The negative plate is similarly supported cadmium powder with an admixture of iron or nickel. The electrolyte is an aqueous solution of potassium hydroxide. With metal containers, the individual cells are insulated by spacers; many are available with plastics containers.

Nickel-cadmium sintered plate cells use a highly porous nickel matrix produced by sintering special types of nickel powder. The porous plate, which is usually strengthened by a reinforcing grid of nickel or nickel-plated steel, contains about 80% voids into which the active materials, nickel and cadmium hydroxides respectively, are introduced. Intimate contact is thus maintained between the active material and a large surface of metallic nickel, to provide the necessary conductivity. Sintered-plate cells are more costly than pocket-plate cells and are therefore used mainly for duties requiring a short-time high discharge rate. They occupy less volume than the equivalent capacity pocket-plate types. Sealed nickel-cadmium cells are only available with relatively low ampere-hour capacities and find wide application for standby duties, particularly emergency lighting. Recharging of these cells without building up dangerous internal pressures requires the negative plate to have a larger capacity than the positive. Both pocket-plate and sintered-plate cells are available in sealed designs.

Characteristics. The basic characteristic of the nickel-cadmium battery is extreme flexibility in regard to electrical and mechanical conditions. The battery will accept high or low rate charges and will deliver high or low currents, or work under any combination of these conditions. It can stand idle for long periods in any state of charge and will operate within very close voltage limits.

Types of battery. Two main types of battery are available:

(i) The normal resistance type which is used where the full ampere-hour capacity of the battery is regularly or occasionally required, e.g. traction and standby lighting.

(ii) The high performance type which is used where ampere-hour capacity is seldom required but where the battery has to supply high currents for short periods, e.g. engine starting and switchgear operation.

Charging and discharging. The 'normal' charge rate for all types of nickel-cadmium battery is C/5 for 7 hours where C is the rated capacity of the battery, which will restore a discharged battery in 7 hours. The charging rate and time can, however, be varied according to the duty.

Because of the all steel construction and the enclosed active materials, batteries may be safely discharged at high rates up to short-circuit without damage. The average discharge voltage is 1.2 V per cell at normal currents.

Maintenance. Only very simple maintenance is required for nickel-cadmium batteries, but for optimum performance it is important that this is regularly and conscientiously carried out. It is only necessary to keep the cells clean and dry and top up as required with distilled water.

Battery Charging

It will be apparent from the foregoing description of battery systems that two basic options for battery charging equipment exist:

(i) Those used to maintain stationary batteries in a satisfactory state of float charge.
(ii) Those required for recharging traction, starter and portable batteries following use.

Float charging of stationary batteries. For this type the selection of the charger is dependent upon the size and type of the load and the size of the battery. A number of suboptions are available within this category. These are:

1. Response mode operation. This is shown diagrammatically in *Figure 21.3*. With this arrangement the load is not normally connected to the battery. When the normal power supply fails the load is connected to the battery with or without interruption of power to the load.

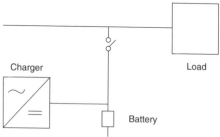

Figure 21.3 Typical response mode operation

In this mode it is only necessary that the charger provides the voltage and current required to maintain the battery on float charge.

2. Parallel mode operation. This is shown in *Figure 21.4*. The load is permanently connected to the charger and the battery so that when the power supply fails the battery supports the load without interruption. Within the

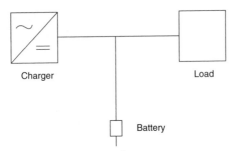

Figure 21.4 Typical parallel mode operation

parallel mode of operation there can be various operating regimes determined by the 'fine tuning' of the system. In the first, normally referred to as 'standby' operation, the charger is able to supply at all times the current required by the battery and the load. The battery is therefore always maintained in a fully charged condition. A variation on the above occurs when the charger is only able to supply *on average* all the current required by the load and battery including the recharge current. In this case the battery is kept fully charged provided the average float current plus average load current remains within the capability of the charger. In the condition referred to as 'buffer' operation high amplitude cycles of load current result in the battery not being able to efficiently accept all the float current during the times that this is available. In this condition the battery is not charged at all times, and this can lead to a 'walkdown' in available capacity. One solution to this problem is to give the battery an occasional offline boost charge.

Charging of traction, starter, and portable batteries. For these types of batteries charging is from a separate dedicated source usually a transformer/rectifier. In the case of electric vehicles this may be carried on the vehicle itself. With lead-acid batteries performance and battery life are very dependent on an efficient charging system. Three forms of charger are in general use, the single step taper charger, the two-step taper charger and the pulse-control charger. With the single step charger, the voltage is held constant just below the battery gassing point, 2.4 volts per cell. When the battery voltage reaches this point, the current is allowed to decay until the battery becomes fully charged. With the two-step process, the first stage follows the same line as with the single step; the current is then reduced to a predetermined value and charging continues for a predetermined period. The pulse type of charger is devised more to keep the battery fully charged. When the main charge is stopped, the battery receives short pulses of charge, controlled by the decay in the battery voltage during the intervening open-circuit period.

Multi-circuit charging. Multi-circuit charging will be considered from its most common application – that of the recharging of battery electric trucks. A typical charging system consists of one or more chargers supplying d.c. to a common busbar and a number of panels, each one being designed to charge a single battery. This system is usually called a 'modified constant potential' system.

Source of power may either be a motor-generator set or a rectifier system. The characteristic must be the provision of a constant voltage throughout the load range within a tolerance of ±3%, and a current capacity equal to the sum of the currents drawn by all the individual charging panels. While there is no limit to the number of batteries which can be charged from one machine it is usual to use several chargers in parallel on one d.c. busbar, or separate the system into several groups, if more than 20 or 30 batteries are to be charged.

When using a single d.c. busbar it is desirable that all batteries consist of the same number of cells and be of the same general type, either lead-acid or alkaline. The ampere-hour rating of the individual batteries, however, may vary.

The basic circuit of a panel is simply a ballast resistor in series with the battery across the d.c. busbar.

Reference Documents

The following reference documents contain much useful information:

BS 6133 Code of Practice for the safe operation of lead acid station-ary batteries

BS 6132 Code of Practice for the safe operation of alkaline secondary cells and batteries

BS EN 50272 Safety requirements for secondary batteries and battery installation, Part 2: Stationary batteries.

22 Cable management systems

Because of the dramatic increase in the electrical and electronic requirements of all types of buildings, domestic, commercial and industrial, has come the demand for the provision of safe yet adaptable wiring systems enabling sections of them to be easily altered without affecting the supply to the remainder. This is essential for those networks supplying so-called electronic offices which have to cater for data, voice and image communications facilities, building services, lighting, power, earthing and lightning protection.

Electronic aids are not only confined to offices but extend to include other buildings like hospitals, retail outlets, hotels and industrial establishments. Each building has different cabling requirements and the density of cabling and degree of versatility needed may vary considerably. This must be reflected in the design of the cable management system enabling it to be easily adapted to meet a wide variety of supply possibilities with the minimum of effort and change of components.

One would expect frequent changes in modern commercial offices, particularly financial establishments containing many computer terminals, telecoms and data facilities but at the other end of the spectrum hotels may only require changes in the offices and reception area.

The growth of cabling systems has been brought about by the rapid increase of information technology (IT). To prevent the generation of masses of loose cables which become unmanageable, manufacturers of electrical distribution equipment have designed integrated systems using cable trunking, conduit, cable trays and ladders to house and support the cabling. These are known generically as 'cable management systems'. They take many forms and provide the user with a versatile facility, well able to cater for frequent changes of the supply network. At the same time they are safe even when being altered.

Integrated Systems

There is no recognized definition of an integrated cable management system but the author suggests the following probably covers it adequately:

> An integrated cable management system is a combination of cable enclosure and or support equipment (i.e. conduit, trunking, floorboxes or other outlets, ladders or trays), so designed as to form an integrated arrangement and being capable of minor changes without having to shut down the entire network or endangering personnel working on it.

One must stress at this point that a most important function is for the system to be 'integrated'. By this is meant that each piece of equipment joins on to the next without any gaps, loose joints, etc. so that the cabling is protected in every part of the network, from the point of supply to the point of use. Furthermore, should the user want to alter any part of the installation it should be a comparatively simple operation: for example, if the position of computer terminals needs to be changed around in an office, or part of an office, this should be possible without disconnecting all the other terminals being served by the same network.

The above definition does not cover some special systems like flat cable and cable harness methods which are being used to perform the same function equally well, and these are described later in the chapter.

Compliance with international, IEC, or British Standards for the products comprising an integrated system is extremely important, because there are many inferior items available on the British market and to introduce any, even as a small part of an installation, will jeopardize the safety of the whole. In particular, an integrated system should meet, wherever relevant, the requirements of BS 7671, formerly the IEE Wiring Regulations.

The different components forming an integrated system are described below.

Conduits

There are three types of conduit systems from which to choose: metal, insulating and composite. Metal conduit is generally steel; insulating conduit has no conductive components and is usually some form of uPVC; composite conduit comprises both metal and insulating material and is generally flexible. We do not highlight all the advantages and disadvantages of these or other integrated systems components here. For a fuller description of all the different types of cable management systems the reader is referred to *Cable Management Systems* by E. A. Reeves, available from Blackwell Science in Oxford, from which the information in this chapter has been extracted.

When considering the purchase of any conduit system, and the same goes for other cable management arrangements, it is wise to ensure that the manufacturer can supply all the necessary accessories required such as bends, junction boxes, outlet boxes, etc.

Metal conduit. This may be steel or stainless steel, the latter being used in sterile areas like food processing. Reputable suppliers generally belong to the British Electrical Systems Association (BESA) and also possibly the British Welded Steel Tube Manufacturers' Association. Where mechanical strength is required many users opt for steel, but there is always the risk of corrosion if the coating is damaged in any way.

Most manufacturers concentrate on supplying Class 2, black stoving enamel on the inside and outside, and Class 4 (*Figure 22.1*) which is hot-dipped galvanized on both inside and outside. Both can be provided in light gauge (about 1 mm thick walls) and heavy gauge (up to 2 mm thick walls).

Insulating conduit. This is often called plastics conduit and covers an enormous range of different materials but here we confine the discussion to conduits made from unplasticized polyvinylchloride (uPVC). Round rigid conduit is available as light or heavy gauge in sizes from 16 to 50 mm diameter. No threading is required as with metallic varieties for joining adjacent sections together a special adhesive being used. White or black finish is available. Being made from plastics there is no danger of corrosion. High impact types withstand the rough handling on site. Oval sections provide a shallower profile where necessary.

Figure 22.1 Class 2 and Class 4 metal conduit has screwed ends, with one end protected by a coupler and the other by a cap (Burn Tubes)

Channelling. Sometimes called capping, it can be employed for switch drops and cable buried in plaster to provide protection against damage by nails knocked in the wall during the lifetime of the installation.

Flexible conduit. Flexible conduit is defined in BS EN 50086 Part 1 as 'A pliable conduit which can be bent by hand with a reasonable small force, but without any assistance, and which is intended to flex frequently throughout its life'. Pliable conduit is defined as a conduit which can be bent by hand with a reasonable force but without other assistance. However, it is important to realize that pliable conduit must never be used where continual flexing is experienced.

Flexible conduit is available in steel or plastics forms, and particularly with steel, in many varieties. Liquidtight flexible conduit is usually of steel core with a PVC sheath, and employing specially designed connectors to provide an IP67 degree of protection.

Cable trunking systems

In this section we are dealing only with trunking that houses and protects cables linking points of an electrical network together. Perimeter, bench, mini, and poles and posts are dealt with under their own headings. Cable trunking under this definition is available in both metal and plastics materials. Three types of metal trunking are marketed, steel, stainless steel and aluminium and where a non-flammable material is specified the choice lies between these three types. Toxic gases are released by some plastics materials when subjected to flame but there are halogen-free designs of insulating trunking. Some plastics are flame retardant while others do not burn once the flame is removed. Heat resistance capability might be very important for trunking housing fire alarm circuit cables which need to operate for as long as possible in the event of a fire. Aluminium trunking is claimed to protect data cables from interference.

In any long vertical trunking runs, particularly where these pass through floors or exceed about 5 m, then some kind of fire barrier must be incorporated inside. This is dealt with in more detail in the book referred to earlier.

Steel trunking is available in 23 sizes from 38×38 mm up to 300×300 mm as laid down in BS 4678 Part 1. Plastics trunking generally follows the same size range. One must bear in mind when comparing plastics and metal cable management systems and their component parts that one big advantage of plastics designs is their lightness compared to metal. Trunking and conduit are also easier to cut to length and join together.

Multi-compartment trunking. Both metal and plastics trunking are available in multi-compartment designs (*Figure 22.2*) to provide complete segregation of cables in the individual compartments. Metal trunking has the added advantage of shielding cables in adjacent compartments.

Figure 22.2 Ega range of uPVC cable trunking (MK Cable Management)

Some designs of multi-compartment trunking can accept outlets on the lids, transforming them to perimeter trunking systems suitable for mounting at either floor or dado level.

Components and fittings. Lids for plastics trunking are usually of the clip-on type while for metal designs some kind of mechanical fixing like a turnbuckle is employed. It is important when ordering trunking to ensure that a full range of fittings is available including angles, tees, corners, crossover units, cable retainers, connectors, flange couplers, reducers, adaptors, etc.

Earth continuity. One of the claimed advantages of metallic cable management systems is that the housing can be used as the earth continuity conductor (e.c.c.) thus saving on cabling space. BS 7671 permits this but there is always the danger of a break in the circuit due to corrosion or a faulty joint at a fitting and many installers prefer to run separate e.c.c. inside the trunking.

Perimeter trunking systems

Perimeter trunking systems include skirting and dado (sometimes called undersill) and are made from plastics (*Figure 22.3*), steel or aluminium. Steel and plastics systems are covered by BS 4678: Parts 1 and 4, respectively, although neither specifically deals with perimeter systems *per se*. Perimeter systems are

Figure 22.3 Sterling multi-compartment plastics perimeter trunking is available in six profile variations and three colours and is suitable for skirting and dado mounting (Marshall-Tufflex)

suitable for domestic, commercial and industrial installations and are ideal for refurbishment applications.

Basically the systems are designed to house power lighting, telecommunications and data cables but there are some hybrid arrangements that incorporate other services as well, such as compressed air or heating pipes. One design combines ducting with a protection rail. Multi-compartment arrangements are the norm, with the different services occupying separate, segregated compartments. This enables work to be carried out on one of the services without interfering with the others. Some multiple socket-outlets can be wall-mounted and are therefore, in a sense, perimeter systems, although they are of limited length and are not in the same category as true perimeter systems.

Generally trunking systems are wired with cable but there are some designs which incorporate busbars used to power the socket-outlets.

Cornice trunking is available and provides an ideal solution to wiring in buildings with concrete floors and ceilings. Manufacturers tend to design it to complement a particular range of their own standard trunking and this needs to be recognized.

Screened compartments. For some computer and data circuits screened compartment systems are available for plastics systems. They may be built into the trunking body or supplied as an extra metallic compartment which fixes on to a standard design. Metallic systems do not need special treatment except where the compartment dividers do not form a complete enclosure.

Socket-outlet positions. One of the advantages of perimeter systems is the facility to fit as many outlet points as required, anywhere along the length, allowing for a minimum distance between adjacent outlets. The position of the outlet is either on the trunking lid itself or on an extension piece mounted above the trunking, which raises it to a level from the floor recommended by BS 7671.

Shapes. Two design shapes are popular, either a rectangular section or one incorporating a chamfered cover plate. The chamfer may be symmetrical, i.e. both the top and bottom of the cover are chamfered or the angle may be confined to the top part of the trunking only. All are of narrow depth with increased cable capacity provided by increasing the height. Three-compartment design is popular with the centre compartment being the largest and the top and bottom compartments generally having the same dimensions.

Miniature trunking

Miniature trunking systems (shortened to mini-trunking) were originally introduced for the refurbishment of domestic dwellings and light commercial establishments. Being surface-mounted and made from plastics material enabled quick installation, it not being unknown to rewire a flat or small house within one day. But since then its application has spread to new buildings, not only for housing power and lighting cable but also telecommunications, alarms and data systems. All mini-trunking is characterized by a low profile, some products being as shallow as 7.5 mm but the usual being of the order of 12.5 to 16 mm. A typical product is shown as *Figure 22.4.*

Figure 22.4 A mini-trunking system showing joints, tee-pieces and other fittings, together with a switchbox accessory. Close-fitting joints provide a high level of safety (Legrand Electric)

Two methods of fixing are available either by screws (or tacks) or by adhesive, the latter having a peel-off backing paper. Once the adhesive design is fixed to the wall or ceiling it cannot be adjusted as is possible with mechanical fixing methods. One design has clip-on components which further speed up installation.

To give some idea of the sizes of this trunking one manufacturer provides eight different sections ranging from 16 mm square up to 50 mm wide by 32 mm deep. Although most are of single-compartment construction there are a limited number of two-compartment designs on the market.

All lids are of the clip-on variety and are either flush with the trunking sides or overlap them. At least one supplier has a double-locking clip-on lid to provide security if it is needed.

Bench trunking systems

Bench trunking is not defined by BS 4678 although the trunking standard is used as a guide by manufacturers. The term describes a triangular-shaped metal

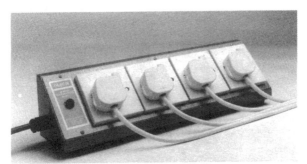

Figure 22.5 From an extensive range suitable for bench, wall or floor mounting, this unit is a steel trunking finished in two-tone colour, with four socket-outlets, 2 m of cable and a 13 A fused plug (Olsen Electronics)

or plastics trunking with the sloping face carrying socket-outlets (*Figure 22.5*), although a rectangular trunking section with multiple socket-outlets, when fitted to a bench, performs the same function. Where necessary the trunking should also comply with BS EN 60529 dealing with ingress of foreign bodies, i.e. an installation may require an IP number.

Bench trunking can be hard wired or may incorporate busbars, and usually comes in specific lengths although these may not be the same from individual manufacturers. Some designs can be compartmented to segregate different services and others may come prewired.

Residual current devices. To protect against earth faults many manufacturers fit residual current devices as standard in their products thus meeting the requirements of BS 7671.

Clean supplies. A special design of bench trunking protects sensitive equipment from transients, spikes, switching surges, radio frequency interference and lightning surges. It can be used to supply computers and other sensitive data equipment. The trunking system may be called a power conditioner and as such is covered by BS EN 60950.

Service poles and posts

Service poles and posts have been developed to meet the demand for concentrated power, telecommunications and data outlet facilities in open-plan offices, isolated workstations, reception areas and financial institutions (*Figure 22.6*). Power requirements are usually quite small and so they may be supplied from a floor outlet box of a platform floor installation. Basically poles and posts are floor-standing multi-compartment trunking sections carrying outlets on one or more faces. Service poles may be fed from the floor (as indicated) or from the ceiling, while posts, being truncated versions of poles, are generally supplied from a floor box.

Virtually all designs consist of an extruded aluminium spine to provide the necessary stiffness, with clip-on covers over formed channels, either of metal or plastics, carrying the various outlets. Some poles may also include one or more of the following: lighting switches, circuit-breakers, luminaires, fire alarm break glass units or other items, increasing their versatility.

Figure 22.6 Single-sided telescopic aluminium multi-compartment service poles (one post also shown), are suitable for mounting against walls. The Prefadis C9000 range is also available in a double-sided version. Most poles are fitted with a 16 A miniature circuit-breaker as standard (Telemecanique)

Although square or rectangular sections are the norm there are part-circular and elliptical designs with special provision for carrying the socket-outlets.

Cavity floor systems

Several terms are used to describe cabling systems that are installed in the space underneath a false floor, or sometimes flush with the floor. Access, raised access, platform, raised and cavity floor are some examples. They may be used to describe the complete system, i.e. cabling and housing, or either the cabling or the trunking. There is no British Standard covering these systems but the Property Services Agency document MOB PF2 PS *Platform floors (raised access floors). Performance Specification* is the standard that reputable suppliers work to. In this section we use the term cavity floor system to describe the complete electrical distribution network serving floor outlet boxes or other facilities providing power, data, telecom or computer services to terminal equipment.

Adjustable jacks are used to support the false floor and provide the space between the screed and the underside of the flooring for the services cabling. This system is often called a pedestal design, but there are flooring systems that do not rely on adjustable jacks to provide the space for the cabling and these are termed non-pedestal designs.

Non-pedestal designs. Floor panels are located close to the structural floor and individual panels are often meant to be permanent fixtures so that service outlet positions are therefore usually decided at the design stage thereby restricting the versatility of the installation. The depth of the floor box is also critical and this will restrict the choice available to the user. Fluted deck and batten type are two arrangements of non-pedestal design, the latter providing some degree of versatility after installation.

One of the biggest advantages of these systems is where there is a limited floor to ceiling space which cannot be altered, i.e. for refurbishment projects.

Pedestal designs. Any depth of floor void can be provided within reason, from 40 mm up to 1000 mm or even greater in special cases, where perhaps other services such as central heating or air conditioning are also to be contained in the space. Rectangular or square panels are generally employed, a popular size being 1200 mm by 600 mm for the fixed type and 600 mm square for the removable design.

Electrical systems. Basically there are three distinct cable management systems used with cavity floors, with quite wide variations in each. The first is where the cables are contained in some form of enclosure fixed to the subfloor (*Figure 22.7*). Alternatively the cable enclosure may be fitted flush with the floor panel. Second the cables may be laid on the subfloor to join components together, sometimes called plug and socket systems. The third method is to suspend the enclosure close up to the underside of the floor panels.

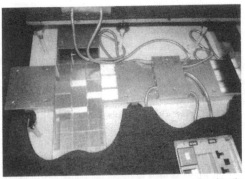

Figure 22.7 This cavity floor metal trunking system shown fitted to the subfloor is the same as the company's surface trunking but usually has conduit knockouts in the sides and dividers for flexible conduit connections between the trunking and the floor outlet-boxes (Cableduct)

Subfloor systems. In some plug and socket systems cables are linked together by multiple outlet feeder boxes to the subfloor and these boxes supply the service outlets fitted in the floor panels. If these feeder boxes have spare capacity it is a simple matter to reposition or install another floor outlet box, by using one of the spare sockets. Segregation of services is maintained within the feeder boxes. A variation is where the cables are replaced by busbar systems and services outlet boxes fed by plugging direct into them.

The most common cavity floor cable management system is where the compartmented trunking (either plastics or metallic) is fixed to the subfloor which carries the services cables. Feeds to outlet boxes are taken out through flexible conduit so preserving the segregation of services. There are many variations of the subfloor systems which are described more fully in the book referenced earlier?

Flushfloor systems. These are usually in the form of metallic trunking, the cover of which is fixed flush with the surface of the floor panels. Outlet boxes fit into the lid at any desired position along the length of the run.

Underpanel systems. The multi-compartment metallic trunking is lifted into position by a jacking system so that the trunking top just touches the underside of the panel, so that no cover for the trunking is needed. Like flushfloor systems they leave the structural slab free for other services and present the cable at a convenient working height.

Outlet boxes. Basically an outlet box consists of three components, the base, an inner section carrying the accessory plates, and a top part usually carrying the lid. Trims are supplied as a separate item. The base is provided with partitions to maintain the segregation between compartments in multi-compartment designs. Up to four compartments are provided as normal although some manufacturers offer more. Knockouts on the sides of the base unit provide facilities for fitting 20 mm or 25 mm flexible conduit. Rectangular or circular boxes are available.

Floor distribution systems

This section covers all types of floor distribution systems except cavity described above, including both metal and plastics flushfloor and undersurface screeded systems, undercarpet wiring and pedestal boxes associated with undersurface systems.

Screeded systems. There are two types of screeded systems. The first is a cable enclosure buried in the screed with access at finished floor level only through junction and services outlet boxes, with interconnection between the junction boxes being by single or multiple ducts through which the cables are drawn. The other has the trunking fixed to the floor with the screed up to the lid which is taken as the datum (*Figure 22.8*). Full access is provided by taking off the lid which is used to house the outlet boxes, or for a modular design the service outlet replaces a section of the lid.

Undercarpet systems. There are basically two ways in which power, telecoms and data services can be taken to equipment, either in a very shallow profile multi-compartment trunking or by very thin specially designed cables. These cables are not to be understood in the conventional sense.

Trunking systems, usually plastics, employ conventional cables. Typically a three-compartment arrangement has dimensions of 60 mm wide by 14 mm deep. Another supplier quotes dimensions of 70 mm wide by 9.6 mm deep. These trunking systems feed shallow profile pedestal boxes which can be fitted with the required outlets, and generally rest on the floor with the carpet brought up to them, i.e. they are visible.

Flat cable designs lie under the carpet and usually have single service capability which can be integrated to give all three services. Special outlet boxes are required.

Pedestal boxes. These are available as standard and non-standard designs, the former being employed for both carpet trunking systems and screeded underfloor arrangements. The non-standard design is for use with flat cables.

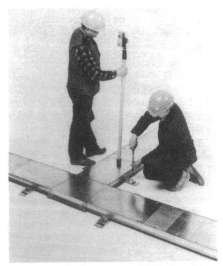

Figure 22.8 Metal flushfloor system being installed on the subfloor prior to screeding. The top of the trunking is taken as the datum line for screeding purposes. Outlet boxes fit direct into the trunking (Walsall Conduits)

Cable trays

Cable trays have traditionally been made from metal but increasingly plastic is being used because of its non-corrosive qualities (*Figure 22.9*). Trays of both materials are characterized by perforations in their bases and sometimes on the sides, these being used to hold the cables in position. Some of the plastics products do not have perforations – they are like trunking without lids. Heights of the longitudinal side members depend on the width and strength of the tray. Different methods are adopted for fastening sections together.

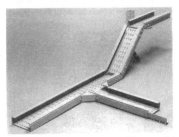

Figure 22.9 Powercomponents PVC cable tray. The special ventilated structure ensures high rigidity combined with light weight. Because the tray cannot corrode it is suitable for use in many hostile environments (Mita)

Perforations in the base are usually in the form of elongated slots although a few circular holes are sometimes included in the pattern. Popular widths start at 50 mm and 75 mm; side depths also range from 24 mm to 100 mm, while standard lengths are 2.44 m, 3 m and 4 m.

Cable ladders

Because both cable ladders and cable trays are used to carry heavy loads of cable over long distances and with large spans they should be treated more like structural support systems than the many forms of cable trunking already described. They find their applications in power stations, chemical and shopping complexes, factories, leisure centres and similar establishments where a great many cables have to be accommodated and routed over long distances.

Cable ladders are generally made of pre-galvanized sheet steel for dry environments but of post-galvanized metal where there is a possibility of corrosion. Post-galvanizing is generally known as hot-dip galvanizing although there are other methods of protection against corrosion (*Figure 22.10*). Stainless steel and extruded aluminium alloys are alternative materials for ladders. There is also a growing interest in plastics designs.

Figure 22.10 KHZ hot-dip galvanized cable ladder having oval rungs and closed rail profile meets Cadbury's stringent requirements at its Bournville factory (Wibe)

There is a great variety in the design of both sides and rungs of ladders, manufacturers claiming advantages with their particular products. Ladder sides carry perforations while some rungs also carry holes. Flat, oval, and T-section rungs indicate the wide variety on the market. Spacing between rungs also varies but 300 mm is a popular choice.

Wire tray system

The wire tray system is neither a cable ladder or cable tray although it performs the function of both to some degree. It is produced from high mechanical strength steel wire to form a welded mesh construction, with sides, to

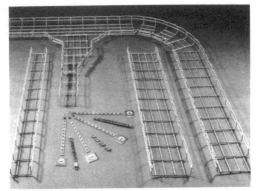

Figure 22.11 The Hi-Way wire tray system can be fabricated on site to the exact size by simply cutting the steel wires, preferably using a bolt cutter with offset cutting jaws. Bends, tees, crosses and level changes can also be formed from the wire tray itself (Vantrunk Engineering)

keep the cable in position. It is therefore much lighter than its counterparts (*Figure 22.11*).

Busbar trunking systems

This section relates to 400 V busbar trunking systems, usually mounted over-head, and fitted with tap-off units to feed equipment on the floor below (*Figure 22.12*). They should comply with BS EN 60439 Parts 1 and 2, particularly Part 2.

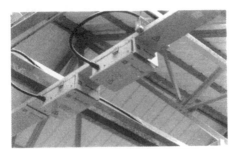

Figure 22.12 Power-Bar 800 busbar trunking has an IP42 enclosure, five tap-off openings per 3 m length and ratings of 250 A, 400 A and 630 A. Fire barriers are positioned to suit (Square D)

Basic components of a busbar trunking system are a set of copper or aluminium busbars complete with supports and insulators, factory assembled within insulated or sheet steel trunking. Ratings range from 25 A up to 5000 A or even greater. Trunking units are provided in straight sections and are usually for three-phase networks although there are some single-phase arrangements at the lower end of the range. Five-busbar designs allow for three-phase and neutral together with a protective conductor. Centre and end feed facilities are available particularly with the higher ratings. Good joints between adjacent sections are essential if hot-spots are to be avoided.

Tap-off units are easily plugged into the busbars and can be altered quite simply should the need arise. They may be of the all-insulated or metal-enclosed design which can be safely connected and disconnected on to live busbars. For busbar systems over 800 A tap-off units are manufactured with ratings from 40 A to 1000 A, and they may be fixed or unpluggable.

There is at least one cast resin design of busbar system with ratings up to 6300 A at 17.5 kV for a.c. networks and 16 000 A at 1.5 kV for d.c. circuits.

Special systems

One should draw attention to cable management systems that do not fit into the normal category but it is not proposed to describe them fully. Mineral insulated cable can be employed using a radial network and junction boxes.

An American modular system, while accommodating cables in trunking systems, can be adapted to provide voice, data and or video systems by utilizing surface-mounted boxes, wall plates and connector modules which snap on to the trunking as required.

Nurse call systems employ trunking at the hospital bedhead to house electrical facilities such as radio, television, sound distribution, power socket-outlets, computer outlets and telecommunications. Zip-up cable jacket and sleeving are ideal for protecting cable supplies to motors, control gear and switchgear.

Wired office furniture enables one to bring the services as close to the point of use as possible, dispensing with any trailing cables on floors.

Harnesses and pre-wired assemblies are made up for specific applications and can be economical where many identical arrangements are required.

23 Hazardous area electrical work

In areas where explosive gas-air mixtures can be present, electrical apparatus that gives off sparks during operation or in the event of a fault, and equipment with hot surfaces constitute a potential source of a hazard. The consequences of an explosion in such an area can be quite horrific, as disasters such as occurred on the Piper 'Alpha' platform, or before that at Flixborough have proved.

In the chemical, mining, and many other branches of industry, and even in everyday life, combustible materials which can form explosive atmospheres and which could under certain circumstances cause an explosion are manufactured, stored and processed. The concept of 'explosion protection' of electrical apparatus has been developed and formalized to prevent accidents occurring in hazardous areas during normal operation of the electrical apparatus.

Historically, the major impetus to the development of special equipment, procedures, standards and codes for particularly hazardous working environments was probably the large-scale development of the coal mining industry in Britain and the rest of the world at the time of the steam power revolution. Davy's miners' safety lamp (*Figure 23.1*) is an example of a piece of equipment designed specifically for use in hazardous areas, and in the early years of the twentieth century, commercial competitions were held for the design of safe electrical equipment for mines. The concept of 'intrinsic safety' is thought to date from a mining accident in South Wales in 1913. An explosion killing 439 men was probably caused by the ignition of methane from the spark created as a miner touched his shovel between the two bare wires which formed the signalling system to call for the mine's haulage winch (*Figure 23.2*).

The legacy of mining's importance in the development of hazardous area electrical equipment Ex protection concepts remains today – the British coal mining industry is governed by slightly different regulations than those applying to other hazardous areas (see sections on gas groupings and certification later in this chapter), but hazardous areas are also found, for example, in offshore oil and gas installations and onshore petrochemical installations and it is quite likely that at some point in his career an electrical engineer will need to know something of the subject of explosion protection.

The numerous sets of standards and codes of practice covering the manufacture, selection, installation and maintenance of electrical equipment in potentially hazardous areas, combined with differences in these standards and codes of practice around the world, make the subject complex and sometimes intimidating. In a single chapter of a book such as this it is impossible to give more than a brief background to the subject and to indicate where full and detailed information can be found. It should also be pointed out that the information contained in this chapter is not intended to replace published standards, codes of practice or other relevant publications.

A worldwide perspective is important because of the increasingly international nature of the electrical system design and building industry. For instance, Britain's technological expertise in the offshore oil and gas industry has meant work for British designers and builders of control systems destined for similar use overseas, but those systems must be suitable for use in the country in which they will be used.

Figure 23.1 Davy's miners' safety lamp (CEAG Crouse Hinds)

An initiative to find common ground worldwide for explosion protected electrical equipment and certification, to relevant IEC (International Electrotechnical Commission) Standards, regardless of the country of origin, has resulted in the launch of a scheme which is identified as IECEx. Member countries of the IEC who participate will have to align their hazardous area standards and test procedures to comply with the IECEx requirements.

The aim therefore is to have acceptance by any IEC member country of Ex certified products regardless of the country of origin manufacture or test authority.

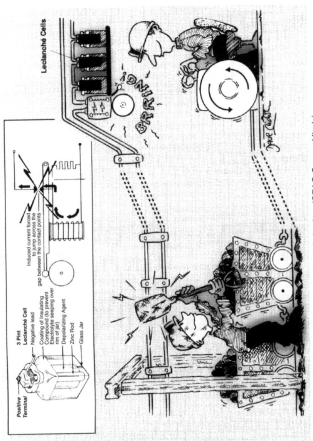

Figure 23.2 Inductance of bell circuit leads to sparking when circuit is broken (CEAG Crouse Hinds)

In common with many areas of trade, steps towards harmonization of standards across the European Union have been and continue to be taken, including the ATEX Directives which harmonize the technical requirements of free trade within the European Union.

ATEX Directives (94/9/EC (ATEX100A) and 99/92/EC (ATEX 137))

ATEX is derived from the French *Atmospheres Explosibles*.

There are two ATEX Directives that affect electrical equipment for use in explosive atmospheres, i.e. 94/9 relative to explosion protected electrical equipment (also non-electrical equipment with ignition potential) manufacture and certification. Directive 99/92 is relative to the workplace and the safety of workers.

Compliance with the Directive 94/9 facilitates free movement of equipment and systems throughout the EU by harmonizing the technical and legal requirements for products intended for use in potentially explosive gas or dust atmospheres.

Equipment that complies with the Directive and has been verified by a Notified Body can be labelled with the distinctive CE mark (*Figure 23.3*).

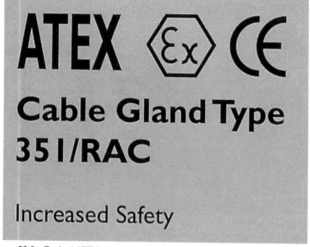

Figure 23.3 Typical ATEX compliance marking (CEAG Crouse Hinds)

Equipment is grouped, i.e. Group I for Mining and Group II for all other Industries. There are two equipment categories for mining applications, M1 and M2, which are dependent on the degree of hazard in the environment. Group II has three categories, which can be broadly described as Category 1 for Zone 0, Category 2 for Zone 1 and Category 3 for Zone 2. (Obviously Category 1 and 2 equipment can be used in Category 3 locations.)

The Directive is legally mandatory from June 2003. All Ex equipment and systems offered for sale in the EU must comply.

The CENELEC EN Standards have been amended to incorporate the essential health and safety requirements of the Directive.

Enforcement of the Directive in the UK is by the Health and Safety Executive and legally implemented by Statutory Instrument 192 'The Equipment and Protective Systems intended for use in Explosive Atmospheres Regulations'.

Directive 99/92 details general safety requirements and assessment of the risks to workers' health and safety. The employer must:

Prevent and provide protection against explosions.

Assess explosion risks.

Ensure that work can be carried out safely.

Classify zones and mark areas where explosive atmospheres may occur.

Draw up an explosion protection document.

Provide adequate training.

The electrical requirements are: the CE marking of equipment, zone classification, gas grouping, T classification, group and category certificates of conformity, installation compliance and maintenance programmes.

Workplaces used for the first time after June 2003 must comply with the Directive. Existing workplaces already in use before June 2003 must comply no later than June 2006. For existing workplaces changes made after 30 June 2003 must comply with the Directive.

This chapter gives information on the definitions of hazardous areas, equipment design concepts for explosion protection and the standards governing the manufacture and certification of electrical equipment designed for use in hazardous areas which will help the electrical engineer in selecting, installing and using such equipment. In addition, information is given on working practices, in general and in specific areas.

Hazardous Areas

The first requirement is to know what a hazardous area is. The principal factors relevant to the classifications of a hazardous area are the nature of the gases or dust present in the potentially explosive atmosphere and the likelihood of that atmosphere being present.

The concept of 'zone classification' has been developed to summarize these factors. The nature of the atmosphere is characterized by the chemical composition of the gas or dust and its auto-ignition temperature. The notions of 'gas grouping' and 'temperature classification' have been developed to formalize this.

Before looking in more detail at these definitions, it is instructive to consider how explosions occur. A useful concept is that of the 'hazard triangle', *Figure 23.4*. The three sides of the triangle represent fuel, oxygen and a source of ignition, all of which are required to create an explosion. The *fuel* considered in this chapter is a flammable gas, vapour or liquid although dust may

476

Ignition source

Figure 23.4 The 'hazard triangle'

also be a potential fuel. *Oxygen* is present in air at a concentration of approximately 21%. The *ignition source* could be a spark or a high temperature. If the potentially flammable atmosphere is between the upper and lower flammable limits for the particular material and an ignition source is introduced then it will explode or burn. Obviously if any side of the hazardous triangle can be removed then a fire or explosion hazard cannot exist. Given that a hazardous area may contain fuel and oxygen, the basis for preventing explosion is ensuring that any ignition source is either eliminated or else does not come into contact with the fuel-oxygen mixture. If there is any possibility of oxygen enrichment, i.e. above 20% by volume, then special consideration is necessary to ensure safety.

Zone classification. *Table 23.1* shows the IEC 79-10 zone classification used in Europe and most other parts of the world. The British Standard BS 5345 Part 2 will become obsolete and replaced by BS/EN/IEC 60079-10. The table also indicates which types of explosion protection are suitable for use within each zone. These explosion protection concepts are described later in the chapter.

The American system of hazardous area classification is structured in a different way, according to the National Electrical Code. In brief, hazardous locations are classified as either Class 1 'Division 1', where ignitable concentrations of flammable gases or vapours may be present during normal operation, or 'Division 2', where flammable gases or vapours occur in ignitable concentrations only in the event of an accident or a failure of a ventilation system.

Class II and Class III Divisions 1 and 2 relate to combustible dust and fibres. The 1999 edition of the National Electric Code (NEC) introduced for the first time in the USA the zone classification concept as an alternative to the class and division definitions of hazardous locations, e.g. Class 1 Zones 0, 1 and 2 for gases and vapour.

In the UK, the Factories Act states that where there is a risk of a flammable dust cloud, explosion protection and measures to reduce the risk of ignition will be required, although this is not dealt with in this chapter. The ATEX Directive legally requires dust hazards to be considered and classified as either Zone 20, 21 or 22.

Gas grouping and temperature classification. Different gases require different amounts of energy (by hot surface or spark) to ignite them and the two concepts of gas grouping and temperature classification are used in

Table 23.1 IEC 79 classification of hazardous area zones

	Suitable protection
Zone 0	
Areas in which hazardous explosive gas atmospheres are present constantly or for long periods, for example in pipes or containers	Ex 'ia' Ex 's' (where specially certified Zone 0)
Zone 1	
Areas in which hazardous explosive gas atmospheres are occasionally present, for example in areas close to pipes or draining stations	Ex 'd'; Ex 'ib'; Ex 'p'; Ex 'e'; Ex 's'; Ex 'o'; Ex 'q'; Ex 'm'; Equipment suitable for Zone 0
Zone 2	
Areas in which hazardous explosive gas atmospheres are rare or only exist for a short time, for example areas close to Zones 0 and 1	Ex 'N'/Ex 'n'; Equipment suitable for Zones 1 and 0

Europe to classify electrical apparatus according to its suitability for use with explosive atmospheres of particular gases.

Table 23.2 lists common industrial gases in their appropriate gas groups. Gas group I is reserved for equipment suitable for use in coal mines, and the differences between electrical equipment certified for use in mines is covered in more detail later. Gas group II – which contains gases found in other

Table 23.2 CENELEC/IEC gas grouping

Group		*Representative gases*
I	(MINING)	Methane
IIA		Acetone, ethane, ethyl acetate, ammonia, benzol, acetic acid, carbon monoxide, methanol, propane, toluene, ethyl alcohol, l-amyl acetate, N-hexane, N-butane, N-butyl alcohol, petrol, diesel, aviation fuel, heating oils, acetaldehyde, ethyl ether
	SURFACE INDUSTRIES	
IIB		Town gas, ethylene (ethene)
IIC		Hydrogen, acetylene (ethyne), hydrogen disulphide

industrial applications – is subdivided IIA, IIB, or IIC according to the relative flammability of the most explosive mixture of the gas with air.

Table 23.3 defines each temperature class according to the maximum allowed apparatus surface temperature exposed to the surrounding atmosphere, and indicates common gases for which these classifications are appropriate.

Table 23.3 CENELEC/IEC temperature classification

Class	Highest permissible surface temperature (°C)	Representative gases
T1	450	Acetone, ethane, ethyl acetate, ammonia, benzol, acetic acid, carbon monoxide, methanol, propane, toluene, town gas, hydrogen
T2	300	Ethyl alcohol, amyl acetate, N-hexane, N-butane, N-butyl alcohol, ethylene
T3	200	Petrol, diesel, aviation fuel, heating oils
T4	135	Acetaldehyde, ethyl ether
T5	100	
T6	85	Carbon disulphide

North American practice defines hazardous materials in classes. Flammable gases and vapours are Class 1 materials, combustible dusts are Class 2 materials and 'flyings' (such as sawdust) are Class 3 materials. Class 1 is subdivided into four groups depending on flammability: A (e.g. acetylene), B (e.g. hydrogen), C (e.g. ethylene) and D (e.g. propane, methane). Note that when compared with the IEC gas groupings, the subgroup letters are in opposite order of flammability.

North American temperature classification is similar to IEC standards, but further subdivides the classes to give more specific temperature data.

Standards and Codes of Practice for UK and Europe

Standards. As in many areas of industry, CENELEC European Norms (EN) exist alongside identical equivalent British Standards (BS). The origin of these norms is the European Directives concerning electrical equipment for use in potentially explosive atmospheres published in 1975 and 1979. A separate directive relating to mines was published in 1982. As a result of moves towards the removal of hindrances to free trade within Europe, CENELEC published European Norms (EN) which constitute harmonized standards. The new European ATEX Directive which came into effect in 1996, covers the conformity assessment procedures for all such electrical and mechanical equipment, protective systems and components. Until the new Directive is fully in place, and until 2003, the structure of standards and codes described in this chapter is likely to be relevant. (The ATEX Directive will be legally enforced by June 2003.)

As with all standards, review and revision is a continual process, and the European Norms that relate to hazardous areas are at present being updated and reissued in their second editions to incorporate the essential health and safety requirements of the ATEX Directive.

It is important for the electrical engineer to be ware of the European Norm standards for the manufacture of explosion protected equipment (EN 50014 and others) and equivalents in their latest editions because they determine the types of equipment available for use, and also impinge on installation procedure. Other relevant standards include those covering cables (see later).

Equipment certified as meeting EN standards carries the distinctive hexagonal conformity mark (*Figure 23.5*).

Figure 23.5 The EN conformity mark for certified explosion-protected equipment

Codes of Practice. Regulations and guidance for the installation and maintenance of electrical equipment in hazardous areas vary from country to country because of historical differences in installation practice. It is essential, of course, to take into account the national Wiring Regulations or general Codes of Practice for electrical installations. In addition in the UK, British Standard BS 5345 covered the selection, installation and maintenance of electrical apparatus for use in potentially explosive atmospheres, although it is not applicable to mining. BS 5345 has nine parts, some of which have already been mentioned (e.g. Part 2, which covers zone classification). This set of Standards is now obsolete (with the exception of Part 2) but should be retained for reference to existing site installations.

The new standard for selection and installation of electrical equipment in explosive atmospheres is BS/EN/IEC 60079-14.

Area assessment procedure. Companies using flammable materials should carry out an area assessment exercise in accordance with either BS 5345 Part 2 (zone classification) or preferably BS/EN/IEC 60079-10. Other relevant industry codes also exist, such as those relating to petroleum or chemical industries. In general, the assessment procedure would result in a written report identifying and listing the flammable materials which are used, recording all potential hazards and their source and type, identifying the extent of the zone by taking factors such as type of potential release of flammable material, ventilation and so on, and including other relevant data. Because the area assessment process is complex, various specialist companies and organizations offer a commercial hazardous area assessment service.

Electrical Equipment

The manufacturing and other processes within designated hazardous areas are similar to those in non-hazardous areas, and so there is a need for electrical power for motors, lighting, control and instrumentation. It is not surprising that as well as complete electrical systems built and certified to Ex standards, all manner of electrical equipment, from motors (*Figure 23.6*) and control stations (*Figure 23.7*), through luminaires (*Figure 23.8*) and switchgear to plugs and sockets (*Figure 23.9*), and even handlamps and torches, can be found in Ex-certified versions from specialist manufacturers.

Figure 23.6 Flameproof motor type EEx 'd', frame size 355 S4, 250 kW 4-pole (ABB Motors)

Equipment designed, manufactured and offered for sale in the EU for use in hazardous areas must comply with the ATEX Directive, incorporate a certification label that provides information about the product's suitability for the intended use and the certifying authority's mark. As an example, we will consider the markings on a twin tube EEx 'ed' fluorescent luminaire unit, certified to harmonized European standards. This is shown in *Figure 23.10*.

Some of these markings are self-explanatory (for example, manufacturer's name), and others have been covered earlier in this chapter (gas group and temperature classification). This section will explain the other markings, and give further information on how electrical equipment is designed and certified for use in hazardous areas.

Apparatus certified as suitable for use in British coal mines will carry different marks. The gas group mark will be I, and there will be no temperature class, because gas group I contains only one hazardous gas and this has a relatively high ignition temperature; thus the T class rating is considered unnecessary. The conformity mark consists of the letters MEx in the distinctive MECS logo. The differences in the certification process for equipment for the mining industry are dealt with later.

Figure 23.7 Explosion protected EEx 'e/d' control station (Enclosure EEX 'e'. Switches and indicator lamp EEx 'de') (CEAG Crouse Hinds)

Types of protection. According to EN 50014, electrical apparatus for use in explosive atmospheres can be designed with various protection concepts, *Table 23.4*. The apparatus is explosion protected (that is, it will not create or transmit the explosion of a surrounding hazardous atmosphere during normal operation) if a Recognized Test Authority has certified it to a relevant CEN-ELEC EN Standard. Equipment is designated EEx if it has been certified to a CENELEC European Norm (EN) harmonized standard and marked as shown in *Figure 23.5*. The more general IEC Ex designation is used in the rest of this chapter.

The electrical system design engineer will have to make a choice as to the method of protection to be used for the system being designed, and to select apparatus and components accordingly. In practice, the majority of electrical equipment for use in hazardous areas will be designated according to Ex 'd', Ex 'e', 'N/n' or Ex 'i' concepts. Detailed descriptions of these Explosion Protection Concepts (Ex) follow, along with brief details on the other explosion protection design concepts.

Ingress protection. The concept of ingress protection ('IP rating') is relevant to the design of explosion protected (Ex) electrical equipment as standards for some Ex apparatus also require a suitable IP rating such as Increased Safety Ex 'e'. Most type tests for certification include more stringent ingress protection tests and will include an impact test, normally of 7 joules. *Table 23.5* defines *ingress protection* ratings.

Installation of the eLLK 92

Whether it is mounted on rails or suspended from the ceiling, the lion's share of the overall costs is taken up by the installation and electrical connection of the light fitting. Here, due to the standardized fixing clearances and the generously dimensioned terminal compartments, the eLLK 92 provides a high saving potential. The terminal compartment can be opened without removing covers or reflectors, thus permitting the easy connection of cables. The standard, single-ended through wiring allows up to 2×6 wires per cable to be connected quickly and economically to the double terminals. Safe and permanent contact is ensured, even for wire cross-sections up to 6 mm^2. Simply push the hinged cover shut and you already have protection against contact to UVV4.

Figure 23.8 Wiring an EEx 'ed' luminaire (CEAG Crouse Hinds)

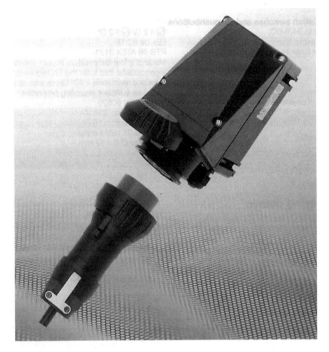

Figure 23.9 EEx 'de' plug and socket designed for hazardous area use (CEAG Crouse Hinds)

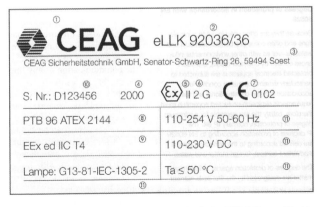

Figure 23.10(a) Label for Eex 'ed' fluorescent luminaire (CEAG Crouse Hinds)

1. Name of manufacturer.
2. Product type code.
3. Address of manufacturer.
4. Year of manufacture.
5. Community marking for explosion protected equipment.
6. Marking of equipment in accordance with ATEX Directive 94/9 for use in hazardous areas, i.e. Group II Zone 1 (or 2) gas atmospheres.
7. CE mark confirming compliance with the ATEX Directive. The number next to the CE mark indicates the Notified Body identity.
8. Test Authority Certificate number, note X indicates special conditions of use.
9. Marking according to CENELEC EN Standards: EEx (Explosion Protection) ed (Increased Safety and Flameproof) IIC (Gas Group) T4 Temp Class.
10. Serial number
11. Ambient temperature.

Figure 23.10(b)   Key to markings on above label

Table 23.4   Types of protection and relevant standards

Protection concept	Designation (European where appropriate)	Relevant standards		
		Harmonized European	Equivalent British Standard	International IEC
General requirements		EN 50014	BS 5501 Pt 1	IEC 79-0
Oil immersion	EEx 'o'	EN 50015	BS 5501 Pt 2	IEC 79-6
Pressurized	EEx 'p'	EN 50016	BS 5501 Pt 3	IEC 79-2
Quartz – sand filled	EEx 'q'	EN 50017	BS 5501 Pt 4	IEC 79-5
Flameproof enclosure	EEx 'd'	EN 50018	BS 5501 Pt 5	IEC 79-1
Increased safety	EEx 'e'	EN 50019	BS 5501 Pt 6	IEC 79-7
Intrinsic safety	EEx 'i'	EN 50020	BS 5501 Pt 7	IEC 79-11
		EN 50039	BS 5501 Pt 9	
Encapsulation	EEx 'm'	EN 50028	BS 5501 Pt 8	
Non-sparking	EEx 'n'		BS/EN 50021	IEC 79-15
Special protection	Ex 's'		SFA 3009	

Flameproof enclosure – Ex 'd'. Those internal parts of electrical apparatus that can ignite an explosive gas-air mixture are contained in an enclosure that can withstand the pressure created in the event of ignition of explosive gases inside the enclosure, and which can prevent the communication of any internal explosion to the external atmosphere surrounding the enclosure.

The rationale of the Ex 'd' concept is the safe containment of any internal explosion which may be created by ingress of an explosive material into the equipment, and thus the concept is applicable to virtually all types of electrical apparatus given that the potential sparking or hot elements can be contained in a suitably sized and strong enclosure. The factors taken into account by equipment manufacturers and system designers include arc and flame path lengths and types, surface temperature, internal temperature with regard to temperature classification, and distance from components in the enclosure to the enclosure wall (12 mm minimum is specified by the standard).

Some component types are unsuitable for use with flameproof enclosures, for example rewireable fuses, and components containing flammable liquids.

Table 23.5 Ingress protection ratings

First digit	Description of protection against solid objects	Second digit	Description of protection against liquids
0	No protection	0	No protection
1	Up to 50 mm, e.g. accidental touch by hands	1	Drops of water, e.g. condensation
2	Up to 12 mm, e.g. fingers	2	Direct sprays of water up to 15° from vertical
3	Up to 2.5 mm (tools and wires)	3	Direct sprays of water up to 60° from vertical
4	Up to 1 mm (tools, wires and small wires)	4	Splashing from all directions
5	Dust (no harmful deposit)	5	Low pressure jets of water from all directions
6	Totally protected against dust	6	Strong pressure jets of water, e.g. on ship decks
		7	Immersion in water
		8	Long periods of immersion under pressure

A major consideration in the use of Ex 'd' enclosures is the making of flameproof joints, which may be flanged, spigoted or screwed. Maximum gaps and minimum widths of any possible flame path through a joint are defined by the CENELEC EN Standard 50018, which also lays down requirements for screw thread pitch, quality and length. It should be noted that the standard states that 'there should be no intentional flamepath gap'.

The certification process for Ex 'd' equipment involves examination of the mechanical strength of the enclosure which must at least withstand 1.5 times the maximum explosion pressure for a specified gas group, also explosion and ignition tests under specified controlled conditions.

Increased safety – Ex 'e'. Measures are taken to prevent the possibility of sparks or an ignition capable temperature rise on internal or external parts of the electrical apparatus during normal and the most onerous operational conditions. The prevention of explosion from normally non-sparking/arcing or hot equipment is the guiding philosophy behind the Ex 'e' concept. Non-sparking devices can be used and various electrical, mechanical and thermal methods are used to increase the level of safety to meet the certification test requirements.

Examples of equipment designed and constructed to the Ex 'e' protection philosophy include luminaires, terminal boxes and motors. One advantage of Ex 'e' as a means of protection is that boxes and enclosures can be made of both metallic and other non-metallic materials that are easier to work with than using Ex 'd' flameproof cast enclosures.

Key design concepts for Ex 'e' equipment are the electrical, physical and thermal stability of the materials, and the compatibility of different materials used for items such as electrical terminations. Manufacturers of specialist explosion protected electrical apparatus have been ingenious in the design of electrical terminations for Ex 'e' equipment to ensure mechanically secure and positive maintenance-free connection of conductors. Ex 'e' certified enclosures require a minimum of IP54 ingress protection following a 7 joule impact test, although in practice IP66 is the industry preferred option, enclosures are available to meet a very wide range of hostile environmental operating conditions.

Non-sparking type N/n protection for Zone 2 hazardous areas. To avoid confusion both upper case (N) and the lower case (n) have been used in the heading above. Type N relates to the British Standard BS 6941 on which the European Norm CENELEC Standard for non-sparking type 'n' for Zone 2 applications are based.

To align with European Standards and comply with the requirements of ATEX Directives the CENELEC Standard EN 50021 is preferred, the protection concept is identified as EEx 'n'. The following text in this section will only be referred to as EEx 'n'.

Definition of EEx 'n' protection. A type of protection applied to electrical apparatus such that in normal operation it is not capable of igniting a surrounding explosive atmosphere or a fault condition capable of causing ignition is not likely to occur. The application of type n protection may require sound engineering judgement, due to the wide range of apparatus and technical aspects it covers. EEx 'n' apparatus is basically a non-sparking design and the allocated temperature rating applies to both internal and external surfaces of the apparatus.

EEx 'n'R. There is also available a restricted breathing version of the protection concept where the temperature rating applies only to the external surfaces exposed to surrounding atmosphere. This concept is in the main applicable to lighting applications where the lamp temperature is very high, but the restricted breathing design prevents direct contact between the lamp surface and any surrounding potentially explosive atmosphere.

Non-sparking electrical apparatus. The concept of flameproof enclosures initially applied to all areas of petrochemical electrical installations. This resulted in heavy and costly apparatus which is both expensive to install and difficult to maintain.

Historically the idea developed that the risk of ignition from electrical apparatus varied from very severe (Zone 0) to areas where flammable mixtures would be present in normal operation (Zone 1) and to much larger areas where a flammable gas or vapour would only be present for short periods under abnormal conditions (Zone 2). This led to the use of non-sparking apparatus such as terminal boxes, motors, luminaries, etc., being used in hazardous areas classified as Zone 2.

Although the Zone 2 non-sparking concept has been safely used for many years in the UK, many other countries in Europe adopted the attitude that certification was not necessary and there was a general reluctance to adopt EEx 'n' certification. This attitude has now changed. In North America good quality non-incendive industrial-type electrical equipment is permitted in Class I Division 2 locations without any formal approval requirements.

General requirements for EEx 'n' apparatus. The apparatus shall not in normal operation produce arcs or sparks; it can also include the following devices that would not normally be capable of providing a source of ignition:

- By use of enclosed break devices.
- By use of non-incendive components.

- By use of hermetically sealed devices.
- By use of encapsulated devices.
- By use of energy limited apparatus and circuits (IS).
- By use of sand-filled EEx 'q' devices.
- By use of EEx 'de' devices.

For definitions of the above devices refer to BS 50021 : 1999 Type of Protection 'n'.

The apparatus shall not in normal operation develop a surface temperature above the specified temperature rating unless the temperature or 'hot-spot' is prevented from causing ignition by the use of restricted breathing in which the temperature rating is only applied to external surfaces of the apparatus.

Although the main applications for type 'n' protection has historically been terminal boxes, lighting and induction motors, with the publication of BS 60021 there are alternative applications for the protection concept such as:

Simplified pressurization which is a technique similar to that allowed under EEx 'p' protection for Zone 1 or 2, which permits the use of standard industrial type components to be fitted internally to the enclosure, provided that an internal pressure is applied to prevent the ingress or migration of explosive atmospheres into the enclosure.

Restricted breathing could also be applied to EEx 'n' type enclosures provided the necessary test requirements have been satisfied.

Energy limited apparatus can be housed in type 'n' enclosures to contain IS devices and terminations such as I/O control systems.

Marking of Ex 'n' apparatus. Certified EEx 'n' apparatus shall be permanently marked in a visible position and include the following information:

The manufacturers' name
The product type identification
The symbol EEx 'n'
The gas group
The T classification
The serial number
The test authority

The symbol EEx 'n' shall include a suffix:

A For non-sparking apparatus.
C For sparking apparatus in which the contacts are suitably protected other than by restricted breathing enclosure, energy limitation and simplified pressurization.
R For restricted breathing enclosures.
L For energy limited apparatus.
P For enclosure with simplified pressurization.

Intrinsic safety – Ex 'i'. An electrical system can only be intrinsically safe (IS) if all the circuits it contains and field devices are intrinsically safe. The EN 50020 standard stipulates that under normal operation and certain specified fault conditions, no sparks or thermal effects are produced which can cause the ignition of a specified gas atmosphere. Sparks or thermal effects are not produced because the energy in the IS circuits is very low; intrinsically safe circuits are control and instrumentation circuits rather than power circuits.

Simple apparatus. The definition of 'simple apparatus' is the maximum stipulated voltage which a field device can generate is less than 1.2 V, with a current not exceeding 0.1 A, an energy of 20 microjoules and power of 25 mW. Simple apparatus does not need to be certified. IS circuitry exceeding these simple apparatus ratings can still meet the requirements of Ex 'i', but does require certification by a Recognized Test Authority.

There are two Ex 'i' protection types. Ex 'ia' equipment will not cause ignition (i) in normal operation, (ii) with a single fault, and (iii) with any two faults; a safety factor of 1.5 applies in normal operation and with one fault, and a safety factor of 1.0 applies with two faults. Ex 'ia' equipment is suitable for use in all zones, including Zone 0.

Ex 'ib' equipment is incapable of causing ignition (i) in normal operation, and (ii) with a single fault; a safety factor of 1.5 applies in normal operation and with one fault, and a safety factor of 1.0 applies with one fault if the apparatus contains no unprotected switch contacts in parts likely to be exposed to the potentially explosive atmosphere, and the fault is self-revealing. Ex 'ib' equipment is suitable for use in all zones except Zone 0.

Components for intrinsically safe circuits contain barriers to prevent excessive electrical energy from entering the circuit. The two principal barrier types are *Zener* barriers (used when an intrinsically safe earth is available) and *galvanically isolated barrier devices* (used where an IS earth *may* not be available, *Figure 23.11*). An IS earth must be provided by a clearly marked conductor of not less than 4 mm^2 with impedance not greater than 1 ohm from the barrier earth to the main power supply earth. The specified ingress protection rating of an enclosure for IS apparatus is IP20 but for hostile exterior environments, IP66 is recommended.

Other types of protection. Ex 'p' protection uses air or inert gas to maintain positive pressure (normally 0.5 mbar) over atmospheric pressure to prevent the entry of flammable gas or vapour into the enclosure or room. Another method is to reduce the volume of gas or vapour within the enclosure or room below the explosive gas-air mixture level by dilution from a clean external source.

Ex 'o' protection refers to apparatus in which ignition of a gas-air mixture is prevented by immersing the live or sparking apparatus in a specified minimum depth of oil, determined by type testing. Ex 'o' equipment is rarely encountered.

Ex 'q' (sand-filled) protected apparatus has the live or sparking elements immersed in granular quartz or other similar material. The required ingress protection rating is IP54.

Ex 'm' encapsulated equipment has potential ignition sources encapsulated to prevent them coming into contact with explosive atmospheres.

Ex 's' special equipment has been shown by test to be suitable for use in the appropriate zone. Although the apparatus may comply with a National Standard, it will not comply with a CENELEC EN Standard of the established protection concepts described above and therefore may not bear the CE mark.

Cables. As in any electrical system, cables are required in hazardous areas for instrumentation, communication, lighting, power and control; they must be selected for the intended purpose and environmental conditions (refer to BS/EN/IEC 60079-14).

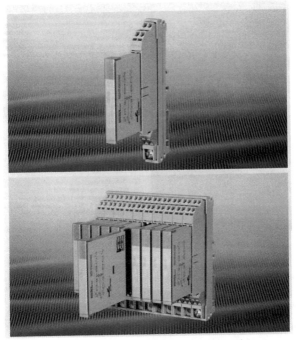

Figure 23.11 The 'intrinsically safe engineering system' – a rack to accommodate various galvanically isolated IS interface cards for control and instrumentation circuits (CEAG Crouse Hinds)

In intrinsically safe (IS) electrical systems, the inductive and capacitive nature of cables means that they can store energy and this energy must be taken into account when designing IS circuits. This is achieved by strict control of capacitance and L/R ratios in conjunction with Zener barriers or galvanic isolators. British Standards relevant to IS circuit cabling include BS 5308 Parts 1 and 2 which cover polyethylene insulated cables for use in petroleum refineries and related applications, and PVC insulated cables widely used in chemical and industrial applications. Although BS 5308 cables are frequently used in IS systems for hazardous area use, the words 'intrinsically safe' do not appear in the standard's title, and the foreword states that the cables 'may be suitable for Group II Intrinsically Safe systems'.

Other cable standards which are relevant to hazardous area work and which may be specified include BS 6883/IEC 92-3 (covering wiring for ships and topside on offshore oil and gas installations), BS 6425/IEC 754-1 (smoke and halogen emission), BS 6387/IEC 331 (fire resistance), BS 4066/IEC 332 (flame retardance) and BS 6207 (mineral insulated cables), *Figure 23.12*.

Cable glands. Cable glands used for the connection of cables to Ex apparatus must be certified and appropriate for the type of Ex protection installation.

Figure 23.12 EEx 'd' cables and cabling components (CEAG Crouse Hinds)

For installations in the EU cable glands shall be marked CE. Unused cable entry holes in Ex electrical equipment must be sealed with a 'stopping device', which is certified for the relevant type of Ex protection.

Certification. Within Europe, there are national test authorities which can issue certification documents for electrical equipment, to verify that equipment meets a specific standard for explosion protection. *Table 23.6* gives details of these authorities in each country. In the UK there are two accredited test authorities: EECS/BASEEFA (which is due to cease operations in its present

Table 23.6 Certifying authorities

Country	Authority
UK	BASEEFA/EECS and SIRA (MECS for mining applications)
Germany	PTB
France	CERCHAR and LCIE
Spain	LOM
Italy	CESI
Belgium	ISSeP
Denmark	DEMKO
Sweden	SEMKO
Norway	NEMKO
USA	FM
Canada	CSA
Netherlands	KEMA

role under the Health and Safety Executive) and SIRA. As well as being accepted throughout the European Union, certification by BASEEFA/SIRA is also recognized in other parts of the world, particularly in the Middle and Far East.

In the UK, certification to the standards relevant for use in mines is carried out by MECS (a unit of EECS). The procedure is similar to that for certification to standards for other hazardous areas, but in addition to explosion protection requirements, electrical equipment must also provide the high degree of electrical, mechanical and operational safety (pitworthiness requirements) demanded by the Mines Inspectorate.

Once a product has been tested by a Recognized Test Authority and certified as meeting a specific standard, the equipment manufacturer is issued with a Certificate of Conformity to verify that that product complies with the specified standards. Individual certified components usually have 'conditions for safe use' attached to their certificate and this is indicated by the suffix 'U'. 'Certificates of conformity' for a complete piece of electrical apparatus allow installation in hazardous areas without further verification but may also have Installation Conditions of Use indicated by the suffix 'X' following the certificate number.

In brief, the certification process involves an assessment of the conformity of the equipment to the specific standard sought. Examination of a prototype to ensure that it complies with the design documents and testing. The certified design is defined in a set of approved drawings listed on the certificate.

Installation, Inspection and Maintenance Practice

Installation. As well as information on selection of electrical apparatus for use in hazardous areas, the British Standard Code of Practice BS 5345 contains guidance on installation, inspection and maintenance of equipment. The code is divided into nine parts: Part 1 gives general guidance, Part 2 covers hazardous area classification, and each remaining part is specific to one of the types of protection concept. The code contains much useful information of a general nature concerning electrical work in hazardous areas. It does not cover work in mines or areas where explosive dusts may be present, and it makes clear that the code is not intended to replace recommendations produced for specific industries or particular applications.

BS 5345 is now obsolete other than Part 2 (Area Classification) but should be retained for reference to relevant existing installations. The BS 5345 Code of Practice is replaced by a single Installation Standard, i.e. BS/EN/IEC 60079-14 'Electrical Installations for explosive gas atmospheres (other than Mines)'. The standard does not replace general electrical wiring regulations or installation 'conditions of use' which may be contained in Ex apparatus certification documentation. Note this standard does not include dust hazards.

BS 5345 Part 2 (Area Classification) will be replaced by BS/EN/ IEC 60079-10.

For installations making use of Ex 'd' flameproof enclosures, the standard recommends preventing solid obstacles such as steelwork, walls or other electrical equipment from approaching near to flanged joints or openings. Minimum clearance distances of up to 40 mm are given.

Flamepath gaps should be protected against the ingress of moisture (if a weatherproof seal is not part of the enclosure design) with approved non-setting agents, e.g. light petroleum-based grease. Extreme care should be taken in the selection of these non-setting agents to avoid potential separation of joint surfaces. The standards specify the type of threads for entry tappings into flameproof enclosures, and draw attention to the requirement to follow any directions contained in the certification documents for cable systems and terminations.

Cable glands and cable entry devices. Cable glands must be certified Ex, marked CE, appropriate to the type of Ex protection being installed, selected according to the cable type and dimension and also have a compatible thread entry form. Earthing requirements must also be considered. Where necessary, barrier-type cable glands may be required for direct entry into flameproof enclosures. When conduits are used for wiring, they must be fitted with a sealing (stopping) device to prevent the migration of liquids or gases through the conduits.

Maintenance (BS/EN/IEC 60078-17). In general, operation and maintenance should be taken into account when designing process equipment and systems in order to minimize the release of flammable gases. For example, the requirement for routine opening and closing of parts of a system should be borne in mind at the design stage. No modifications should be made to plant without reference to those responsible for hazardous area classification, who should be knowledgeable in such matters. Whenever equipment is reassembled, it should be carefully examined. The standard gives recommended inspection schedules for equipment of each type of protection concept, which sets out what should be inspected on commissioning and at periodic intervals. For all equipment, the protection type, surface temperature class and gas group should all be checked to ensure that the equipment is suitable for its zone of use, and also that no unauthorized modifications have been carried out. Circuit identification should also be checked. Ex 'd' apparatus with integral cables or sealed terminal bushings or cable terminations which are encapsulated should not be tampered with but returned to the manufacturer if maintenance is required.

The section of the standard on inspection, maintenance and testing of increased safety Ex 'e'/Ex 'n' equipment also includes a recommendation to check that ratings of lamps used which may have been replaced are correct. Also gaskets or encapsulation material used in Ex 'e'/Ex 'n' equipment is satisfactory. Ex 'e' cage motors and associated protection equipment gives recommendations intended to ensure that all parts of the motor do not rise above a safe temperature when stalled (locked rotor).

The section of the standard BS/EN/IEC 60079-14 dealing with Ex 'i' equipment pays particular attention to the interconnecting cables used in IS systems, recommending, for instance, minimum conductor sizes to ensure temperature compliance in fault conditions, specific separations between individual IS circuits and earth, insulation thicknesses, screening and mechanical protection of cables. The use of multicore cable is considered, as is the siting of cables to avoid potential induction problems. In general, cable entries should be designed to minimize mechanical damage to cables. With all IS equipment, the need to follow the specific requirements of certification documents is emphasized, and during inspection it is recommended that attention be paid to lamps, fuses, earthing and screens, barriers and cabling. Certain specified on-site testing and maintenance of energized IS circuits is permitted inside the hazardous area provided any test equipment used is certified as intrinsically safe, and that conditions on certification documents are followed.

It is impossible in this section to give detailed, specific guidance on the installation, inspection and maintenance of electrical equipment in hazardous areas. The electrical engineer is advised to consult the BS/EN/IEC Standards relative to gas and dust hazardous areas or other codes and regulations relevant to specific industries or applications.

Sources of Further Information

BSI Enquiries, 389 Chiswick High Road. London, telephone 020 8996 7000.

SIRA Test and Certification, Saighton Lane, Saighton, Chester, telephone (01244) 332200.

EECS (BASEEFA and MECS), Harpur Hill, Buxton, Derbyshire, telephone (01298) 28000.

CEAG telephone 024 7630 8930

CEAG Crouse Hinds of Coventry supplied this chapter.

Index